AF251981

The

Geometric Process
and Its Applications

The
Geometric Process
and Its Applications

Yeh Lam

The University of Hong Kong & Northeastern University at Qinhuangdao, China

World Scientific

NEW JERSEY · LONDON · SINGAPORE · BEIJING · SHANGHAI · HONG KONG · TAIPEI · CHENNAI

Published by

World Scientific Publishing Co. Pte. Ltd.

5 Toh Tuck Link, Singapore 596224

USA office: 27 Warren Street, Suite 401-402, Hackensack, NJ 07601

UK office: 57 Shelton Street, Covent Garden, London WC2H 9HE

British Library Cataloguing-in-Publication Data
A catalogue record for this book is available from the British Library.

THE GEOMETRIC PROCESS AND ITS APPLICATIONS

ISBN-13 978-981-270-003-2
ISBN-10 981-270-003-X

Printed in Singapore by B & JO Enterprise

To my parents:
Kai Cheung Lam and Lai Ying Yip

Preface

The motivation of introducing geometric process is due to many practical problems in different fields, including science, engineering, medicine and social science. In maintenance problems, as most systems are deteriorating, the successive operating times of a system after repair will be decreasing, while the consecutive repair times after failure will be increasing. In epidemiology study, the number of infected cases usually shows an increasing trend in the early stage, and is stable in the middle stage, but has a decreasing trend in the late stage. In economics research, the economic development of a country or a region often shows a periodic cycle, so that its gross national product (GNP) will be increasing in the early stage of a cycle, and stable in the middle stage of the cycle, then decreasing in the late stage of the cycle. Many models such as minimal repair model, nonlinear regression model, nonlinear time series model, were developed for treating problems of the above phenomena with trend. However, it seems that a more direct approach is to introduce a monotone process model. As a simple monotone process, Lam (1988a, b) first introduced the geometric process.

Definition　A stochastic process $\{X_i, i = 1, 2, \ldots\}$ is called a geometric process if there exists a real $a > 0$ such that $\{a^{i-1}X_i, i = 1, 2, \ldots\}$ form a renewal process. The real number a is called the ratio of the geometric process.

Clearly, a geometric process is stochastically decreasing, if $a \geq 1$; it is stochastically increasing, if $0 < a \leq 1$. It will become a renewal process, if $a = 1$. Therefore, the geometric process is a generalization of a renewal process.

So far, much research work has been done in developing the theory and applications of geometric process. Many fruitful theoretical results and interesting applications have attracted more and more attention. As a result, the second edition of Encyclopedia of Statistical Science has added "Geometric Process" (see Lam (2006) for reference) as one new item.

In this book "Geometric Process and Its Application", we shall summarize the research work on theory and applications of geometric process developed since 1988. We expect that the publication of this book will stimulate further research in this topic and strengthen the practical applications. We sincerely hope more new results will follow after the publication of this book. We also hope the penetration of geometric process into different application fields will be accelerated after this book comes out.

Chapter 1 is a chapter of preliminaries. It contains a brief review of the Poisson process and renewal process. Stochastic order and some classes of lifetime distribution are then studied. In addition, a new class of lifetime distribution, namely earlier preventive repair better (worse) than later in expectation, is introduced. In Chapter 1, some martingale theory is reviewed. Furthermore, a formula for the rate of occurrence of failure is also studied.

In Chapter 2, the definition of geometric process is introduced, followed by the study of its probability properties.

In Chapter 3, like the renewal function in a renewal process, the expected number of events occurred by time t in a geometric process is defined as the geometric function of the geometric process. As the renewal function, the geometric function is also very important in application such as in warranty problem and block replacement model. In Chapter 3, several methods for the evaluation of the geometric function are suggested, these methods are then compared with a simulation method.

In Chapter 4, the statistical inference of a geometric process is studied. To do this, two statistics are introduced for testing if a data set agrees with a geometric process. Afterward, some nonparametric and parametric estimators for the parameters in a geometric process are proposed. The asymptotic normality of these estimators are then studied.

In Chapter 5, a geometric process model is applied to analysis of data with trend. Many real data sets are analyzed. In comparison with the results obtained by other three models, the geometric process model is the best model among four models. Furthermore, a threshold geometric process model is introduced for analyzing data with multiple trends. As an example, the threshold geometric process model is used to study the data sets of daily infected cases in SARS.

In Chapter 6, the application of geometric process to the maintenance problem of a one-component system is studied. The optimal replacement policy is determined analytically. Furthermore, the monotonicity of the optimal replacement policy is discussed. Afterward, a monotone process model for a multiple state system is considered, and a threshold geometric process maintenance model is introduced for a system with multiple trends, in particular for a system with a bathtub failure rate function. Moreover, a shock geometric process model and a δ-shock geometric process model are considered, both models study the effect of environment on the system. Then as a preventive repair is very useful in increasing the availability and reliability of a system, at the end of Chapter 6, a geometric process model with preventive repair is considered.

In Chapter 7, the reliability analysis for a two-component series, parallel and cold-standby systems respectively, is discussed. A geometric process maintenance model for a two-component cold standby system is also investigated.

In Chapter 8, the applications of geometric process to other topics in operational research is discussed. We consider here the applications in queueing theory and warranty problem.

Thus, Chapters 2-4 concentrate on the theoretical research of the probability theory and statistical inference of geometric process. Therefore, Chapters 2-4 form the foundation of the application of the geometric process. Nevertheless, Chapters 5-8 demonstrate the applications of geometric process to statistics, applied probability, operational research and management science.

The prerequisite for reading this book is an undergraduate calculus course plus a probability and statistics course. Only a few sections or results which are marked with an asterisk $*$ may require more advanced mathematics. However, the readers can skip over these parts without affecting the understanding of the main contents in the manuscript.

On the one hand, this book is a reference book for the researchers who are interested in geometric process theory or its application; on the other hand, it is a handbook for practitioners such as maintenance managers or reliability engineers or data analysts. Moreover, this book can be used as a postgraduate textbook or senior undergraduate textbook for statistics, applied probability, operational research, management science and reliability courses for one semester.

I take this opportunity to thank Professors A. C. Atkinson, M. D.

Cheng, D. R. Cox, S. S. Ding, L. C. Thomas and H. Tong, who were my teachers respectively during my undergraduate and postgraduate studies, for their continuing instruction and advice. I would also like to thank Professor Y. L. Zhang for our long-term cooperation in research, especially in the joint research of geometric process. I am grateful to Dr. Jennifer S. K. Chan and Dr. Yayong Tang for their assistance in preparation of the manuscript. I should express my gratitude to all of my friends and students for their help, concern and encouragement. I should also express my thanks to the colleagues of World Scientific for their effort and work, so that this book can be prosperously published.

Last but not least, I would like to thank my wife Chiu Bing Tan and my sons Fong Lam and Ying Lam for their continuous concern and support in order that I can devote time and energy to teaching, research and preparation of this book.

Yeh Lam
The University of Hong Kong and
Northeastern University at Qinhuangdao
ylam@saas.hku.hk
May 2007

Contents

Chapter 1

Preliminaries

1.1 Introduction

A fundamental concept in probability theory is random experiment. An random experiment is an experiment whose outcome cannot be determined in advance. The set of all possible outcomes of an random experiment is called the sample space of the experiment and is denoted by S.

An event E is a subset of sample space S. An event E is occurred if the outcome of the random experiment is an element of E. A random variable X is a real function defined on S.

A stochastic process $\{X(t), t \in T\}$ is a family of random variables so that for each $t \in T, X(t)$ is a random variable, where T is called the index set. We may interpret t as time, and then $X(t)$ is called the state of the stochastic process at time t. If the index set T is a countable set, the process $\{X(t), t \in T\}$ is called a discrete time stochastic process; if T is a continuum such, an interval for example, the process $\{X(t), t \in T\}$ is called a continuous time stochastic process.

Now, we introduce two concepts.

Definition 1.1.1. A stochastic process $\{X(t), t \in T\}$ is said to have independent increments if for all $t_0 < t_1 < \cdots < t_n$ and $t_i \in T, i = 0, 1, \ldots, n$, the random variables

$$X(t_1) - X(t_0), X(t_2) - X(t_1), \ldots, X(t_n) - X(t_{n-1})$$

are independent.

According to Definition 1.1.1, a stochastic process has independent increments if the changes of the process over nonoverlapping time intervals are independent.

1

Definition 1.1.2. A stochastic process $\{X(t), t \in T\}$ is said to have stationary increments if for all $t, t+s \in T$, the distribution of $X(t+s) - X(t)$ is the same for all t.

Thus, a stochastic process has stationary increments if the distribution of the change of the process between two times depends only on their time difference.

Chapter 1 introduces some preliminaries. In Section 1.2, we shall study the Poisson process that is a simple counting process. Section 1.3 will introduce the renewal process that is a more general counting process. In Section 1.4, we shall consider the stochastic order and discuss some classes of lifetime distribution. A new class of lifetime distribution, namely earlier preventive repair better (worse) than later in expectation, is introduced. Section 1.5 briefly studies the concept of martingale, then introduces martingale convergence theorem and the Doob, Riesz and Krickeberg composition theorems. In Section 1.6, we shall study the rate of occurrence of failures that is an important concept in reliability.

1.2 The Poisson Process

First of all, we define the counting process.

Definition 1.2.1. A stochastic process $\{N(t), t \geq 0\}$ is said to be a counting process if $N(t)$ is the total number of events occurred by time t.

Properties of the counting process.

(1) $N(t) \geq 0$.
(2) $N(t)$ is integer valued.
(3) If $s < t$, then $N(s) \leq N(t)$.
(4) For $s < t$, $N(t) - N(s)$ represents the number of events occurred in $(s, t]$.

It follows from Definition 1.1.1 that a counting process has independent increments if the numbers of events occurred in disjoint time intervals are independent. On the other hand, a counting process has a stationary increments if the distribution of the number of events occurred in a time interval depends only on the length of interval.

Now, we are available to define the Poisson process.

Definition 1.2.2. A counting process $\{N(t), t \geq 0\}$ is called a Poisson process with rate $\lambda > 0$, if

(1) $N(0) = 0$.
(2) The counting process has independent increments.
(3) For all $s, t \geq 0$,

$$P\{N(t+s) - N(s) = n\} = \frac{(\lambda t)^n}{n!} e^{-\lambda t} \quad n = 0, 1, \ldots \qquad (1.2.1)$$

Note that (1.2.1) means that the number of events occurred in an time interval of length t is a Poisson random variable with mean λt.

Definition 1.2.3 gives an alternative definition of a Poisson process.

Definition 1.2.3. A counting process $\{N(t), t \geq 0\}$ is called a Poisson process with rate $\lambda > 0$, if
(1) $N(0) = 0$.
(2) The counting process has independent and stationary increments.
(3) $P\{N(h) = 1\} = \lambda h + o(h)$.
(4) $P\{N(h) \geq 2\} = o(h)$.

Theorem 1.2.4. Definitions 1.2.2 and 1.2.3 are equivalent.
Proof.

It is trivial to show that Definition 1.2.2 implies Definition 1.2.3. To show that Definition 1.2.3 implies Definition 1.2.2, let

$$p_n(t) = P\{N(t) = n\}.$$

By classical probability analysis, $p_n(t)$ will satisfy the following differential equations

$$p_0'(t) = -\lambda p_0(t), \qquad (1.2.2)$$

$$p_n'(t) = -\lambda p_n(t) + \lambda p_{n-1}(t), \quad n = 1, 2, \ldots \qquad (1.2.3)$$

with initial condition

$$p_n(0) = \begin{cases} 1 & n = 0, \\ 0 & n \neq 0. \end{cases} \qquad (1.2.4)$$

In fact, for $n > 0$,

$$\begin{aligned} p_n(t+h) &= P\{N(t+h) = n\} \\ &= P\{N(t) = n, N(t+h) - N(t) = 0\} \\ &\quad + P\{N(t) = n - 1, N(t+h) - N(t) = 1\} \\ &\quad + \sum_{k=2}^{n} P\{N(t) = n - k, N(t+h) - N(t) = k\} \\ &= p_n(t)p_0(h) + p_{n-1}(t)p_1(h) + o(h) \qquad (1.2.5) \\ &= (1 - \lambda h)p_n(t) + \lambda h p_{n-1}(t) + o(h), \qquad (1.2.6) \end{aligned}$$

where (1.2.5) is due to conditions (2) and (4), while (1.2.6) is in virtue of conditions (3) and (4) in Definition 1.2.3 respectively, so that

$$p_0(h) = 1 - P\{N(h) = 1\} - P\{N(h) \geq 2\} = 1 - \lambda h + o(h).$$

Therefore

$$\frac{p_n(t+h) - p_n(t)}{h} = -\lambda p_n(t) + \lambda p_{n-1}(t) + \frac{o(h)}{h}.$$

Letting $h \to 0$ yields (1.2.3). The proof of (1.2.2) is similar. Now we can solve the differential equations with initial conditions. In fact, from (1.2.2) and (1.2.4) it is easy to see that

$$p_0(t) = e^{-\lambda t}.$$

Then assume that

$$p_{n-1}(t) = \frac{(\lambda t)^{n-1}}{(n-1)!} e^{-\lambda t}.$$

From (1.2.3) with the help of (1.2.4), we have

$$p_n(t) = \frac{(\lambda t)^n}{n!} e^{-\lambda t}. \tag{1.2.7}$$

Thus by induction, (1.2.7) holds for all integers $n = 0, 1, \ldots$. This proves that Definition 1.2.3 implies Definition 1.2.2. Hence, the proof of Theorem 1.2.4 is completed.

Definition 1.2.5. A continuous random variable X is said to have an exponential distribution $Exp(\lambda)$, if its density is given by

$$f(x) = \begin{cases} \lambda e^{-\lambda x} & x > 0, \\ 0 & \text{elsewhere.} \end{cases} \tag{1.2.8}$$

An important property of exponential distribution is the memoryless property . To explain this, consider

$$P\{X > s + t \mid X > s\} = \frac{P\{X > s+t, X > s\}}{P\{X > s\}} = \frac{P\{X > s+t\}}{P\{X > s\}}$$

$$= \frac{e^{-\lambda(s+t)}}{e^{-\lambda s}} = e^{-\lambda t} = P\{X > t\}. \tag{1.2.9}$$

This means that the conditional probability that a system survives for at least $s + t$ hours, given that it has survived for s hours, is the same as the unconditional probability that it survives for at least t hours. In other words, the system cannot remember how long it has survived. It can be shown that a continuous distribution is memoryless if and only if it is an exponential distribution.

A discrete random variable X is said to have a geometric distribution $G(p)$ with parameter p, if the probability mass function is given by

$$p(x) = P\{X = x\} = pq^{x-1} \quad x = 1, 2, \ldots \qquad (1.2.10)$$

with $q = 1 - p$. Then $E\{X\} = 1/p$. An integer valued random variable X is memoryless if

$$P\{X > m + n \mid X > n\} = P\{X > m\} \;\text{ for } m, \; n = 0, 1, \ldots \quad (1.2.11)$$

It can be shown that an integer valued distribution is memoryless if and only if it is a geometric distribution.

Given a Poisson process with rate λ, let X_1 be the time of the first event. In general, for $n \geq 1$, let X_n be the interarrival time between the $(n-1)$th and the nth events. Then we have the following theorem.

Theorem 1.2.6. Given a Poisson process with rate λ, the interarrival times $X_n, n = 1, 2, \ldots$, are independent and identically distributed (i.i.d.) random variables each having an exponential distribution $Exp(\lambda)$.
Proof.
It is clear that

$$P\{X_1 > t\} = P\{N(t) = 0\} = e^{-\lambda t}.$$

Then X_1 has an exponential distribution $Exp(\lambda)$. Now consider

$$
\begin{aligned}
P\{X_1 > s, X_2 > t\} &= \int_0^\infty P\{X_1 > s, X_2 > t \mid X_1 = x\} \lambda e^{-\lambda x} dx \\
&= \int_s^\infty P\{\text{no event in}(x, t + x] \mid X_1 = x\} \lambda e^{-\lambda x} dx \\
&= \int_s^\infty P\{\text{no event in}(x, t + x]\} \lambda e^{-\lambda x} dx \qquad (1.2.12) \\
&= e^{-\lambda t} \int_s^\infty \lambda e^{-\lambda x} dx \qquad (1.2.13) \\
&= e^{-\lambda s} e^{-\lambda t},
\end{aligned}
$$

where (1.2.12) is due to independent increments, and (1.2.13) is in virtue to (1.2.1). Therefore, X_1 and X_2 are independent each having $Exp(\lambda)$ distribution. Then Theorem 1.2.4 follows by induction.

Now define $S_0 = 0$ and

$$S_n = \sum_{i=1}^{n} X_i,$$

S_n is the arrival time of the nth event. Clearly, we have

$$N(t) \geq n \Leftrightarrow S_n \leq t. \tag{1.2.14}$$

Because $X_i, i = 1, 2, \ldots, n$, are i.i.d. random variables each having exponential distribution $Exp(\lambda)$. Then (1.2.14) yields that

$$P\{S_n \leq t\} = P\{N(t) \geq n\} = \sum_{i=n}^{\infty} \frac{(\lambda t)^i}{i!} e^{-\lambda t}. \tag{1.2.15}$$

Consequently, by differentiating with respect to t, S_n will have a gamma distribution $\Gamma(n, \lambda)$ with density function

$$f_n(x) = \begin{cases} \frac{\lambda^n}{\Gamma(n)} x^{n-1} e^{-\lambda x} & x > 0, \\ 0 & \text{elsewhere.} \end{cases} \tag{1.2.16}$$

Assume that N is a Poisson random variable with parameter λ, and $\{X_i, i = 1, 2, \ldots\}$ are i.i.d. random variables each having a distribution F. Assume further $\{X_i, i = 1, 2, \ldots\}$ are independent of N. Then random variable

$$X = X_1 + \cdots + X_N$$

is called a compound Poisson random variable with Poisson parameter λ and component distribution F.

Definition 1.2.7. Assume that $\{N(t), t \geq 0\}$ is a Poisson process, and $\{X_i, i = 1, 2, \ldots\}$ are i.i.d. random variables and are independent of process $\{N(t), t \geq 0\}$. Let

$$X(t) = X_1 + \cdots + X_{N(t)}.$$

Then stochastic process $\{X(t), t \geq 0\}$ is called a compound Poisson process.

Clearly, if $\{X(t), t \geq 0\}$ is a compound Poisson process, then $X(t)$ is a compound Poisson random variable with Poisson parameter λt.

In Definition 1.2.2 or 1.2.3, the rate λ of a Poisson process is a constant, hence the Poisson process is called a homogeneous Poisson process. Now, we shall consider the case that the rate is a function of time t. This is a nonhomogeneous Poisson process.

Definition 1.2.8. A counting process $\{N(t), t \geq 0\}$ is called a nonhomogeneous Poisson process with intensity function $\lambda(t), t \geq 0$, if
(1) $N(0) = 0$.
(2) The counting process has independent increments.
(3) $P\{N(t+h) - N(t) = 1\} = \lambda(t)h + o(h)$.
(4) $P\{N(t+h) - N(t) \geq 2\} = o(h)$.

Then we have the following theorem.

Theorem 1.2.9. Given a nonhomogeneous Poisson process $\{N(t), t \geq 0\}$ with intensity function $\lambda(t), t \geq 0$, then

$$P\{N(t+s) - N(s) = n\} = \frac{(\int_{s}^{t+s} \lambda(x)dx)^n}{n!} \exp\{-\int_{s}^{t+s} \lambda(x)dx\}$$

$$n = 0, 1, \ldots \tag{1.2.17}$$

The proof is similar to that of Theorem 1.2.4, see Ross (1996) for details.

Now, let X be the lifetime until the first failure occurs. Then from (1.2.17), we have

$$\bar{F}(t) = P\{X > t\} = P\{N(t) = 0\} = \exp\{-\int_{0}^{t} \lambda(x)dx\}, \tag{1.2.18}$$

and $\bar{F}(t)$ is called the survival function . Thus the distribution F and the density f of X are respectively given by

$$F(x) = 1 - \exp\{-\int_{0}^{x} \lambda(u)du\}, \tag{1.2.19}$$

and

$$\lambda(x) = \frac{f(x)}{\bar{F}(x)}. \tag{1.2.20}$$

In other words, the intensity function $\lambda(t)$ and the distribution F are uniquely determined each other.

In practice, two particular nonhomogeneous Poisson processes are important.

(1) The Cox-Lewis model
   The intensity function is

$$\lambda(t) = \exp(\alpha_0 + \alpha_1 t) \qquad -\infty < \alpha_0, \alpha_1 < \infty, \ \ t > 0. \qquad (1.2.21)$$

(2) The Weibull process model
   The intensity function is

$$\lambda(t) = \alpha\theta t^{\theta-1} \qquad \alpha, \theta > 0, t > 0. \qquad (1.2.22)$$

1.3   The Renewal Process

Now, let $\{X_n, n = 1, 2, \ldots\}$ be a sequence of nonnegative i.i.d. random variables each having a distribution F with $F(0) = P\{X_n = 0\} < 1$. Denote

$$\mu = E(X_n) = \int_0^\infty x dF(x).$$

Obviously $0 < \mu \leq \infty$. If we interpret X_n as the interarrival time between the $(n-1)$th and nth events (or renewals), then as in Poisson process, we can define the arrival time of the nth event by

$$S_n = \sum_{i=1}^n X_i \qquad (1.3.1)$$

with $S_0 = 0$. Thus, the number of events occurred by time t is then given by

$$N(t) = \sup\{n : S_n \leq t\}. \qquad (1.3.2)$$

Then, $\{N(t), t \geq 0\}$ is a counting process.

Definition 1.3.1.   The counting process $\{N(t), t \geq 0\}$ is called a renewal process.

   If the common distribution F is an exponential distribution, then the renewal process becomes a Poisson process. Therefore, renewal process is a generalization of Poisson process. Now define

$$M(t) = E[N(t)],$$

$M(t)$ is called the renewal function, it is the expected number of events occurred by time t.

As in Poisson process, we have

$$N(t) \geq n \iff S_n \leq t. \tag{1.3.3}$$

Then

$$P\{N(t) = n\} = P\{N(t) \geq n\} - P\{N(t) \geq n+1\}$$
$$= P\{S_n \leq t\} - P\{S_{n+1} \leq t\} = F_n(t) - F_{n+1}(t),$$

where F_n is the distribution function of S_n. Thus we have the following result.

Theorem 1.3.2.

$$M(t) = \sum_{n=1}^{\infty} F_n(t). \tag{1.3.4}$$

Proof.

$$M(t) = E[N(t)] = \sum_{n=0}^{\infty} nP\{N(t) = n\}$$
$$= \sum_{n=1}^{\infty} n\{F_n(t) - F_{n+1}(t)\} = \sum_{n=1}^{\infty} F_n(t).$$

This completes the proof.

Theorem 1.3.3.

$$M(t) < \infty \quad \text{for all } 0 \leq t < \infty. \tag{1.3.5}$$

Proof.

Since $P\{X_n = 0\} < 1$, there exists an $\alpha > 0$ such that $p = P\{X_n \geq \alpha\} > 0$. Now define

$$\tilde{X}_n = \begin{cases} 0 & \text{if } X_n < \alpha, \\ \alpha & \text{if } X_n \geq \alpha. \end{cases} \tag{1.3.6}$$

Let

$$\tilde{S}_n = \tilde{X}_1 + \cdots + \tilde{X}_n, \tilde{N}(t) = \sup\{n : \tilde{S}_n \leq t\}.$$

It is clear that $\{\tilde{N}(t), t \geq 0\}$ forms a renewal process whose events can only take place at times $t = k\alpha, k = 0, 1, \ldots$. The number of events at these times $t = k\alpha$ are i.i.d. random variables each having a geometric distribution $G(p)$. Thus

$$E[\tilde{N}(t)] = \sum_{k=0}^{[t/\alpha]} E[\text{number of events occurred at } k\alpha]$$

$$= \frac{[t/\alpha] + 1}{P(X_n \geq \alpha)} < \infty,$$

where $[x]$ is the largest integer no more than x. As $\tilde{X}_n \leq X_n$, then we have $N(t) \leq \tilde{N}(t)$, and (1.3.5) follows.

Note that the above proof also shows that $E[(N(t))^r] < \infty$ for all $t \geq 0, r \geq 0$.

By conditional on the value of X_1, we have

$$M(t) = E[N(t)] = \int_0^\infty E[N(t) \mid X_1 = x] dF(x). \tag{1.3.7}$$

Because

$$N(t) \mid X_1 = x = \begin{cases} 0 & x > t, \\ 1 + N_1(t - x) & x \leq t, \end{cases} \tag{1.3.8}$$

where

$$N_1(t) = \sup\{n : \sum_{i=2}^{n+1} X_i \leq t\}.$$

Then by substituting (1.3.8) to (1.3.7), it follows that

$$M(t) = \int_0^t \{1 + E[N(t - x)]\} dF(x),$$

since $N_1(t)$ and $N(t)$ have the same distribution. Thus we have

$$M(t) = F(t) + \int_0^t M(t - x) dF(x). \tag{1.3.9}$$

Equation (1.3.9) is called the renewal equation, it is an integral equation satisfied by renewal function $M(t)$.

Then assume that $\{X_n, n = 1, 2, \ldots\}$ are continuous with common density function f. Let the density of S_n be f_n. Thus, Theorem 1.3.2 gives

$$M'(t) = \sum_{n=1}^\infty f_n(t). \tag{1.3.10}$$

Denote the Laplace transform of $M(t)$ by

$$M^*(s) = \int_0^\infty e^{-st} M(t) dt,$$

and the Laplace transform of $f(t)$ by

$$f^*(s) = \int_0^\infty e^{-st} f(t) dt.$$

Because $S_n = \sum_{i=1}^{n} X_i$ is the sum of $X_1, \ldots, X_n$, the density of S_n is the n-fold convolution of f. Then the Laplace transform of $f_n(t)$ is given by

$$f_n^*(s) = [f^*(s)]^n.$$

Now, let the Laplace transform of $M'(t)$ be $M'^*(s)$. By taking Laplace transforms on both sides of (1.3.10), we have

$$M'^*(s) = \sum_{n=1}^{\infty} f_n^*(s) = \sum_{n=1}^{\infty} [f^*(s)]^n = \frac{f^*(s)}{1 - f^*(s)}.$$

However, as

$$M'^*(s) = sM^*(s) - M(0) = sM^*(s).$$

It follows from the above two equalities that

$$M^*(s) = \frac{f^*(s)}{s[1 - f^*(s)]}. \tag{1.3.11}$$

Thus, to determine renewal function $M(t)$, we can either use Theorem 1.3.2, or solve renewal equation (1.3.9). Alternatively, we can determine the Laplace transform $M^*(s)$ by (1.3.11) and then obtain $M(t)$ by inverting.

In many practical situations, an approximate formula for $M(t)$ is convenience. To derive an approximate formula, we need to introduce the concept of lattice.

A random variable X is lattice if it only takes on integral multiples of some $d \geq 0$. In other words, X is lattice, if there exists $d \geq 0$ such that

$$\sum_{n=-\infty}^{\infty} P\{X = nd\} = 1.$$

The largest d having this property is called the period of X. If X is lattice and F is the distribution of X, then we also say F is lattice.

Theorem 1.3.4. If F is not a lattice, $E[X] = \mu$ and $Var(X) = \sigma^2$, then

$$M(t) = \frac{t}{\mu} + \frac{\sigma^2 - \mu^2}{2\mu^2} + o(1) \quad \text{as} \quad t \to \infty. \tag{1.3.12}$$

Proof.

First of all, let the common density be f. Then the Laplace transform of f is given by

$$f^*(s) = E[e^{-sX}] = E[1 - sX + \frac{1}{2}(sX)^2 - \cdots]$$

$$= 1 - s\mu + \frac{s^2}{2}(\sigma^2 + \mu^2) + O(s^3). \tag{1.3.13}$$

By substitution, (1.3.11) becomes

$$M^*(s) = \frac{1 - s\mu + s^2(\sigma^2 + \mu^2)/2 + O(s^3)}{s[s\mu - s^2(\sigma^2 + \mu^2)/2 + O(s^3)]}$$

$$= \frac{1}{\mu s^2}[1 - s\mu + s^2(\sigma^2 + \mu^2)/2 + O(s^3)][1 + \frac{\sigma^2 + \mu^2}{2\mu}s + O(s^2)]$$

$$= \frac{1}{\mu s^2} + \frac{\sigma^2 - \mu^2}{2\mu^2 s} + O(1) \quad \text{as} \quad s \to 0. \tag{1.3.14}$$

Inverting (1.3.14) yields

$$M(t) = \frac{t}{\mu} + \frac{\sigma^2 - \mu^2}{2\mu^2} + g(t) \quad \text{as} \quad t \to \infty. \tag{1.3.15}$$

Let $g^*(s)$ be the Laplace transform of $g(t)$. Then from (1.3.14) and (1.3.15), by using the Tauberian theorem, we have

$$\lim_{t \to +\infty} g(t) = \lim_{s \to 0} sg^*(s)$$

$$= \lim_{s \to 0} s[M^*(s) - \frac{1}{\mu s^2} - \frac{\sigma^2 - \mu^2}{2\mu^2 s}] = 0.$$

This completes the proof of Theorem 1.3.4.

Now, let $N(\infty) = \lim_{t \to \infty} N(t)$, then we have

Lemma 1.3.5.

$$P\{N(\infty) = \infty\} = 1. \tag{1.3.16}$$

Proof.

From (1.3.3), we have

$$P\{N(\infty) < \infty\} = P\{X_n = \infty, \text{for some } n\}$$
$$= P\{\bigcup_{n=1}^{\infty} (X_n = \infty)\}$$
$$\leq \sum_{n=1}^{\infty} P\{X_n = \infty\} = 0.$$

Lemma 1.3.5 means that with probability 1, $N(t)$ tends to infinity as t tends to infinity. The following lemma gives the limiting rate of $N(t)$ as t tends to infinity.

Lemma 1.3.6. With probability 1,

$$\frac{N(t)}{t} \to \frac{1}{\mu} \quad \text{as } t \to \infty. \tag{1.3.17}$$

Proof.

Because $S_{N(t)} \leq t < S_{N(t)+1}$, then

$$\frac{S_{N(t)}}{N(t)} \leq \frac{t}{N(t)} < \frac{S_{N(t)+1}}{N(t)}.$$

It follows from Lemma 1.3.5 that with probability 1, $N(t) \to \infty$ as $t \to \infty$. Then the strong law of large numbers shows that

$$\frac{S_{N(t)}}{N(t)} \to \mu \quad \text{as } t \to \infty.$$

On the other hand, we have

$$\frac{S_{N(t)+1}}{N(t)} = \{\frac{S_{N(t)+1}}{N(t)+1}\}\{\frac{N(t)+1}{N(t)}\}$$
$$\to \mu \quad \text{as } t \to \infty.$$

As $t/N(t)$ is between two numbers, both of them converge to μ as $t \to \infty$, by principle of squeezing, so does $t/N(t)$. This completes the proof of Lemma 1.3.6.

Definition 1.3.7. Given a sequence of random variables $\{X_n, n = 1, 2, \ldots\}$, an integer-valued random variable N is called a stopping time for $\{X_n, n = 1, 2, \ldots\}$, if for all $n = 1, 2, \ldots$, event $\{N = n\}$ is independent of $X_{n+1}, X_{n+2}, \ldots$.

Assume that N is a stopping time for stochastic process $\{X_n, n = 1, 2, \ldots\}$. Then we can observe the process in sequential order and let N be the number observed before stopping. If $N = n$, then we shall stop after observing $X_1, \ldots, X_n$ and before observing $X_{n+1}, X_{n+2}, \ldots$ Obviously, if N is a stopping time for $\{X_n, n = 1, 2, \ldots\}$, then for all $n = 1, 2, \ldots$ events $\{N \leq n\}$ and $\{N > n\}$ will be determined by $X_1, \ldots, X_n$ only. Now, we have the following theorem.

Theorem 1.3.8.   **(Wald's Equation)**

If $\{X_n, n = 1, 2, \ldots\}$ is a sequence of i.i.d. random variables each having a common expectation $E[X] < \infty$, and if N is a stopping time for $\{X_n, n = 1, 2, \ldots\}$ with $E[N] < \infty$, then

$$E[\sum_{n=1}^{N} X_n] = E[N]E[X]. \tag{1.3.18}$$

Proof.

Let the indicator of event $\{N \geq n\}$ be

$$I_n = \begin{cases} 1 & \text{if } N \geq n, \\ 0 & \text{if } N < n. \end{cases} \tag{1.3.19}$$

Then

$$E[\sum_{n=1}^{N} X_n] = E[\sum_{n=1}^{\infty} X_n I_n] = \sum_{n=1}^{\infty} E[X_n I_n]$$

$$= E[X] \sum_{n=1}^{\infty} E[I_n] \tag{1.3.20}$$

$$= E[X] \sum_{n=1}^{\infty} P\{N \geq n\} = E[N]E[X], \tag{1.3.21}$$

where (1.3.20) is because N is a stopping time, then event $\{N \geq n\}$ is determined by $X_1, \ldots, X_{n-1}$ and hence independent of X_n. This completes the proof.

Now, assume that $\{N(t), t \geq 0\}$ is a renewal process whose interarrival times are $\{X_n, n = 1, 2, \ldots\}$. Then $N(t)+1$ is a stopping time for $\{X_n, n = 1, 2, \ldots\}$. This is because

$$N(t) + 1 = n \Leftrightarrow N(t) = n - 1$$

$$\Leftrightarrow X_1 + \cdots + X_{n-1} \leq t, \quad X_1 + \cdots + X_n > t.$$

Thus event $\{N(t) + 1 = n\}$ is determined by $X_1, \cdots, X_n$. Consequently, $N(t) + 1$ is a stopping time. Therefore Theorem 1.3.8 yields the following result.

Corollary 1.3.9. If $\mu < \infty$, then

$$E[S_{N(t)+1}] = \mu\{M(t) + 1\}. \tag{1.3.22}$$

The following theorems are important in renewal process.

Theorem 1.3.10. (The Elementary Renewal Theorem)

$$\frac{M(t)}{t} \to \frac{1}{\mu} \quad \text{as} \ \ t \to \infty, \tag{1.3.23}$$

where $1/\infty = 0$.

See Ross (1996) for the proof.

Theorem 1.3.11. Let μ and σ^2 be the mean and variance of the common distribution of the interarrival time in a renewal process $\{N(t), t \geq 0\}$. Then

$$P\{\frac{N(t) - t/\mu}{\sigma\sqrt{t/\mu^3}} < y\} \to \frac{1}{\sqrt{2\pi}} \int_{-\infty}^{y} e^{-x^2/2} dx \quad \text{as} \ \ t \to \infty. \tag{1.3.24}$$

Proof.

Let $r_t = t/\mu + y\sigma\sqrt{t/\mu^3}$. Then by using (1.3.3)

$$P\{\frac{N(t) - t/\mu}{\sigma\sqrt{t/\mu^3}} < y\} = P\{N(t) < r_t\}$$

$$= P\{S_{r_t} > t\} = P\{\frac{S_{r_t} - r_t\mu}{\sigma\sqrt{r_t}} > \frac{t - r_t\mu}{\sigma\sqrt{r_t}}\}$$

$$= P\{\frac{S_{r_t} - r_t\mu}{\sigma\sqrt{r_t}} > -y(1 + \frac{y\sigma}{\sqrt{t\mu}})^{-1/2}\}$$

$$\to \frac{1}{\sqrt{2\pi}} \int_{-y}^{\infty} e^{-x^2/2} dx \tag{1.3.25}$$

$$= \frac{1}{\sqrt{2\pi}} \int_{-\infty}^{y} e^{-x^2/2} dx \quad \text{as} \ \ t \to \infty,$$

(1.3.25) is due to the central limit theorem so that $(S_{r_t} - r_t\mu)/\sigma\sqrt{r_t}$ converges to standard normal distribution $N(0,1)$ as t tends to infinity, and because

$$-y(1 + \frac{y\sigma}{\sqrt{t\mu}})^{-1/2} \to -y \quad \text{as} \ t \to \infty.$$

This completes the proof.

Theorem 1.3.11 states that if t is large, then $N(t)$ will approximately be normal with mean t/μ and variance $t\sigma^2/\mu^3$.

Theorem 1.3.12.   (Blackwell's Theorem)
Let F be the common distribution of interarrival times in a renewal process.
(1) If F is not lattice, then for all $a \geq 0$

$$M(t+a) - M(t) \to a/\mu \ \text{ as } \ t \to \infty. \tag{1.3.26}$$

(2) If F is lattice with period d, then

$$E[\text{number of renewals at } nd] \ \to \ d/\mu \ \text{ as } \ t \to \infty. \tag{1.3.27}$$

For the proof of Theorem 1.3.12, see e.g. Feller (1970).

Given a function h defined on $[0, \infty)$, for any $a > 0$, let $\overline{m}_n(a)$ and $\underline{m}_n(a)$ be the supremum and infimum of $h(t)$ on interval $[(n-1)a, na]$ respectively. Then h is said to be directly Riemann integrable if for any $a > 0$, $\overline{m}_n(a)$ and $\underline{m}_n(a)$ are finite and

$$\lim_{a \to 0} \{a \sum_{n=1}^{\infty} \overline{m}_n(a)\} = \lim_{a \to 0} \{a \sum_{n=1}^{\infty} \underline{m}_n(a)\}.$$

A sufficient condition for a function h to be directly Riemann integrable is that
(1) $h(t) \geq 0$ for all $t \geq 0$,
(2) $h(t)$ is nonincreasing,
(3) $\int_0^{\infty} h(t)dt < \infty$.
Then we have the following theorem.

Theorem 1.3.13.   (The Key Renewal Theorem)
Let F be the common distribution of the interarrival times in a renewal process. If F is not lattice, and h is directly Riemann integrable, then

$$\lim_{t \to \infty} \int_0^t h(t - x)dM(x) = \frac{1}{\mu} \int_0^{\infty} h(t)dt, \tag{1.3.28}$$

where

$$M(x) = \sum_{n=1}^{\infty} F_n(x) \quad \text{and} \quad \mu = \int_0^{\infty} \bar{F}(t)dt.$$

For the proof of Theorem 1.3.13, see e.g. Feller (1970).

Given a renewal process $\{N(t), t \geq 0\}$ with interarrival times $\{X_n, n = 1, 2, \ldots\}$, we can define the age at t as

$$A(t) = t - S_{N(t)}, \tag{1.3.29}$$

the residual life at t as

$$B(t) = S_{N(t)+1} - t, \tag{1.3.30}$$

and the total life at t as

$$X_{N(t)+1} = S_{N(t)+1} - S_{N(t)} = A(t) + B(t). \tag{1.3.31}$$

Then we have the following theorem.

Theorem 1.3.14. Let F be the common distribution of the interarrival times in a renewal process. If F is not lattice with mean $\mu < \infty$, then

$$(1) \quad \lim_{t \to \infty} P\{A(t) \leq x\} = \lim_{t \to \infty} P\{B(t) \leq x\} = \tfrac{1}{\mu} \int_0^x \bar{F}(y) dy, \tag{1.3.32}$$

$$(2) \qquad \lim_{t \to \infty} P\{X_{N(t)+1} \leq x\} = \tfrac{1}{\mu} \int_0^x y dF(y). \tag{1.3.33}$$

See Ross (1996) for the proof of Theorem 1.3.14.

Now, consider a renewal process $\{N(t), t \geq 0\}$ with interarrival times $\{X_n, n = 1, 2, \ldots\}$ having a common distribution F. Assume that each time a renewal occurs we shall receive a reward. Let R_n be the reward earned at the time of the nth renewal. Usually, R_n will depend on X_n, but we assume that the pairs $\{(X_n, R_n), n = 1, 2, \ldots\}$ are i.i.d. random vectors. Then $\{(X_n, R_n), n = 1, 2, \ldots\}$ is called a renewal reward process. Thus, the total reward earned by time t is given by

$$R(t) = \sum_{n=1}^{N(t)} R_n.$$

Let

$$E[R] = E[R_n], \quad E[X] = E[X_n].$$

Afterward, we have the following theorem.

Theorem 1.3.15. Assume that $E[R] < \infty$ and $E[X] < \infty$. Then with probability 1,

$$\frac{R(t)}{t} \to \frac{E[R]}{E[X]} \quad \text{as } t \to \infty, \tag{1.3.34}$$

and

$$\frac{E[R(t)]}{t} \rightarrow \frac{E[R]}{E[X]} \quad \text{as } t \rightarrow \infty. \tag{1.3.35}$$

See Ross (1996) for the proof of Theorem 1.3.15. This is an important result and has a lot of applications in practice.

Remarks

If we say a cycle is completed whenever a renewal occurs, then (1.3.35) states that the long-run average reward per unit time (or simply average reward) is just the expected reward earned during a cycle, divided by the expected time of a cycle.

The reward can be negative. In this case, we may define R_n as the cost incurred at the time of the nth renewal. Then (1.3.35) states that the long-run average cost per unit time (or simply average cost) is given by

$$\text{average cost} = \frac{\text{the expected cost incurred in a cycle}}{\text{the expected time of a cycle}}. \tag{1.3.36}$$

Although we have assumed that the reward is earned all at once at the end of the renewal cycle, Theorem 1.3.15 still holds when the reward is earned gradually during the renewal cycle.

1.4   Stochastic Order and Class of Lifetime Distributions

At first, we introduce the concept of stochastic order here.

Definition 1.4.1. A random variable X is said to be stochastically larger (less) than a random variable Y, if

$$P\{X > t\} \geq (\leq) P\{Y > t\} \quad \text{for all } t. \tag{1.4.1}$$

It is denoted by $X \geq_{st} (\leq_{st})Y$. Let F and G be the distributions of X and Y respectively, then (1.4.1) is equivalent to

$$\bar{F}(t) \geq (\leq) \bar{G}(t) \quad \text{for all } t. \tag{1.4.2}$$

A sequence of random variables $\{X_n, n = 1, 2, \ldots\}$ is stochastically increasing (decreasing) if

$$X_{n+1} \geq_{st} (\leq_{st})X_n, n = 1, 2, \ldots$$

Lemma 1.4.2. If $X \geq_{st} Y$, then $E[X] \geq E[Y]$.

Proof.

Assume first that X and Y are nonnegative random variables. Then

$$E[X] = \int_0^\infty P\{X > t\}dt \geq \int_0^\infty P\{Y > t\}dt = E[Y]. \qquad (1.4.3)$$

In general, we have

$$X = X^+ - X^-,$$

where

$$X^+ = \begin{cases} X & \text{if } X \geq 0, \\ 0 & \text{if } X < 0, \end{cases} \qquad (1.4.4)$$

and

$$X^- = \begin{cases} 0 & \text{if } X \geq 0, \\ -X & \text{if } X < 0. \end{cases} \qquad (1.4.5)$$

Similarly, $Y = Y^+ - Y^-$. It is easy to show that

$$X \geq_{st} Y \Rightarrow X^+ \geq_{st} Y^+, \quad X^- \leq_{st} Y^-. \qquad (1.4.6)$$

Consequently, it follows from (1.4.3) and (1.4.6) that

$$E[X] = E[X^+] - E[X^-] \geq E[Y^+] - E[Y^-] = E[Y]. \qquad (1.4.7)$$

The following theorem gives a sufficient and necessary condition so that X is stochastically larger than Y.

Theorem 1.4.3.

$$X \geq_{st} Y \iff E[f(X)] \geq E[f(Y)]$$
$$\text{for any nondecreasing function } f. \qquad (1.4.8)$$

Proof.

Suppose that $X \geq_{st} Y$ and f is a nondecreasing function. For any t, let

$$f^{-1}(t) = \inf\{x : f(x) > t\}.$$

Then

$$P\{f(X) > t\} = P\{X > f^{-1}(t)\} \geq P\{Y > f^{-1}(t)\} = P\{f(Y) > t\}.$$

Consequently, $F(X) \geq_{st} f(Y)$. Then, from Lemma 1.4.2, $E[f(X)] \geq E[f(Y)]$.

Now assume that $E[f(X)] \geq E[f(Y)]$ for any nondecreasing function f. For any t, define

$$f_t(x) = \begin{cases} 1 & x > t, \\ 0 & \text{elsewhere.} \end{cases}$$

Then

$$E[f_t(X)] = P\{X > t\}, \quad E[f_t(Y)] = P\{Y > t\}. \tag{1.4.9}$$

Because $E[f_t(X)] \geq E[f_t(Y)]$, (1.4.9) implies that $X \geq_{st} Y$.

Then assume that X is the lifetime of a system with distribution F. The residual life of the system at age t given that the system has survived up time t will be given by

$$X_t = X - t \mid X > t. \tag{1.4.10}$$

A very popular class of life distribution is studied on the basis of the monotonicity of X_t.

Definition 1.4.4. A life distribution F has an increasing failure rate (IFR) if X_t is decreasing in t, i.e.

$$X_s \geq_{st} X_t \quad \text{for} \quad s < t. \tag{1.4.11}$$

Or for any $x \geq 0$, we have

$$P(X - s > x \mid X > s) \geq P(X - t > x \mid X > t) \quad \text{for} \quad s < t. \tag{1.4.12}$$

It has a decreasing failure rate (DFR) if X_t is increasing in t, i.e.

$$X_s \leq_{st} X_t \quad \text{for} \quad s < t. \tag{1.4.13}$$

Or for any $x \geq 0$, we have

$$P(X - s > x \mid X > s) \leq P(X - t > x \mid X > t) \quad \text{for} \quad s < t. \tag{1.4.14}$$

If F has an IFR, it is denoted by $F \in$ IFR, if F has a DFR, it is denoted by $F \in$ DFR. Now define

$$\bar{F}_t(x) = P(X - t > x \mid X > t) = \frac{\bar{F}(t + x)}{\bar{F}(t)}.$$

Then we have the following result.

Theorem 1.4.5. $F \in$ IFR (DFR) if and only if $\bar{F}_t(x)$ is decreasing (increasing) in t.

The class of IFR (DFR) distributions is closely related to the concept of failure rate. Consider a continuous random variable X, let the distribution and density be F and f respectively. Then the failure rate function is defined by

$$\lambda(t) = \frac{f(t)}{\bar{F}(t)}. \tag{1.4.15}$$

Let X be the lifetime of a system. Given that the system has survived for time t, the conditional probability that it will not survive for an additional time dt is given by

$$P\{t < X < t + dt \mid X > t\} = \frac{P\{t < X < t + dt, X > t\}}{P\{X > t\}}$$

$$= \frac{P\{t < X < t + dt\}}{P\{X > t\}} = \frac{\int\limits_{t}^{t+dt} f(u)du}{\bar{F}(t)}$$

$$\doteq \frac{f(t)dt}{\bar{F}(t)} = \lambda(t)dt. \tag{1.4.16}$$

Thus $\lambda(t)$ is the probability intensity that a t-year-old system will fail in $(t, t + dt)$.

Now, assume that X is the lifetime until the first failure in a nonhomogeneous Poisson process with intensity function $\lambda(t)$. Then it follows from (1.2.19), (1.2.20) and (1.4.15) that intensity function $\lambda(t)$ is the failure rate function of X.

In particular, if X is the lifetime before the first failure in a Poisson process with rate λ, then lifetime X will have an exponential distribution $Exp(\lambda)$ with density function given by (1.2.8). As a result, the failure rate is

$$\lambda(t) = \frac{\lambda e^{-\lambda t}}{e^{-\lambda t}} = \lambda.$$

Therefore, the failure rate of an exponential distribution $Exp(\lambda)$ is a constant λ that is the rate of the Poisson process.

Let

$$d = \sup\{t : t \geq 0, \ \bar{F}(t) > 0\}.$$

It can be shown that if $F \in \text{IFR(DFR)}$, then F is absolutely continuous in $(0, d)$, i.e. F has a probability density in $(0, d)$. See Barlow and Proschan (1975) or Cheng (1999) for reference. Now we have the following equality.

$$\frac{d\bar{F}_t(x)}{dt} = \frac{1}{\{\bar{F}(t)\}^2}\{-f(t + x)\bar{F}(t) + f(t)\bar{F}(t + x)\}$$

$$= \frac{\bar{F}(t + x)}{\bar{F}(t)}\{\lambda(t) - \lambda(t + x)\}.$$

Then we have the following theorem.

Theorem 1.4.6. A continuous life distribution F has an IFR (DFR) if and only if its failure rate function $\lambda(t)$ is increasing (decreasing).

From Theorem 1.4.6, we can see intuitively that X is IFR (DFR) means that the older the system is, the more (less) likely to fail in a small time dt is.

The following results are well known, see Barlow and Proschan (1975) for reference.

Theorem 1.4.7.
(1) A gamma distribution $\Gamma(\alpha, \beta)$ with density function

$$f(x) = \begin{cases} \frac{\beta^\alpha}{\Gamma(\alpha)} x^{\alpha-1} e^{-\beta x} & x > 0, \\ 0 & \text{elsewhere,} \end{cases} \tag{1.4.17}$$

is IFR if $\alpha \geq 1$ and DFR if $\alpha \leq 1$.
(2) A Weibull distribution $W(\alpha, \beta)$ with density function

$$f(x) = \begin{cases} \alpha\beta x^{\alpha-1} e^{-\beta x^\alpha} & x > 0, \\ 0 & \text{elsewhere,} \end{cases} \tag{1.4.18}$$

is IFR if $\alpha \geq 1$ and DFR if $\alpha \leq 1$.

In particular, an exponential distribution $Exp(\lambda)$ with density function (1.2.8) is IFR as well as DFR.

The set of all IFR (DFR) distributions forms a class of life distribution. Now, we consider another class of life distribution.

Definition 1.4.8. A life distribution F with finite mean μ is new better than used in expectation (NBUE) if

$$\bar{F}(t) \geq \frac{1}{\mu} \int_t^\infty \bar{F}(x)dx \quad \text{for} \quad t \geq 0. \tag{1.4.19}$$

It is new worse than used in expectation (NWUE) if

$$\bar{F}(t) \leq \frac{1}{\mu} \int_t^\infty \bar{F}(x)dx \quad \text{for} \quad t \geq 0. \tag{1.4.20}$$

They are denoted respectively by $F \in$ NBUE and $F \in$ NWUE.

The set of all NBUE (NWUE) distributions forms a class of life distributions. The following lemma gives the relationship between IFR (DFR) and NBUE (NWUE) classes.

Lemma 1.4.9. If a lifetime distribution $F \in$ IFR (DFR), then $F \in$ NBUE (NWUE).

See Barlow and Proschan (1975) for the proof.

Let the lifetime of a system be X. Given that $X > t$, then $X_t = X - t$ is the residual life. Thus from Definition 1.4.8, we can see that a life distribution F is NBUE (NWUE) if the expected residual life of a system which has survived for time t is no larger (less) than the expected lifetime of a new and identical system. In fact, if $F \in$ NBUE (NWUE), then from Definition 1.4.8, it yields that

$$E[X_t] = \int_0^\infty x dF_t(x) = \frac{1}{\bar{F}(t)} \int_0^\infty x dF(t+x)$$

$$= \frac{1}{\bar{F}(t)} \int_t^\infty \bar{F}(x)dx \leq (\geq)\mu.$$

Besides IFR (DFR) and NBUE (NWUE) classes, many other classes of life distributions have been considered (see Barlow and Proschan (1975) and Cheng (1999) for reference). However, all the classes of life distributions studied so far are defined on the basis of the life distribution of a system itself, i.e. the ageing effect on the system itself. In practice, for improving the system reliability and implementing the system more economically, besides the failure repair, we may also adopt a preventive repair that is a repair during the operating time of the system. Then, it is interesting to study the effect of preventive repair on the distribution of the total operating time of the system. For this purpose, we shall introduce a new class of lifetime distribution by taking into account the effect of preventive repair.

Let $F(x)$ be the distribution of the operating time X of a system. Assume that a preventive repair is taken when the operating time of the system reaches T, and the system after preventive repair is as good as new. Let M be the number of preventive repairs before system failure. Then, $Y = M + 1$ will have a geometric distribution $G(p)$ with probability mass function given by (1.2.10), where

$$p = P\{X \leq T\} = F(T)$$

and $q = P(X > t) = 1 - p$. Moreover,

$$E[M] = E(Y) - 1 = \frac{1}{p} - 1 = \frac{q}{p}.$$

Now, the total operating time with preventive repair at T is given by

$$X_T = \sum_{i=1}^{M} T + X_{(M+1)} = MT + X_{(M+1)},$$

where $X_{(M+1)} < T$ is the $(M+1)$th operating time of the system. Thus, the expected total operating time with preventive repair at T is given by

$$\begin{aligned}
E[X_T] &= E\{E[X_T \mid M]\} \\
&= E\{E[MT + X_{(M+1)}] \mid M)\} \\
&= E\{MT + \frac{1}{F(T)} \int_0^T t\,dF(t)\} \\
&= \frac{1}{p}\{qT + \int_0^T t\,dF(t)\} \\
&= \frac{1}{F(T)} \int_0^T \{1 - F(t)\}dt.
\end{aligned} \qquad (1.4.21)$$

Then, Lam (2006b) introduced the following definition.

Definition 1.4.10. A life distribution F is called earlier preventive repair better than later in expectation (ERBLE) if the expected total operating time with preventive repair at t

$$\mu(t) = \frac{1}{F(t)} \int_0^t \{1 - F(x)\}dx \qquad (1.4.22)$$

is nonincreasing in t, this is denoted by $F \in$ ERBLE. It is called earlier preventive repair worse than later in expectation (ERWLE) if $\mu(t)$ is nondecreasing in t, and is denoted by $F \in$ ERWLE.

The following results are also due to Lam (2006b).

Lemma 1.4.11. If a life distribution $F \in$ IFR (DFR), then $F \in$ ERBLE (ERWLE).

Proof.

If $F \in$ IFR, F will be absolutely continuous. Let the density function be f, then the failure rate function is given by $\lambda(t) = f(t)/\{1 - F(t)\}$. Thus it follows from (1.4.22) that

$$\mu'(t) = \frac{1}{F(t)^2}\{F(t)[1 - F(t)] - f(t)\int_0^t [1 - F(x)]dx\}$$

$$= \frac{f(t)}{F(t)^2}\{\frac{F(t)}{\lambda(t)} - \int_0^t \frac{f(x)}{\lambda(x)}dx\}$$

$$\leq \frac{f(t)}{\lambda(t)F(t)^2}\{F(t) - \int_0^t f(x)dx\} = 0.$$

Thus, $F \in$ ERBLE. For the case $F \in$ DFR, the proof is similar.

Then, from Theorem 1.4.7, we have

Corollary 1.4.12.
(1) A gamma distribution $\Gamma(\alpha, \beta)$ with density function (1.4.17) is ERBLE if $\alpha \geq 1$ and ERWLE if $\alpha \leq 1$.
(2) A Weibull distribution $W(\alpha, \beta)$ with density function (1.4.18) is ERBLE if $\alpha \geq 1$ and ERWLE if $\alpha \leq 1$.

In particular, an exponential distribution $Exp(\lambda)$ with density function (1.2.8) is ERBLE as well as ERWLE.

To see the relation between the classes of ERBLE (ERWLE) and IFR (DFR). we can consider the following two examples.

Example 1.4.13. Let

$$F(x) = \begin{cases} 1 - e^{-x} & 0 \leq x < 1, \\ 1 - e^{-3x} & x \geq 1. \end{cases} \tag{1.4.23}$$

It is easy to check that for $s < 1 < t$ and $1 < s + x < t + x$, then

$$\bar{F}_s(x) = \frac{\bar{F}(s + x)}{\bar{F}(s)} = e^{-(2s+3x)}$$

$$< \bar{F}_t(x) = \frac{\bar{F}(t + x)}{\bar{F}(t)} = e^{-3x}.$$

Thus $F \notin$ IFR. However, for $0 \leq t < 1, \mu(t) = 1$, it is a constant. Moreover, for $t \geq 1$

$$\mu(t) = \frac{1}{1 - e^{-3t}}(1 - e^{-1} + \frac{1}{3}e^{-3} - \frac{1}{3}e^{-3t}),$$

it is decreasing. This implies that $F \in$ ERBLE.

Example 1.4.14. Let

$$F(x) = \begin{cases} 1 - e^{-1} & 0 \leq x < 1, \\ 1 - e^{-x} & x \geq 1. \end{cases} \tag{1.4.24}$$

By a similar argument, it is also easy to show that $F \in$ ERWLE, but $F \notin$ DFR.

From Lemma 1.4.11 and two examples above, we can conclude that the ERBLE class (ERWLE) is greater than the IFR (DFR) class. The class of ERBLE (ERWLE) is a new class of life distributions by taking into account the effect of preventive repair. This new class of life distributions should have its own theoretical interest and practical application.

1.5   * Martingales

Consider a stochastic process $\{Z_n, n = 1, 2, \ldots\}$, let

$$\mathcal{F}_n = \sigma\{Z_1, \ldots, Z_n\}$$

be the σ-algebra generated by $\{Z_1, \ldots, Z_n\}$. If we interpret Z_n as the system state at time n, then $\mathcal{F}_n$ will represent the history of the system up to time n.

A stochastic process $\{X_n, n = 1, 2, \ldots\}$ is called a martingale with respect to $\{\mathcal{F}_n, n = 1, 2, \ldots\}$ if for all n

$$E[|X_n|] < \infty, \tag{1.5.1}$$

$$X_n \in \mathcal{F}_n \ \text{ i.e. } \ X_n \ \text{ is } \ \mathcal{F}_n \ \text{ measurable}, \tag{1.5.2}$$

$$E[X_{n+1} \mid \mathcal{F}_n] = X_n. \tag{1.5.3}$$

If $\{X_n, n = 1, 2, \ldots\}$ is a martingale, then taking expectation on both sides of (1.5.3) yields

$$E[X_{n+1}] = E[X_n],$$

and hence

$$E[X_n] = E[X_1] \ \text{ for all } \ n. \tag{1.5.4}$$

A stochastic process $\{X_n, n = 1, 2, \ldots\}$ is called a supermartingale (submartingale) with respect to $\{\mathcal{F}_n, n = 1, 2, \ldots\}$ if for all n, conditions (1.5.1), (1.5.2) and

$$E[X_{n+1} \mid \mathcal{F}_n] \leq (\geq) X_n, \tag{1.5.5}$$

hold. If $\{X_n, n = 1, 2, \ldots\}$ is a supermartingale (submartingale), then taking expectation of (1.5.5) yields

$$E[X_{n+1}] \leq (\geq) E[X_n],$$

and hence

$$E[X_n] \leq (\geq)E[X_1] \quad \text{for all} \quad n. \tag{1.5.6}$$

Theorem 1.5.1. (Martingale Convergence Theorem)
Assume that $\{X_n, n = 1, 2, \ldots\}$ is a supermartingale (submartingale) with

$$\sup_{1 \leq n < \infty} E[|X_n|] < \infty.$$

Then with probability 1, random variable

$$X = \lim_{n \to \infty} X_n$$

exists and $E[|X|] < \infty$. See e. g. Stout (1974) for the proof of Theorem 1.5.1.

Theorem 1.5.2. (The Doob Decomposition)
Assume that $\{X_n, n = 1, 2, \ldots\}$ is a submartingale with respect to $\{\mathcal{F}_n, n = 1, 2, \ldots\}$. Then $\{X_n, n = 1, 2, \ldots\}$ has a decomposition such that

$$X_n = M_n - A_n, \tag{1.5.7}$$

where $M_n, n = 1, 2, \ldots$ is a martingale, $\{A_n, n = 1, 2, \ldots\}$ is an decreasing sequence of random variables such that $A_{n+1} \leq A_n$ with probability 1, $A_1 = 0$ and $A_n \in \mathcal{F}_{n-1}$. Moreover, such a decomposition is unique.
Proof.
Let $M_1 = X_1$ and $A_1 = 0$. Then define M_n, A_n for $n \geq 2$ as follows.

$$M_n = M_{n-1} + (X_n - E[X_n \mid \mathcal{F}_{n-1}]), \tag{1.5.8}$$
$$A_n = A_{n-1} + (X_{n-1} - E[X_n \mid \mathcal{F}_{n-1}]). \tag{1.5.9}$$

From (1.5.8) and (1.5.9), we have

$$M_n - A_n = \sum_{i=2}^{n} (X_i - X_{i-1}) + X_1 - A_1 = X_n,$$

and (1.5.7) follows. It is then easy to check $M_n, n = 1, 2, \ldots$ and $\{A_n, n = 1, 2, \ldots\}$ satisfy the requirements. To show such a decomposition is unique, suppose $X_n = M_n' - A_n'$ is another decomposition. Then

$$M_n - M_n' = A_n - A_n'.$$

Thus $M_1 = M_1'$, since $A_1 = A_1' = 0$. Moreover, because $M_2 - M_2' = A_2 - A_2' \in \mathcal{F}_1$, therefore,

$$M_2 - M_2' = E[M_2 - M_2' \mid \mathcal{F}_1] = M_1 - M_1' = 0.$$

Then by induction, we can show that $M_n = M_n'$ and hence $A_n = A_n'$ for all n. This completes the proof of Theorem 1.5.2.

Theorem 1.5.3. (The Riesz Decomposition)

Assume that $\{X_n, n = 1, 2, \ldots\}$ is a submartingale with respect to $\{\mathcal{F}_n, n = 1, 2, \ldots\}$. Then the following two conditions are equivalent.

$$(1) \qquad \lim_{n \to \infty} E[X_n] < \infty, \qquad (1.5.10)$$

(2) $\{X_n, n = 1, 2, \ldots\}$ has a decomposition such that

$$X_n = Y_n + Z_n, \qquad (1.5.11)$$

where $\{Y_n, n = 1, 2, \ldots\}$ is a martingale and $\{Z_n, n = 1, 2, \ldots\}$ is a nonpositive submartingale with $\lim_{n \to \infty} E[Z_n] = 0$. Moreover, such a decomposition is unique. If in addition, $\{X_n, n = 1, 2, \ldots\}$ are nonnegative, then $\{Y_n, n = 1, 2, \ldots\}$ are also nonnegative.

Proof.

$(1) \Rightarrow (2)$ By Theorem 1.5.2, $\{X_n, n = 1, 2, \ldots\}$ has a Doob decomposition $X_n = M_n - A_n$, then

$$E[X_n] = E[M_n] - E[A_n] = E[M_1] - E[A_n].$$

As A_n is nonpositive, then $E[\|A_n\|] = E[X_n] - E[M_1] < \infty$, and $A_\infty = \lim_{n \to \infty} A_n$ is integrable. Define

$$Y_n = M_n - E[A_\infty \mid \mathcal{F}_n], \qquad (1.5.12)$$

$$Z_n = E[A_\infty \mid \mathcal{F}_n] - A_n. \qquad (1.5.13)$$

Now $\{Y_n, n = 1, 2, \ldots\}$ is a martingale, since $\{E[A_\infty \mid \mathcal{F}_n], n = 1, 2, \ldots\}$ is a martingale. As $A_n, n = 1, 2, \ldots$, is decreasing and $A_n \in \mathcal{F}_{n-1}$, then $\{Z_n, n = 1, 2, \ldots\}$ is nonpositive submartingale since

$$E[Z_n \mid \mathcal{F}_{n-1}] = E[E(A_\infty \mid \mathcal{F}_n) - A_n \mid \mathcal{F}_{n-1}]$$
$$= E[A_\infty \mid \mathcal{F}_{n-1}] - A_n$$
$$\geq E[A_\infty \mid \mathcal{F}_{n-1}] - A_{n-1} = Z_{n-1}.$$

On the other hand, by monotone convergence theorem, (1.5.13) gives

$$\lim_{n \to \infty} E[Z_n] = \lim_{n \to \infty} E[E(A_\infty \mid \mathcal{F}_n) - A_n]$$
$$= E[A_\infty] - E[A_\infty] = 0.$$

To show that such a composition is unique, suppose $X_n = Y_n' + Z_n'$ is another decomposition. Now let

$$W_n = Y_n - Y_n' = Z_n' - Z_n.$$

Then $\{W_n, n = 1, 2, \ldots\}$ is a martingale and

$$E[|W_n|] \leq E[|Z_n'|] + E[|Z_n|]$$
$$= -E[Z_n'] - E[Z_n] \to 0 \quad \text{as} \quad n \to 0.$$

This implies that $\{W_n, n = 1, 2, \ldots\}$ is uniformly integrable and $W_\infty = \lim_{n \to \infty} W_n = 0$. Consequently,

$$Y_n - Y_n' = W_n = \lim_{k \to \infty} E[W_{n+k} \mid \mathcal{F}_n] = E[W_\infty \mid \mathcal{F}_n] = 0.$$

Thus such a decomposition is unique. Now if in addition, $\{X_n, n = 1, 2, \ldots\}$ are nonnegative, then (1.5.12) yields

$$\begin{aligned}
Y_n &= M_n - E[A_\infty \mid \mathcal{F}_n] \\
&= \lim_{k \to \infty} E[M_{n+k} \mid \mathcal{F}_n] - \lim_{k \to \infty} E[A_{n+k} \mid \mathcal{F}_n] \\
&= \lim_{k \to \infty} E[M_{n+k} - A_{n+k} \mid \mathcal{F}_n] = \lim_{k \to \infty} E[X_{n+k} \mid \mathcal{F}_n] \geq 0.
\end{aligned}$$

Thus, $\{Y_n, n = 1, 2, \ldots\}$ are also nonegative.

$(2) \Rightarrow (1)$ Now

$$E[X_n] = E[Y_n] + E[Z_n] = E[Y_1] + E[Z_n] \leq E[Y_1] < \infty.$$

This completes the proof of Theorem 1.5.3.

Theorem 1.5.4.   (The Krickeberg Decomposition)
Assume that $\{X_n, n = 1, 2, \ldots\}$ is a submartingale with respect to $\{\mathcal{F}_n, n = 1, 2, \ldots\}$. Then the following two conditions are equivalent.

(1) $$\sup_n E[X_n^+] < \infty. \tag{1.5.14}$$

(2) $\{X_n, n = 1, 2, \ldots\}$ has a decomposition such that
$$X_n = L_n - M_n, \tag{1.5.15}$$

where $\{L_n, n = 1, 2, \ldots\}$ is a nonpositive submartingale and $\{M_n, n = 1, 2, \ldots\}$ is a nonpositive martingale. Moreover, such a decomposition has the maximality such that for any other decomposition $X_n = L_n' - M_n'$, where L_n', M_n' are nonpositive submartingale and nonpositive martingale respectively, then

$$L_n \geq L_n', \quad M_n \geq M_n'.$$

Proof.
$(1) \Rightarrow (2)$ Because $\{X_n, n = 1, 2, \ldots\}$ is a submartingale, then

$\{X_n^+, n = 1, 2, \ldots\}$ is a nonnegative submartingale and $\{E[X_n^+], n = 1, 2, \ldots\}$ is nondecreasing. As a result,

$$\lim_{n \to \infty} E[X_n^+] = \sup_n E[X_n^+] < \infty. \tag{1.5.16}$$

By Theorem 1.5.3, $\{X_n^+, n = 1, 2, \ldots\}$ has a Riesz decomposition

$$X_n^+ = Y_n + Z_n,$$

where $\{Y_n, n = 1, 2, \ldots\}$ is a nonnegative martingale and $\{Z_n, n = 1, 2, \ldots\}$ is nonpositive submartingale. Now for $n = 1, 2, \ldots$, let $M_n = -Y_n$ and define

$$L_n = X_n + M_n \leq X_n^+ + M_n = X_n^+ - Y_n = Z_n \leq 0.$$

Then $\{M_n, n = 1, 2, \ldots\}$ is a nonpositive martingale and $\{L_n, n = 1, 2, \ldots\}$ is a nonpositive submartingale. To show the maximality, suppose that $X_n = L_n' - M_n'$ is another decomposition, then

$$M_n' = L_n' - X_n \leq 0 \wedge (-X_n) = -X_n^+ = -Y_n - Z_n = M_n - Z_n.$$

Therefore

$$\begin{aligned}
M_n' &= E[M_{n+k}' \mid \mathcal{F}_n] \\
&\leq E[-X_{n+k}^+ \mid \mathcal{F}_n] = E[M_{n+k} - Z_{n+k} \mid \mathcal{F}_n] \\
&= M_n - E[Z_{n+k} \mid \mathcal{F}_n].
\end{aligned} \tag{1.5.17}$$

Because $\{Z_n, n = 1, 2, \ldots\}$ is nonpositive and

$$\lim_{n \to \infty} E[Z_n] = 0,$$

then $\{Z_n, n = 1, 2, \ldots\}$ is uniformly integrable and $Z_\infty = \lim_{n \to \infty} Z_n = 0$. Thus, $\lim_{k \to \infty} E[Z_{n+k} \mid \mathcal{F}_n] = 0$. Consequently, letting $k \to \infty$, (1.5.17) yields

$$M_n' \leq M_n,$$

hence

$$L_n' - L_n = M_n' - M_n \leq 0.$$

$(2) \Rightarrow (1)$ Since $M_n = L_n - X_n$, we have $M_n \leq -X_n^+$. Then

$$E[X_n^+] \leq -E[M_n] = -E[M_0] < \infty.$$

This completes the proof of Theorem 1.5.4.

1.6 * The Rate of Occurrence of Failures

Given a repairable system, let $N_f(t)$ be the number of failures of the system that have occurred by time t. Then the expected number of failures by time t is $M_f(t) = E[N_f(t)]$. Its derivative $m_f(t)$ is called the rate of occurrence of failures (ROCOF) at time t. Obviously, if $m_f(t)$ is increasing, the system is deteriorating, if $m_f(t)$ is decreasing, the system is improving. Therefore, the ROCOF is an important index in reliability theory.

Lam (1995, 1997) introduced a simple formula for the determination of the ROCOF $m_f(t)$ for a Markov chain with infinite state space. Suppose the state of a system at time t is $X(t)$. Assume that $\{X(t), t \geq 0\}$ is a continuous time homogeneous Markov chain with a finite or infinite state space $S = \{0, 1, 2, \ldots\}$. Let the infinitesimal matrix of the process be $A = (q_{ij})$. Thus

$$p_{ij}(\Delta t) = P(X(t + \Delta t) = j \mid X(t) = i)$$

$$= \begin{cases} q_{ij}\Delta t + o(\Delta t) & j \neq i, \\ 1 - q_i\Delta t + o(\Delta t) & j = i, \end{cases} \tag{1.6.1}$$

where $q_i = -q_{ii} \geq 0$ and $q_{ij} \geq 0$ for $j \neq i$.

Assume that the system has two kinds of state: up state and down state say. Denote the set of up states by W and set of down states by F. Then obviously $S = W \cup F$. At the beginning, suppose the system is in an up state. Let the number of transitions of the Markov chain by time t be $N(t)$ and the number of failures by time t be $N_f(t)$ respectively. Denote $N(t, t + \Delta t] = N(t + \Delta t) - N(t)$ and $N_f(t, t + \Delta t] = N_f(t + \Delta t) - N_f(t)$. Then clearly

$$N_f(t, t + \Delta t] \leq N(t, t + \Delta t]. \tag{1.6.2}$$

Moreover, let $p_i(t) = P(X(t) = i)$. Now, we make the following two assumptions.

Assumption 1. The Markov chain $\{X(t), t \geq 0\}$ is conservative, i.e.

$$q_i = \sum_{j \neq i} q_{ij}.$$

Assumption 2.

$$q = \sup_i q_i < \infty.$$

Note that if the Markov chain has a finite state space S, Assumptions 1 and 2 will automatically hold. Thus Assumptions 1 and 2 are actually made for the Markov chain with an infinite state space S.

It is well known that for a Markov chain $\{X(t), t \geq 0\}$, the sojourn time at state i has an exponential distribution $\text{Exp}(q_i)$. Therefore, Assumption 2 means that the expected sojourn time will have a positive lower bound, i.e.

$$\frac{1}{q} = \inf_i \frac{1}{q_i} > 0.$$

To derive the formula for the determination of the ROCOF $m_f(t)$ for the Markov chain $\{X(t), t \geq 0\}$, we start with time $T_0 = 0$, for $n > 1$, let T_n be the nth transition time of the process $\{X(t), t \geq 0\}$. If $X(T_n) = i$, then $T_{n+1} - T_n$ is the sojourn time of the process in state i. Now we can prove the following lemmas.

Lemma 1.6.1. For all integer n, we have

$$1. \quad P(T_n - T_0 \leq \Delta t \mid X(T_n) = j, X(T_0) = i) \leq (q\Delta t)^n, \qquad (1.6.3)$$

$$2. \qquad P(T_n - T_0 \leq \Delta t \mid X(T_0) = i) \leq (q\Delta t)^n. \qquad (1.6.4)$$

Proof.

For $n = 1$, it is well known that the sojourn time in state i has exponential distribution $Exp(q_i)$ with density function $f(x) = q_i \exp(-q_i x)$ for $x > 0$ and 0 otherwise, and T_1 is independent of the state $X(T_1) = j$. Therefore

$$P(T_1 - T_0 \leq \Delta t \mid X(T_1) = j, X(T_0) = i)$$
$$= P(T_1 - T_0 \leq \Delta t \mid X(T_0) = i) = 1 - \exp(-q_i \Delta t)$$
$$\leq q_i \Delta t \leq q \Delta t, \qquad (1.6.5)$$

where (1.6.5) is due to the inequality $1 - \exp(-x) \leq x$ for $x \geq 0$ and Assumption 2. Thus, (1.6.3) and (1.6.4) are true for $n = 1$. Now, assume that (1.6.3) and (1.6.4) hold for n. For $n + 1$,

$$P(T_{n+1} - T_0 \leq \Delta t \mid X(T_{n+1}) = j, X(T_0) = i)$$
$$\leq \sum_k P(T_{n+1} - T_n \leq \Delta t, T_n - T_0 \leq \Delta t, X(T_n) = k$$
$$\mid X(T_{n+1}) = j, X(T_0) = i)$$
$$= \sum_k P(T_{n+1} - T_n \leq \Delta t$$
$$\mid T_n - T_0 \leq \Delta t, X(T_n) = k, X(T_{n+1}) = j, X(T_0) = i)$$
$$\times P(T_n - T_0 \leq \Delta t \mid X(T_n) = k, X(T_{n+1}) = j, X(T_0) = i)$$
$$\times P(X(T_n) = k \mid X(T_{n+1}) = j, X(T_0) = i). \qquad (1.6.6)$$

Then

$$P(T_{n+1} - T_0 \leq \Delta t \mid X(T_{n+1}) = j, X(T_0) = i)$$
$$\leq \sum_k P(T_{n+1} - T_n \leq \Delta t \mid X(T_{n+1}) = j, X(T_n) = k)$$
$$\times P(T_n - T_0 \leq \Delta t \mid X(T_n) = k, X(T_0) = i)$$
$$\times P(X(T_n) = k \mid X(T_{n+1}) = j, X(T_0) = i) \tag{1.6.7}$$
$$\leq (q\Delta t)^{n+1} \sum_k P(X(T_n) = k \mid X(T_{n+1}) = j, X(T_0) = i) \tag{1.6.8}$$
$$= (q\Delta t)^{n+1},$$

where (1.6.7) is because of the Markov property and (1.6.8) is due to the homogeneity. Consequently by induction, (1.6.3) holds for all integers. A similar argument shows that (1.6.4) is also true for all integers.

Lemma 1.6.2.

$$P(T_1 - T_0 \leq \Delta t, T_2 - T_0 \geq \Delta t, X(T_1) = j \mid X(T_0) = i) = q_{ij}\Delta t + o(\Delta t).$$
$$\tag{1.6.9}$$

Proof.

$$P(T_1 - T_0 \leq \Delta t, T_2 - T_0 \geq \Delta t, X(T_1) = j \mid X(T_0) = i)$$
$$= \int_0^\infty P(T_1 - T_0 \leq \Delta t, T_2 - T_0 > \Delta t, X(T_1) = j \mid T_1 - T_0 = x, X(T_0) = i)$$
$$\times q_i \exp(-q_i x) dx$$
$$= \int_0^{\Delta t} P(T_2 - T_1 > \Delta t - x \mid X(T_1) = j, T_1 - T_0 = x, X(T_0) = i)$$
$$\times P(X(T_1) = j \mid T_1 - T_0 = x, X(T_0) = i) q_i \exp(-q_i x) dx$$
$$= \int_0^{\Delta t} P(T_2 - T_1 > \Delta t - x \mid X(T_1) = j)[(q_{ij}/q_i) + o(1)]q_i \exp(-q_i x) dx$$
$$= \int_0^{\Delta t} \exp[-q_j(\Delta t - x)][(q_{ij}/q_i) + o(1)]q_i \exp(-q_i x) dx$$
$$= q_{ij}\Delta t + o(\Delta t).$$

Lemma 1.6.3.

$$P(N_f(t, t + \Delta t] = 1) = \sum_{i \in W, j \in F} p_i(t) q_{ij} \Delta t + o(\Delta t).$$

Proof.

Because of (1.6.2), we have

$$P(N_f(t, t + \Delta t] = 1)$$

$$= \sum_{i \in W} P(N_f(t, t + \Delta t] = 1 \mid X(t) = i) P(X(t) = i)$$

$$+ \sum_{i \in F} P(N_f(t, t + \Delta t] = 1 \mid X(t) = i) P(X(t) = i)$$

$$= \sum_{i \in W} \{ P(N_f(t, t + \Delta t] = 1, N(t, t + \Delta t] = 1 \mid X(t) = i) P(X(t) = i)$$

$$+ P(N_f(t, t + \Delta t] = 1, N(t, t + \Delta t] \geq 2 \mid X(t) = i) P(X(t) = i) \}$$

$$+ \sum_{i \in F} P(N_f(t, t + \Delta t] = 1, N(t, t + \Delta t] \geq 2 \mid X(t) = i) P(X(t) = i)$$

$$= \sum_{i \in W} P(N_f(t, t + \Delta t] = 1, N(t, t + \Delta t] = 1 \mid X(t) = i) P(X(t) = i)$$

$$+ \sum_{i} P(N_f(t, t + \Delta t] = 1, N(t, t + \Delta t] \geq 2 \mid X(t) = i) P(X(t) = i)$$

$$= I_1 + I_2. \tag{1.6.10}$$

On the one hand, it follows from Lemma 1.6.2 that

$$I_1 = \sum_{i \in W, j \in F} P(T_1 - T_0 \leq \Delta t, T_2 - T_0 \geq \Delta t, X(T_1) = j \mid X(T_0) = i) p_i(t)$$

$$= \sum_{i \in W, j \in F} p_i(t) q_{ij} \Delta t + o(\Delta t). \tag{1.6.11}$$

On the other hand, Lemma 1.6.1 gives

$$I_2 \leq \sum_{i} P(N(t, t + \Delta t] \geq 2 \mid X(T_0) = i) p_i(t)$$

$$= \sum_{i} P(T_2 - T_0 \leq \Delta t \mid X(T_0) = i) p_i(t)$$

$$\leq (q \Delta t)^2 \sum_{i} p_i(t) = (q \Delta t)^2. \tag{1.6.12}$$

The combination of (1.6.11) and (1.6.12) yields Lemma 1.6.3.

Lemma 1.6.4.

$$\sum_{k=2}^{\infty} k P(N_f(t, t + \Delta t] = k) = o(\Delta t).$$

Proof.

$$\sum_{k=2}^{\infty} kP(N_f(t,t+\Delta t] = k)$$

$$= \sum_{k=2}^{\infty}\sum_{i} kP(N_f(t,t+\Delta t] = k \mid X(t) = i)P(X(t) = i)$$

$$\leq \sum_{k=2}^{\infty}\sum_{i} kP(N(t,t+\Delta t] \geq k \mid X(t) = i)P(X(t) = i)$$

$$= \sum_{k=2}^{\infty}\sum_{i} kP(T_k - T_0 \leq \Delta t \mid X(T_0) = i)P(X(t) = i)$$

$$\leq \sum_{k=2}^{\infty} k(q\Delta t)^k = o(\Delta t), \qquad (1.6.13)$$

where (1.6.13) follows from Lemma 1.6.1.

Now, since

$$M_f(t + \Delta t) - M_f(t) = E[N_f(t,t+\Delta T)]$$

$$= P(N_f(t,t+\Delta t] = 1) + \sum_{k=2}^{\infty} kP(N_f(t,t+\Delta t] = k)$$

$$= \sum_{i\in W, j\in F} p_i(t)q_{ij}\Delta t + o(\Delta t),$$

by Lemmas 1.6.3 and 1.6.4. Thus we have proven the following theorem.

Theorem 1.6.5. Assume that a Markov chain is conservative, i.e.

$$q_i = \sum_{j\neq i} q_{ij}$$

and

$$q = \sup_{i} q_i < \infty,$$

then the ROCOF at time t is given by

$$m_f(t) = \sum_{i\in W, j\in F} p_i(t)q_{ij}. \qquad (1.6.14)$$

In particular, if the state space S of a Markov chain is finite, Assumptions 1 and 2 are clearly true, then Theorem 1.6.5 holds. This special case was considered by Shi (1985).

1.7 Notes and References

For reading this monograph, we assume that the readers should have learned Calculus and an undergraduate course in Probability and Statistics. Of course, some other preliminaries are needed. Chapter 1 contains almost all additional knowledge needed in this monograph. Sections 1.2 to 1.4 are mainly based on Ross (1996), Barlow and Proschan (1975), Ascher and Feingold (1984). However, in Section 1.4, the class of life distributions ERBLE (ERWLE) that takes into account the effect of preventive repair is new, it was introduced by Lam (2007a). We expect that ERBLE (ERWLE) will have important theoretical interest and wide practical application. In Section 1.5, we introduce martingales and the martingale convergence theorem. Moreover, the Doob, Riesz and Krickberg decomposition theorems are also studied. For reference of these three decomposition theorems, see Dellacherie and Meyer (1982) or He et al. (1995). As in many reference books, they just highlight on the supermartingale case. Here we state the martingale convergence theorem for both the supermartingale and submartingale cases. Then we state these three decomposition theorems for submartingale case and give detailed proof for convenience of application in geometric process. In Section 1.6, we study the rate of occurrence of failures (ROCOF). If the expected sojourn time in a state of a conservative continuous time Markov chain has a positive lower bound, Lam (1995, 1997) gave a formula for the evaluation of ROCOF. In the case that a process is not a Markov chain but can be reduced to a Markov process after introducing some supplementary variables, Lam (1995, 1997) also gave a formula for the determination of the ROCOF. Section 1.6 is based on Lam (1997) that was published in Journal of Applied Probability by Applied Probability Trust.

Chapter 2

Geometric Process

2.1   Introduction

In Chapter 1, we mention that for the purpose of modelling a deteriorating
or improving system, a direct approach is to introduce a monotone process.
On the other hand, in analysis of data from a series of events with trend,
a natural approach is also to apply a monotone process. In this chapter,
as a simple monotone process, geometric process is introduced. Then, we
shall study the probability properties of geometric process. Furthermore, a
threshold geometric process is also proposed.

The structure of Chapter 2 is as follows. In Section 2.2, the definition of
geometric process is introduced, some properties of the geometric process
are discussed. In Section 2.3, as in renewal process, the age, residual life and
total life of a geometric process are considered. Then, some limit theorems
in geometric process are studied in Section 2.4. In Section 2.5, a special
geometric process with exponential distribution is considered.

2.2   Geometric Process

As a simple monotone process, Lam (1988a, b) introduced the geometric
process.

Definition 2.2.1. A sequence of nonnegative random variables $\{X_n, n = 1, 2, \ldots\}$ is said to be a geometric process (GP), if they are independent
and the distribution function of X_n is given by $F(a^{n-1}x)$ for $n = 1, 2, \ldots$,
where $a > 0$ is called the ratio of the GP.

To make sense, in practice we should assume $F(0) = P(X_1 = 0) < 1$.

37

If $\{X_n, n = 1, 2, \ldots\}$ is a GP and the density function of X_1 is f, then from Definition 2.2.1, the probability density function of X_n will be given by $a^{n-1}f(a^{n-1}x)$. Note that throughout this book, we shall use GP as an abbreviation of geometric process.

Definition 2.2.2. A stochastic process $\{X_n, n = 1, 2, \ldots\}$ is said to be a GP, if there exists a real $a > 0$ such that $\{a^{n-1}X_n, n = 1, 2, \ldots\}$ forms a renewal process. The positive number a is called the ratio of the GP.

Clearly, Definitions 2.2.1 and 2.2.2 are equivalent. Furthermore, a GP is stochastically increasing if the ratio $0 < a \leq 1$; it is stochastically decreasing if the ratio $a \geq 1$. A GP will become a renewal process if the ratio $a = 1$. Therefore, GP is a simple monotone process and is a generalization of the renewal process.

Assume that $\{X_n, n = 1, 2, \ldots\}$ is a GP with ratio a. Let the distribution function and density function of X_1 be F and f respectively, and denote $E(X_1) = \lambda$ and $\mathrm{Var}(X_1) = \sigma^2$. Then

$$E[X_n] = \frac{\lambda}{a^{n-1}}, \tag{2.2.1}$$

and

$$\mathrm{Var}(X_n) = \frac{\sigma^2}{a^{2(n-1)}}. \tag{2.2.2}$$

Thus, a, λ and σ^2 are three important parameters of a GP. Note that if $F(0) < 1$, then $\lambda > 0$. Now define $S_0 = 0$ and

$$S_n = \sum_{i=1}^{n} X_i.$$

Then let $\mathcal{F}_n = \sigma(X_1, X_2, \cdots, X_n)$ be the σ-algebra generated by $\{X_i, i = 1, \cdots, n\}$. Note that $\{S_n, n = 1, 2, \cdots\}$ is a sequence of nonnegative increasing random variables with

$$E[S_{n+1} \mid \mathcal{F}_n] = S_n + E[X_{n+1}] \geq S_n.$$

If ratio $a \leq 1$, then it is straightforward that

$$S_n \xrightarrow{a.s.} \infty \quad \text{as} \quad n \to \infty. \tag{2.2.3}$$

However, if ratio $a > 1$, then we have

$$\sup_{n \geq 0} E[|S_n|] = \lim_{n \to \infty} E[S_n] = \lambda \lim_{n \to \infty} \frac{1 - a^{-n}}{1 - a^{-1}} = \frac{a\lambda}{a - 1} < \infty. \tag{2.2.4}$$

This implies that $\{S_n, n = 1, 2, \cdots\}$ is a nonnegative submartingale with respect to $\{\mathcal{F}_n, n = 1, 2, \cdots\}$. Furthermore, we have the following result.

Theorem 2.2.3. If $a > 1$, there exists a random variable S such that

$$S_n \xrightarrow{a.s.} S \quad \text{as} \quad n \to \infty, \tag{2.2.5}$$

$$S_n \xrightarrow{m.s.} S \quad \text{as} \quad n \to \infty, \tag{2.2.6}$$

and

$$E[S] = a\lambda/(a-1), \tag{2.2.7}$$

$$Var[S] = \frac{a^2\sigma^2}{a^2 - 1}. \tag{2.2.8}$$

Proof.

Because $\{S_n, n = 1, 2, \cdots\}$ is a nonnegative submartingale with respect to $\{\mathcal{F}_n, n = 1, 2, \cdots\}$, then from Theorem 1.5.1, there exists a nonnegative random variable S such that

$$S_n \xrightarrow{a.s.} S \quad \text{as} \quad n \to \infty.$$

Thus (2.2.5) follows. Afterward, we can write $S = \sum\limits_{i=1}^{\infty} X_i$. It follows from (2.2.2) that

$$E[(S_n - S)^2] = E[(\sum_{i=n+1}^{\infty} X_i)^2]$$

$$= \text{Var}[\sum_{i=n+1}^{\infty} X_i] + \{E[\sum_{i=n+1}^{\infty} X_i]\}^2$$

$$= \frac{1}{a^{2(n-1)}} \left\{ \frac{\sigma^2}{a^2 - 1} + \frac{\lambda^2}{(a-1)^2} \right\} \longrightarrow 0, \quad n \to \infty.$$

Then (2.2.6) follows. Thus, by using monotone convergence theorem, (2.2.5) yields that

$$E[S_n] \longrightarrow E[S] \quad \text{as} \quad n \to \infty,$$

and (2.2.7) holds. On the other hand, we have

$$E[S_n^2] = Var[S_n] + \{E[S_n]\}^2 = \frac{a^2\sigma^2(1 - a^{-2n})}{a^2 - 1} + \frac{a^2\lambda^2(1 - a^{-n})^2}{(a-1)^2}.$$

By using monotone convergence theorem again, we obtain

$$E[S^2] = \frac{a^2\sigma^2}{a^2 - 1} + \frac{a^2\lambda^2}{(a-1)^2}.$$

Then (2.2.8) follows from $Var[S] = E[S^2] - \{E[S]\}^2$. This completes the proof of Theorem 2.2.3.

Therefore, according to Theorem 2.2.3, S_n converges to S not only almost surely but also in mean squares. Furthermore, Theorem 2.2.3 implies that if $a > 1$, then for any integer n, because event $\{S < \infty\}$ implies that $\{S_n < \infty\}$. Therefore

$$P(S_n < \infty) = 1. \tag{2.2.9}$$

Now, by applying Theorems 1.5.2 to the nonnegative submartingale $\{S_n, n = 1, 2, \ldots\}$, we have the following theorem.

Theorem 2.2.4. (The Doob Decomposition for GP)
If $a > 1$, the process $\{S_n, n = 1, 2, \ldots\}$ has a unique Doob decomposition such that

$$S_n = M_n - A_n, \tag{2.2.10}$$

where $\{M_n, n = 1, 2, \ldots\}$ is a martingale, $\{A_n, n = 1, 2, \ldots\}$ is decreasing with $A_1 = 0$ and $A_n \in \mathcal{F}_{n-1}$.
Note that if $a > 1$, then $\{S_n, n = 1, 2, \ldots\}$ is a nonnegative submartingale and

$$\lim_{t \to \infty} E[S_n] = \sup_n E[S_n^+] = \sup_n E[S_n] = \frac{a\lambda}{a-1} < \infty.$$

Then Theorems 1.5.3 and 1.5.4 yield respectively the following two results.

Theorem 2.2.5. (The Riesz Decomposition for GP)
If $a > 1$, the process $\{S_n, n = 1, 2, \ldots\}$ has a unique Riesz decomposition such that

$$S_n = Y_n + Z_n, \tag{2.2.11}$$

where $\{Y_n, n = 1, 2, \ldots\}$ is a nonnegative martingale and $\{Z_n, n = 1, 2, \ldots\}$ is a nonpositive submartingale with $\lim_{n \to \infty} E[Z_n] = 0$.

Theorem 2.2.6.   (The Krickeberg Decomposition for GP)
If $a > 1$, the process $\{S_n, n = 1, 2, \ldots\}$ has a Krickeberg decomposition so that

$$S_n = L_n - M_n, \tag{2.2.12}$$

where $\{L_n, n = 1, 2, \ldots\}$ is a nonpositive submartingale and $\{M_n, n = 1, 2, \infty\}$ is a nonpositive martingale. Moreover, such a decomposition has the maximality such that for any other decomposition $X_n = L_n^{'} - M_n^{'}$, where $L_n^{'}, M_n^{'}$ are nonpositive submartingale and nonpositive martingale respectively, then

$$L_n \geq L_n^{'}, \quad M_n \geq M_n^{'}.$$

In reliability engineering, the failure rate or hazard rate of a system expresses the propensity of the system to fail in a small time interval after t, given that it has survived for time t. In practice, many systems demonstrate that their failure rate has the shape of a bathtub curve. As the early failures of a system are due to quality-related defects, the failure rate is decreasing at the early stage or the infant mortality phase. During the middle stage or the useful life phase of the system, the failure rate may be approximately constant because failures are caused by external shocks that occur at random. In the late stage or the wearout phase of the system, the late-life failures are due to wearout and the failure rate is increasing. Therefore, in practice, many systems are improving in the early stage, then will be steady in the middle stage, and will be deteriorating in the late stage. In data analysis, many data sets show the existence of multiple trends. Consequently, we need to introduce a threshold GP for modelling a system with bathtub shape failure rate or analysing data with multiple trends.

Definition 2.2.7. A stochastic process $\{Z_n, \ n = 1, 2, \ldots\}$ is said to be a threshold geometric process (threshold GP) , if there exists real numbers $\{a_i > 0, i = 1, 2, \ldots, k\}$ and integers $\{1 = M_1 < M_2 < \ldots < M_k < M_{k+1} = \infty\}$ such that for each $i = 1, \ldots, k$,

$$\{a_i^{n-M_i} Z_n, \ M_i \leq n < M_{i+1}\}$$

forms a renewal process. The real numbers $\{a_i, i = 1, 2, \ldots, k\}$ are called the ratios and $\{M_1, M_2, \ldots, M_k\}$ are called the thresholds and k is called the number of thresholds of the threshold GP. Moreover, $\{Z_n, \ M_i \leq n < M_{i+1}\}$ is called the ith piece of the threshold GP.

Note that in Definition 2.2.7, it is not necessary that the renewal processes for different i are the same. Let the common mean and variance of $\{a_i^{n-M_i} Z_n, \ M_i \leq n < M_{i+1}\}$ be λ_i and σ_i^2 respectively, then

$$E[Z_n] = \frac{\lambda_i}{a_i^{n-M_i}} \tag{2.2.13}$$

and

$$\text{Var}(Z_n) = \frac{\sigma_i^2}{a_i^{2(n-M_i)}} \quad M_i \leq n < M_{i+1}, \ i = 1, \ldots, k. \tag{2.2.14}$$

Therefore number of thresholds k, threshold M_i, ratio a_i, mean λ_i and variance σ_i^2 for $i = 1, \ldots, k$ are important parameters for a threshold GP. Clearly, if $k = 1$, the threshold GP reduces to a GP defined by Definition 2.2.1 or 2.2.2.

2.3 Age, Residual Life and Total Life

Given a GP $\{X_n, n = 1, 2, \ldots\}$, as in renewal process, we can define the age at t as

$$A(t) = t - S_{N(t)}, \tag{2.3.1}$$

the residual life at t as

$$B(t) = S_{N(t)+1} - t, \tag{2.3.2}$$

and the total life at t as

$$X_{N(t)+1} = S_{N(t)+1} - S_{N(t)} = A(t) + B(t). \tag{2.3.3}$$

Let F_n be the distribution of S_n. Then we have the following result.

Theorem 2.3.1.

$$(1) \quad P(A(t) > x) = \begin{cases} \bar{F}(t) + \sum_{n=1}^{\infty} \int_0^{t-x} \bar{F}(a^n(t-u))dF_n(u) & 0 < x < t, \\ 0 & x \geq t. \end{cases}$$

$$(2) \quad P(B(t) > x) = \begin{cases} \bar{F}(t+x) + \sum_{n=1}^{\infty} \int_0^{t} \bar{F}(a^n(x+t-y))dF_n(y) & x > 0, \\ 1 & x \leq 0. \end{cases}$$

$$(3) \quad P(X_{N(t)+1} > x) = \begin{cases} \bar{F}(t \vee x) + \sum_{n=1}^{\infty} \int_0^{t} \bar{F}(a^n(x \vee (t-y)))dF_n(y) & x > 0, \\ 1 & x \leq 0. \end{cases}$$

$$(4) \quad P(S_{N(t)} \leq x) = \begin{cases} \bar{F}(t) + \sum_{n=1}^{\infty} \int_0^{x} \bar{F}(a^n(t-y))dF_n(y) & 0 < x \leq t, \\ 1 & x > t, \end{cases}$$

where F is the distribution of X_1 and $\bar{F}(x) = 1 - F(x)$, while F_n is the distribution of S_n and $\bar{F}_n(x) = 1 - F_n(x)$.

Proof.

For $x \geq t$, part (1) is trivial. Now, assume that $0 < x < t$,

$$P(A(t) > x) = P(S_{N(t)} < t - x)$$

$$= \sum_{n=0}^{\infty} P(S_{N(t)} < t - x, \ N(t) = n)$$

$$= \sum_{n=0}^{\infty} P(S_n < t - x, \ S_{n+1} > t)$$

$$= \bar{F}(t) + \sum_{n=1}^{\infty} \int_0^{t-x} P(S_{n+1} > t \mid S_n = y) dF_n(y)$$

$$= \bar{F}(t) + \sum_{n=1}^{\infty} \int_0^{t-x} P(X_{n+1} > t - y) dF_n(y)$$

$$= \bar{F}(t) + \sum_{n=1}^{\infty} \int_0^{t-x} \bar{F}(a^n(t - y)) dF_n(y).$$

Thus, part (1) follows. The proof of part (2) is similar to part (1). To prove part (3), we note that

$$P(X_{N(t)+1} > x) = \sum_{n=0}^{\infty} P(X_{N(t)+1} > x, N(t) = n)$$

$$= \sum_{n=0}^{\infty} \int_0^t P(X_{n+1} > x, S_n \leq t < S_{n+1} \mid S_n = y) dF_n(y)$$

$$= \sum_{n=0}^{\infty} \int_0^t P(X_{n+1} > \max(x, t - y)) dF_n(y) \qquad (2.3.4)$$

$$= \bar{F}(t \vee x) + \sum_{n=1}^{\infty} \int_0^t \bar{F}(a^n(x \vee (t - y))) dF_n(y).$$

Then part (3) follows. Moreover, the proof of part (4) is similar to part (1). This completes the proof of Theorem 2.3.1.

Furthermore, we can also obtain an upper and lower bounds for $P(X_{N(t)+1} > x)$.

Theorem 2.3.2.

$$\sum_{n=0}^{\infty} \bar{F}(a^n x) F_n(t) P(N(t) = n) \le P(X_{N(t)+1} > x)$$

$$\le \sum_{n=0}^{\infty} \bar{F}(a^n x) F_n(t) \quad t > 0, \ x > 0,$$

where F is the distribution of X_1 and $\bar{F}(x) = 1 - F(x)$, while F_n is the distribution of S_n and $\bar{F}_n(x) = 1 - F_n(x)$.

Proof.

It follows from (2.3.4) that

$$P(X_{N(t)+1} > x) = \sum_{n=0}^{\infty} \int_0^t P(X_{n+1} > \max(x, t - y)) dF_n(y)$$

$$\le \sum_{n=0}^{\infty} \int_0^t P(X_{n+1} > x) dF_n(y)$$

$$= \sum_{n=0}^{\infty} \bar{F}(a^n x) F_n(t). \tag{2.3.5}$$

On the other hand,

$$P(X_{N(t)+1} > x) = \sum_{n=0}^{\infty} P(X_{N(t)+1} > x \mid N(t) = n) P(N(t) = n)$$

$$= \sum_{n=0}^{\infty} P(X_{n+1} > x \mid S_n \le t < S_{n+1}) P(N(t) = n)$$

$$= \sum_{n=0}^{\infty} \int_0^t P(X_{n+1} > x \mid S_n \le t < S_{n+1}, S_n = y) dF_n(y) P(N(t) = n)$$

$$= \sum_{n=0}^{\infty} \int_0^t P(X_{n+1} > x \mid X_{n+1} > t - y) dF_n(y) P(N(t) = n)$$

$$\ge \sum_{n=0}^{\infty} P(X_{n+1} > x) \int_0^t dF_n(y) P(N(t) = n) \tag{2.3.6}$$

$$= \sum_{n=0}^{\infty} \bar{F}(a^n x) F_n(t) P(N(t) = n), \tag{2.3.7}$$

where (2.3.6) is due to

$$P(X_{n+1} > x \mid X_{n+1} > t - y) = \begin{cases} 1 & y < t - x, \\ \dfrac{P(X_{n+1}>x)}{P(X_{n+1}>t-y)} & y \geq t - x, \end{cases}$$
$$\geq P(X_{n+1} > x).$$

Integrating (2.3.5) and (2.3.7) from 0 to ∞ with respect to x gives the following corollary.

Corollary 2.3.3.

$$\lambda \sum_{n=0}^{\infty} a^{-n} F_n(t) P(N(t) = n) \leq E[X_{N(t)+1}] \leq \lambda \sum_{n=0}^{\infty} a^{-n} F_n(t) \quad t > 0.$$

2.4 Limit Theorems for Geometric Process

It is well known that Wald's equation plays an important role in renewal process. The following theorem is a generalization of the Wald's equation to a GP, it is called as Wald's equation for GP.

Theorem 2.4.1. (Wald's Equation for GP)
Suppose that $\{X_n, n = 1, 2, \ldots\}$ forms a GP with ratio a, and $E[X_1] = \lambda < \infty$, then for $t > 0$, we have

$$E[S_{N(t)+1}] = \lambda E\left[\sum_{n=1}^{N(t)+1} a^{-n+1} \right]. \tag{2.4.1}$$

If $a \neq 1$, then

$$E[S_{N(t)+1}] = \frac{\lambda}{1 - a} \{ E[a^{-N(t)}] - a \}. \tag{2.4.2}$$

Proof.
Let I_A be the indicator function of event A. Then X_n and $I_{\{S_{n-1} \leq t\}} = I_{\{N(t)+1 \geq n\}}$ are independent. Consequently, for $t > 0$, we have

$$E[S_{N(t)+1}] = E[\sum_{n=1}^{N(t)+1} X_n] = \sum_{n=1}^{\infty} E[X_n I_{\{N(t)+1 \geq n\}}]$$

$$= \sum_{n=1}^{\infty} E[X_n] P(N(t) + 1 \geq n) = \lambda \sum_{n=1}^{\infty} a^{-n+1} P(N(t) + 1 \geq n)$$

$$= \lambda \sum_{j=1}^{\infty} (\sum_{n=1}^{j} a^{-n+1}) P(N(t) + 1 = j)$$

$$= \lambda E[\sum_{n=1}^{N(t)+1} a^{-n+1}]. \tag{2.4.3}$$

Then (2.4.1) holds. Moreover, if $a \neq 1$, then (2.4.2) follows from (2.4.3) directly. This completes the proof of Theorem 2.4.1.

Corollary 2.4.2.

$$E[a^{-N(t)}] \begin{cases} > a + \frac{(1-a)t}{\lambda} & 0 < a < 1, \\ = 1 & a = 1, \\ < a + \frac{(1-a)t}{\lambda} & a > 1,\ t \leq \frac{a\lambda}{a-1}. \end{cases}$$

Proof.

For $a = 1$, the result is trivial. Now, assume that $a \neq 1$, then for all $t > 0$, Theorem 2.4.1 yields

$$t < E[S_{N(t)+1}] = \frac{\lambda}{1-a}(E[a^{-N(t)}] - a).$$

Consequently, Corollary 2.4.2 follows.

Moreover, from (2.4.1), we have the following inequality.

Corollary 2.4.3

$$E[S_{N(t)+1}] \begin{cases} > \lambda\{E[N(t)] + 1\} & 0 < a < 1, \\ = \lambda\{E[N(t)] + 1\} & a = 1, \\ < \lambda\{E[N(t)] + 1\} & a > 1. \end{cases}$$

Note that if $a = 1$, the GP reduces to a renewal process, while Corollary 2.4.3 gives $E[S_{N(t)+1}] = \lambda\{E[N(t)] + 1\}$. This is Wald's equation for the renewal process.

The following theorem follows Theorem 2.4.1 directly.

Theorem 2.4.4. If a stochastic process $\{X_n, \ n = 1, 2, \ldots\}$ is a GP with ratio $a > 1$, then we have

$$(1) \quad \lim_{t \to \infty} \tfrac{1}{t} E[a^{-N(t)}] = 0;$$

$$(2) \quad \lim_{t \to \infty} \tfrac{1}{t} E[\sum_{n=1}^{N(t)} a^{-n+1}] = 0;$$

$$(3) \quad \lim_{t \to \infty} \tfrac{1}{t} E[S_{N(t)+1}] = 0. \qquad\qquad (2.4.4)$$

We can see (2.4.4) from different way. Because from (2.2.5), we have

$$P(N(\infty) \geq n) = P(S_n < \infty) = 1 \ \text{ for all } \ n.$$

Thus

$$P(N(\infty) = \infty) = 1.$$

Therefore, $\lim\limits_{t \to \infty} E[S_{N(t)+1}] = E[S]$, and (2.4.4) follows.

However, the limit properties for the case of $0 < a \leq 1$ is completely different. For example, if $a = 1$, the GP reduces to a renewal process. The elementary renewal theorem in renewal process yields that

$$\lim_{t \to \infty} \frac{E[S_{N(t)+1}]}{t} = \lim_{t \to \infty} \frac{\lambda\{E[N(t)] + 1\}}{t} = 1.$$

Then from Theorem 2.4.1 and (2.3.2), we have

Theorem 2.4.5. Assume that $\{X_n, \ n = 1, 2, \ldots\}$ is a GP with ratio $a = 1$, then we have

$$(1) \quad \lim_{t \to \infty} \tfrac{1}{t} E[a^{-N(t)}] = 0;$$

$$(2) \quad \lim_{t \to \infty} \tfrac{1}{t} E[\sum_{n=1}^{N(t)} a^{-n+1}] = \tfrac{1}{\lambda};$$

$$(3) \quad \lim_{t \to \infty} \tfrac{1}{t} E[S_{N(t)+1}] = 1,$$

$$(4) \quad \lim_{t \to \infty} \tfrac{1}{t} B(t) = 0.$$

Now, we shall introduce the following lemma for NBUE (NWUE) class of the life distribution.

Lemma 2.4.6. The NBUE (NWUE) class is close under convolution operation.

Proof.

To prove Lemma 2.4.6, assume that two nonnegative random variables

X and Y are independent having distributions F and G respectively. Assume further that $X \in$ NBUE and $Y \in$ NBUE. Let $E[X] = \mu$ and $E[Y] = \nu$. Denote the distribution of $Z = X + Y$ by H. Then

$$
\begin{aligned}
\bar{H}(x+y) &= P(Z > x + y) \\
&= P(X > x, Y > y, X + Y > x + y) \\
&\quad + P(X > x, Y \leq y, X + Y > x + y) + P(X \leq x, X + Y > x + y) \\
&= \bar{F}(x)\bar{G}(y) + \int_0^y \bar{F}(x + y - u)dG(u) + \int_0^x \bar{G}(x + y - u)dF(u).
\end{aligned}
\qquad (2.4.5)
$$

Integrating (2.4.5) from 0 to ∞ with respect to y gives

$$
\begin{aligned}
\int_x^\infty \bar{H}(u)du &= \int_0^\infty \bar{H}(x+y)dy \\
&= \bar{F}(x)\int_0^\infty \bar{G}(y)dy + \int_0^\infty \int_0^y \bar{F}(x+y-u)dG(u)dy \\
&\quad + \int_0^\infty \int_0^x \bar{G}(x+y-u)dF(u)dy \\
&= \nu\bar{F}(x) + \int_0^\infty \int_x^\infty \bar{F}(v)dvdG(u) + \int_0^x \int_{x-u}^\infty \bar{G}(v)dvdF(u) \\
&\leq \nu\bar{F}(x) + \mu\bar{F}(x) + \nu\int_0^x \bar{G}(x-u)dF(u) \qquad (2.4.6) \\
&= \mu\bar{F}(x) + \nu\bar{H}(x) \qquad (2.4.7) \\
&\leq (\mu + \nu)\bar{H}(x), \quad x \geq 0, \qquad (2.4.8)
\end{aligned}
$$

where (2.4.6) is because X and $Y \in$ NBUE, while (2.4.7) and (2.4.8) are due to the fact

$$
\bar{H}(x) = \bar{F}(x) + \int_0^x \bar{G}(x-u)dF(u) \geq \bar{F}(x).
$$

Therefore, $Z = X + Y \in$ NBUE. For the case of NWUE, the proof is similar. This completes the proof of Lemma 2.4.6.

Now, we are able to show that for a GP $\{X_n, n = 1, 2, \cdots\}$, if $X_1 \in$ NBUE (NWUE), then for any integer n, $X_n \in$ NBUE (NWUE) and $S_n \in$ NBUE

(NWUE). This is the following lemma.

Lemma 2.4.7. Given a GP $\{X_n, n = 1, 2, \cdots\}$ with ratio a, if $X_1 \in$NBUE (NWUE), then X_n and $S_n \in$NBUE (NWUE) for $n = 1, 2, \cdots$

Proof.

Let the distribution function of X_1 be $F(x)$ and $E[X_1] = \lambda$. Moreover, let the distribution function of X_n be $F_n(x)$ and $E[X_n] = \lambda_n = \lambda/a^{n-1}$. Then

$$\frac{1}{\lambda_n} \int_t^\infty \bar{F}_n(x)dx = \frac{a^{n-1}}{\lambda} \int_t^\infty \bar{F}(a^{n-1}x)dx$$

$$= \frac{1}{\lambda} \int_{a^{n-1}t}^\infty \bar{F}(x)dx \leq \bar{F}(a^{n-1}t) = \bar{F}_n(t),$$

since F is NBUE. Now, because $\{X_i, i = 1, 2, \ldots\}$ are all NBUE, then by Lemma 2.4.6, $S_n = \sum_{i=1}^n X_i$ is also NBUE. For the case of NWUE, the proof is similar.

Afterward, we are available to study the limit theorem for $0 < a \leq 1$.

Theorem 2.4.8. If $\{X_n, \ n = 1, 2, \cdots\}$ is a GP with ratio $0 < a \leq 1$ and $X_1 \in$ NBUE, then

$$(1) \qquad 1 \leq \varliminf_{t \to \infty} \frac{E[S_{N(t)+1}]}{t} \leq \varlimsup_{t \to \infty} \frac{E[S_{N(t)+1}]}{t} \leq \frac{1}{a}.$$

$$(2) \qquad 0 \leq \varliminf_{t \to \infty} \frac{E[B(t)]}{t} \leq \varlimsup_{t \to \infty} \frac{E[B(t)]}{t} \leq \frac{1-a}{a};$$

$$(3) \qquad \frac{1-a}{\lambda} \leq \varliminf_{t \to \infty} \frac{1}{t} E[a^{-N(t)}] \leq \varlimsup_{t \to \infty} \frac{1}{t} E[a^{-N(t)}] \leq \frac{1-a}{a\lambda};$$

$$(4) \ (0 \vee \tfrac{2a-1}{a\lambda}) \leq \varliminf_{t \to \infty} \frac{1}{t} E[\sum_{n=1}^{N(t)} a^{-n+1}] \leq \varlimsup_{t \to \infty} \frac{1}{t} E[\sum_{n=1}^{N(t)} a^{-n+1}] \leq \frac{1}{\lambda}.$$

Proof.

Consider $B(t)$, the residual life at time t for the GP, then

$$E[B(t)] = E[S_{N(t)+1} - t] = E\{E[S_{N(t)+1} - t|N(t)]\}$$
$$\leq E\{E[X_{N(t)+1}|N(t)]\} = \lambda E[a^{-N(t)}]. \tag{2.4.9}$$

Therefore

$$t \leq E[S_{N(t)+1}] \leq t + \lambda E[a^{-N(t)}]. \tag{2.4.10}$$

The combination of (2.4.2) and (2.4.10) gives

$$t \leq \frac{\lambda}{1-a}(E[a^{-N(t)}] - a) \leq t + \lambda E[a^{-N(t)}]. \tag{2.4.11}$$

Inequality (2.4.11) implies that

$$\frac{(1-a)t}{\lambda} + a \leq E[a^{-N(t)}] \leq \frac{(1-a)t}{a\lambda} + 1. \tag{2.4.12}$$

Then part (3) follows. Now based on (2.4.9) and (2.4.12), part (2) is trivial. Moreover, part (1) follows from (2.4.10) and (2.4.12).

On the other hand, because $S_{N(t)} \leq t$, then

$$E[\sum_{n=1}^{N(t)} a^{-n+1}] \leq \frac{t}{\lambda}. \tag{2.4.13}$$

Furthermore, Theorem 2.4.1 yields

$$t \leq E[S_{N(t)+1}] = \lambda E[\sum_{n=1}^{N(t)} a^{-n+1}] + \lambda E[a^{-N(t)}]. \tag{2.4.14}$$

Consequently, it follows from (2.4.12)-(2.4.14) that

$$\frac{t}{\lambda} \geq E[\sum_{n=1}^{N(t)} a^{-n+1}] \geq \frac{t}{\lambda} - E[a^{-N(t)}] \geq \frac{t}{\lambda} - \frac{(1-a)t}{a\lambda} - 1. \tag{2.4.15}$$

Hence part (4) follows. Note that if $0 < a < 1/2$, the left hand side of part (3) should be replaced by 0. This completes the proof of Theorem 2.4.8.

2.5   A Geometric Process with Exponential Distribution

Now, we study the properties of a GP in which X_1 has an exponential distribution $Exp(1/\lambda)$ with density function.

$$f(x) = \begin{cases} \frac{1}{\lambda}\exp(-\frac{x}{\lambda}) & x > 0, \\ 0 & x \leq 0. \end{cases} \tag{2.5.1}$$

Then, we have

Theorem 2.5.1.   Given a GP $\{X_n, n = 1, 2, \ldots\}$ with ratio a, assume that X_1 has an exponential distribution $Exp(1/\lambda)$. Then for $0 < a \leq 1$ or $a > 1, t < \frac{a\lambda}{a-1}$, we have

$$1. \quad E[a^{-N(t)}] = 1 + \frac{(1-a)t}{a\lambda}, \tag{2.5.2}$$

$$2. \quad E[S_{N(t)+n+1}] = \frac{a\lambda}{a-1} + \frac{1}{a^{n+1}}(t - \frac{a\lambda}{a-1}). \tag{2.5.3}$$

Proof.

Let $p_i(t) = P(N(t) = i)$. Then by a classical probability analysis, we have

$$p_0'(t) = -\frac{1}{\lambda}p_0(t), \tag{2.5.4}$$

$$p_i'(t) = -\frac{a^i}{\lambda}p_i(t) + \frac{a^{i-1}}{\lambda}p_{i-1}(t) \quad i = 1, 2, \ldots. \tag{2.5.5}$$

Equations (2.5.4) and (2.5.5) are in fact the Kolmogorov forward equations. As an example, we shall derive equation (2.5.5) here.

$$p_i(t + \Delta t) = P(\text{no event occurs in } (t, t + \Delta t]|N(t) = i)P(N(t) = i)$$

$$+P(\text{one event occurs in } (t, t + \Delta t]|N(t) = i - 1)P(N(t) = i - 1)$$

$$+P(\text{two or more event occur in } (t, t + \Delta t], N(t, t + \Delta t] = i)$$

$$= (1 - \frac{a^i}{\lambda}\Delta t)p_i(t) + \frac{a^{i-1}}{\lambda}\Delta t p_{i-1}(t) + o(\Delta t) \quad i = 1, 2, \ldots$$

Then (2.5.5) follows by letting $\Delta t \to 0$. Thus

$$E[a^{-N(t)}] = \sum_{i=0}^{\infty} E[a^{-N(t)} \mid N(t) = i]P(N(t) = i) = \sum_{i=0}^{\infty} a^{-i}p_i(t).$$

Consequently

$$\frac{dE[a^{-N(t)}]}{dt} = \sum_{i=0}^{\infty} a^{-i}p_i'(t)$$

$$= -\frac{1}{\lambda}p_0(t) + \sum_{i=1}^{\infty} a^{-i}\{-\frac{a^i}{\lambda}p_i(t) + \frac{a^{i-1}}{\lambda}p_{i-1}(t)\}$$

$$= -\frac{1}{\lambda}\sum_{i=0}^{\infty} p_i(t) + \frac{1}{a\lambda}\sum_{i=1}^{\infty} p_{i-1}(t)$$

$$= \frac{1 - a}{a\lambda}.$$

Then (2.5.2) follows. Now, by conditional on the numbers of events occurred by time t, it follows that

$$E[S_{N(t)+n+1}] = \sum_{k=0}^{\infty} E[S_{N(t)+n+1} \mid N(t) = k]P(N(t) = k)$$

$$= \sum_{k=0}^{\infty} E[\sum_{i=1}^{k+n+1} X_i]P(N(t) = k)$$

$$= \frac{a\lambda}{a-1}\left[1 - \sum_{k=0}^{\infty}\frac{1}{a^{k+n+1}}P(N(t) = k)\right]$$

$$= \frac{a\lambda}{a-1}\left[1 - \frac{1}{a^{n+1}}E\{a^{-N(t)}\}\right]$$

$$= \frac{a\lambda}{a-1} + \frac{1}{a^{n+1}}(t - \frac{a\lambda}{a-1}), \qquad a \leq 1 \ \text{ or } \ a > 1, \ t < \frac{a\lambda}{a-1}.$$

Therefore (2.5.3) follows. This completes the proof of Theorem 2.5.1. In particular, we have the following corollary.

Corollary 2.5.2.

$$1. \ E[S_{N(t)}] = t, \tag{2.5.6}$$

$$2. \ E[S_{N(t)+1}] = \lambda + \frac{t}{a}. \tag{2.5.7}$$

By recalling that the age $A(t)$, residual life $B(t)$ and total life $X_{N(t)+1}$ at t defined in Section 2.3, Corollary 2.5.3 is a direct conclusion from Corollary 2.5.2.

Corollary 2.5.3.

$$1. \ E[A(t)] = 0, \tag{2.5.8}$$

$$2. \ E[B(t)] = \lambda + (\frac{1}{a} - 1)t, \tag{2.5.9}$$

$$3. \ E[X_{N(t)+1}] = \lambda + (\frac{1}{a} - 1)t. \tag{2.5.10}$$

Now suppose $0 < a < 1$, we can use Theorem 2.4.8 since exponential distribution is NBUE, while for $a = 1$, Theorem 2.4.5 is applicable. However, the following theorem due to Lam et al. (2003) is a better result.

Theorem 2.5.4. Assume that $\{X_n, \ n = 1, 2, \cdots\}$ is a GP with ratio $0 < a \leq 1$, and X_1 has an exponential distribution $Exp(1/\lambda)$. Then

$$(1) \quad \lim_{t\to\infty} \tfrac{1}{t} E[S_{N(t)+1}] = \tfrac{1}{a};$$

$$(2) \quad \lim_{t\to\infty} \frac{E[B(t)]}{t} = \frac{1-a}{a};$$

$$(3) \quad \lim_{t\to\infty} \tfrac{1}{t} E[a^{-N(t)}] = \frac{1-a}{a\lambda};$$

$$(4) \quad \lim_{t\to\infty} \tfrac{1}{t} E\Big[\sum_{n=1}^{N(t)} a^{-n+1}\Big] = \tfrac{1}{\lambda}.$$

In comparison Theorem 2.4.8 with Theorem 2.5.4, we can see that for the exponential distribution case, the limits always exist and equal the upper bounds in the inequalities of Theorem 2.4.8 respectively.

2.6 Notes and References

In this chapter, we introduce the GP and study its probability properties. As GP is a generalization of renewal process, most results in this chapter generalize the corresponding results in renewal process. Lam (1988a, b) first introduced the definition of GP and discussed its simple properties that form the basis of Section 2.2. However, Theorem 2.2.3 is based on Lam et al. (2003), while Theorems 2.2.4-2.2.6 are direct applications of the Doob, Riesz and Krickeberg decompositions to the process $\{S_n, n = 1, 2, \ldots\}$. On the other hand, Sections 2.3 and 2.4 are based on Lam et al. (2003), in which part 2 of Theorem 2.3.1 is originally due to Zhang (1991). In Section 2.5, we study a particular GP with exponential distribution, in which Theorem 2.5.1 and Corollaries 2.5.2 and 2.5.3 are new, but Theorem 2.5.4 is from Lam et al. (2003).

Chapter 3

Geometric Function

3.1 Introduction

Given a GP $\{X_n, n = 1, 2, \ldots\}$ with ratio a, let

$$S_n = \sum_{i=1}^{n} X_i.$$

Then we can define a counting process

$$N(t) = \sup\{n \mid S_n \leq t\}, \quad t \geq 0. \tag{3.1.1}$$

Now let

$$M(t, a) = E[N(t)]. \tag{3.1.2}$$

Function $M(t, a)$ is called the geometric function of GP $\{X_n, n = 1, 2, \ldots\}$.

In practice, if X_n is the interarrival time between the $(n-1)$th event and the nth event, then $M(t, a)$ will be the expected number of events occurred by time t. If X_n is the operating time after the $(n-1)$th repair, then $M(t, a)$ will be the expected number of failures by time t. Obviously, if the ratio $a = 1$, the GP reduces to a renewal process, and the geometric function $M(t, 1)$ becomes the renewal function $M(t)$ of the renewal process. Therefore, geometric function $M(t, a)$ is a natural generalization of the renewal function. As the renewal function plays an important role in renewal process, the geometric function will also play an important role in GP.

In this chapter, we shall study the properties of the geometric function. In Section 3.2, an integral equation called geometric equation for $M(t, a)$ will be derived. The existence of $M(t, a)$ for a GP is studied in Section 3.3. In Section 3.4, the Laplace transform of $M(t, a)$ is determined. Then in Section 3.5, an analytic solution of the geometric equation is studied. A numerical solution to the geometric equation is introduced in Section 3.6.

55

An approximate solution to the geometric equation is derived in Section 3.7. Then, the solutions obtained by above three methods are compared with the solution by a simulation method in Section 3.8. Finally, as a particular case, the geometric function of a GP with exponential distribution is considered in Section 3.9.

3.2 Geometric Equation

Now, we shall derive an integral equation for the geometric function $M(t, a)$. First of all, it follows from (3.1.1) that

$$N(t) \geq n \iff S_n \leq t. \tag{3.2.1}$$

Let $F_n(x)$ be the distribution function of S_n with

$$F_0(t) = \begin{cases} 1 & t \geq 0, \\ 0 & \text{elsewhere.} \end{cases}$$

The following theorem is a generalization of Theorem 1.3.2. The proof is exactly the same.

Theorem 3.2.1.

$$M(t, a) = \sum_{n=1}^{\infty} F_n(t). \tag{3.2.2}$$

Furthermore, Let F be the distribution of X_1, then similar to (1.3.9), we have the following integral equation for $M(t, a)$.

$$M(t, a) = F(t) + \int_0^t M(a(t - u), a)dF(u), \tag{3.2.3}$$

To prove (3.2.3), we shall first derive a result by induction.

$$F_n(t) = \int_0^t F_{n-1}(a(t - u))dF(u). \tag{3.2.4}$$

In fact, for n $= 1$, (3.2.4) is trivial. Assume that (3.2.4) holds for n. For $n + 1$, because S_{n+1} is the sum of two independent random variables S_n and X_{n+1}, we have

$$S_{n+1} = S_n + X_{n+1}.$$

Then

$$F_{n+1}(t) = \int_0^t F_n(t-u)dF(a^n u)$$

$$= \int_0^t \left\{ \int_0^{t-u} F_{n-1}(a(t-u-v))dF(v) \right\} dF(a^n u) \qquad (3.2.5)$$

$$= \int_0^t \left\{ \int_0^{t-v} F_{n-1}(a(t-v-u))dF(a^n u) \right\} dF(v)$$

$$= \int_0^t \left\{ \int_0^{a(t-v)} F_{n-1}(a(t-v)-y)dF(a^{n-1}y) \right\} dF(v)$$

$$= \int_0^t F_n(a(t-v))dF(v), \qquad (3.2.6)$$

where (3.2.5) is due to the induction assumption, while (3.2.6) is because S_n is the sum of S_{n-1} and X_n. Therefore (3.2.4) holds for any integer n. As a result, (3.2.3) follows by substituting (3.2.4) into (3.2.2).

As equation (3.2.3) is an integral equation satisfied by the geometric function $M(t,a)$, (3.2.3) is called the geometric equation. If the density of X_1 is f, then (3.2.3) becomes

$$M(t,a) = F(t) + \int_0^t M(a(t-u),a)f(u)du. \qquad (3.2.7)$$

3.3 Existence of Geometric Function

To start with, we shall study the existence of the geometric function of a GP.

Theorem 3.3.1. If $\{X_n, n = 1, 2, \ldots\}$ is a GP with ratio $0 < a \leq 1$, then for all $t \geq 0, N(t)$ is finite with probability 1, and the geometric function $M(t,a)$ is also finite.

Proof.

Given a GP $\{X_n,\ n = 1, 2, \ldots\}$ with ratio a, define a renewal process $\{\tilde{X}_n = a^{n-1} X_n,\ n = 1, 2, \ldots\}$. Now let

$$\tilde{S}_n = \tilde{X}_1 + \tilde{X}_2 + \cdots + \tilde{X}_n,$$

and

$$\tilde{N}(t) = \sup\{n \mid \tilde{S}_n \leq t\}.$$

From Theorem 1.3.3, it is clear that for all $t \geq 0$, $\tilde{N}(t) < \infty$ with probability 1 and $E[\tilde{N}(t)] < \infty$. However,

$$\tilde{S}_n \leq S_n,$$

then $N(t) \leq \tilde{N}(t)$. This implies that

$$P(N(t) < \infty) \geq P(\tilde{N}(t) < \infty) = 1,$$

and

$$M(t, a) = E[N(t)] \leq E[\tilde{N}(t)] < \infty.$$

Then Theorem 3.3.1 follows.

Consequently, for a GP with ratio $0 < a \leq 1$, a finite geometric function always exists. However, for $a > 1$, Theorem 2.2.3 gives

$$S_n \xrightarrow{a.s.} S \quad \text{as} \quad n \to \infty$$

with

$$E(S) = \frac{a\lambda}{a - 1} < \infty.$$

Therefore, $P(\lim_{n \to \infty} S_n = \infty) = 0$. This implies that for $a > 1$,

$$P(N(\infty) = \infty) = 1, \tag{3.3.1}$$

where $N(\infty) = \lim_{t \to \infty} N(t)$. Note that from Lemma 1.3.5, (3.3.1) also holds for $a = 1$. Therefore, using the monotone convergence theorem yields that for $a > 1$,

$$\lim_{t \to \infty} M(t, a) = \infty.$$

Moreover, Lam (1988b) gave the following example showing that if $a > 1$, then $M(t, a)$ is not finite for all $t > 0$. Let $\{X_n, n = 1, 2, \ldots\}$ be a GP with $a > 1$. Assume that X_1 is a degenerate random variable with distribution given by

$$F(x) = \begin{cases} 1 & x \geq \theta, \\ 0 & \text{elsewhere.} \end{cases} \tag{3.3.2}$$

Let the distribution function of S_n be F_n. Then define

$$M_n(t, a) = \sum_{i=1}^{n} F_i(t).$$

By induction, it is easy to verify that

$$M_n(t, a) = \begin{cases} 0 & t < \theta, \\ k & \theta \sum_{i=1}^{k} \frac{1}{a^{i-1}} \le t < \theta \sum_{i=1}^{k+1} \frac{1}{a^{i-1}}, \quad k = 1, 2, \ldots, n-1, \\ n & t \ge \theta \sum_{i=1}^{n} \frac{1}{a^{i-1}}. \end{cases}$$

Letting $n \to \infty$, then we have

$$M(t, a) = \begin{cases} 0 & t < \theta, \\ k & \theta \sum_{i=1}^{k} \frac{1}{a^{i-1}} \le t < \theta \sum_{i=1}^{k+1} \frac{1}{a^{i-1}}, \quad k = 1, 2, \ldots \quad (3.3.3) \\ \infty & t \ge \frac{a\theta}{a-1}. \end{cases}$$

Thus $M(t, a) = \infty$ for $t \ge \frac{a\theta}{a-1}$. Actually, for $a > 1$, a more meticulous result for the geometric function $M(t, a)$ is given by the following theorem.

Theorem 3.3.2. Given a GP $\{X_n, n = 1, 2, \ldots\}$ with ratio $a > 1$, let the distribution of X_1 be F and assume that

$$\theta = \inf\{x \mid F(x) > F(0)\}, \tag{3.3.4}$$

then

$$M(t, a) = \infty \quad \text{for} \quad t > \frac{a\theta}{a-1}. \tag{3.3.5}$$

Proof.
According to Theorem 2.2.3, we have

$$\lim_{n \to \infty} S_n = S$$

and

$$s_0 = E[S] = a\lambda/(a-1). \tag{3.3.6}$$

Then, there exists $\delta > 0$ such that $P\{S \le E[S]\} > \delta$. If otherwise, $P\{S \le E[S]\} = 0$. Then

$$E[S] = E\{S \mid S \le E[S]\}P\{S \le E[s]\} + E\{S \mid S > E[S]\}P\{S > E[s]\}$$
$$> E[S].$$

This is impossible. Now, because $S_n, n = 0, 1, \ldots,$ is nondecreasing in n, it follows from (3.2.1) that for any integer n, we have

$$P\{N(a\lambda/(a-1)) \geq n\} = P\{S_n \leq a\lambda/(a-1)\}$$
$$\geq P\{S \leq E(S)\} > \delta.$$

This implies that for $t \geq s_0 = E(S) = a\lambda/(a-1)$,

$$M(t, a) = E\{N(t)\} \geq E\{N(a\lambda/(a-1))\}$$
$$\geq nP\{N(\lambda a/(a-1)) \geq n\} > n\delta.$$

Letting $n \to \infty$ yields that

$$M(t, a) = \infty \quad \text{for} \quad t \geq s_0 = a\lambda/(a-1). \tag{3.3.7}$$

Thus if $\theta \geq \lambda$, then (3.3.5) follows from (3.3.7) directly. Otherwise, if $\theta < \lambda$, let

$$s_1 = s_0/a + \theta.$$

Then (3.3.6) yields

$$s_1 = \frac{\lambda}{a-1} + \theta < \frac{\lambda}{a-1} + \lambda = E(S) = s_0.$$

Therefore, for $s_1 \leq t < s_0$, we can write

$$t = \frac{\lambda}{a-1} + s = \frac{s_0}{a} + s$$

such that $\theta \leq s < \lambda$ and $s < t$. Now (3.2.3) with the help of (3.3.7) gives

$$M(t, a) = F(t) + \int_0^t M(a(t-u), a)dF(u)$$
$$\geq F(t) + \int_0^s M(a(t-u), a)dF(u)$$
$$= F(t) + \int_0^s M(s_0 + as - au, a)dF(u)$$
$$= \infty \quad \text{for} \quad s_1 \leq t < s_0. \tag{3.3.8}$$

The combination of (3.3.7) and (3.3.8) shows that

$$M(t, a) = \infty \quad \text{for} \quad t \geq s_1. \tag{3.3.9}$$

Then by induction, it is straightforward to prove that $M(t,a) = \infty$ for $t \geq s_n$, where

$$s_n = \frac{s_{n-1}}{a} + \theta = \frac{s_0}{a^n} + \sum_{i=0}^{n-1} \frac{\theta}{a^i}$$

$$\rightarrow \frac{a\theta}{a-1} \quad \text{as} \quad n \rightarrow \infty.$$

This completes the proof of Theorem 3.3.2.

Applying Theorem 3.3.2 to Lam's example, we can see that $M(t,a) = \infty$, for $t > a\theta/(a-1)$, this result agrees with (3.3.3). On the other hand, if the distribution of X_1 is increasing at 0, then (3.3.4) implies that $\theta = 0$. Therefore, Theorem 3.3.2 concludes that $M(t,a) = \infty$ for $t > 0$. Then we have the following corollary.

Corollary 3.3.3. Given a GP $\{X_n, n = 1, 2, \ldots\}$ with ratio $a > 1$, if the distribution function of X_1 is increasing at 0, then

$$M(t,a) = \infty \quad \text{for} \quad t > 0.$$

In other words, a finite geometric function does not exist for $t > 0$. As a result, we can concentrate the study of geometric function on the case of $0 < a \leq 1$.

3.4 General Solution to Geometric Equation

To solve geometric equation (3.2.3) for the geometric function $M(t,a)$, we can start with $M_0(t,a) = F(t)$. Then by iteration, it follows that

$$M_n(t,a) = F(t) + \int_0^t M_{n-1}(a(t-u),a)dF(u). \qquad (3.4.1)$$

By induction, we can see that $M_n(t,a)$ is nondecreasing in n for all $t > 0$. Consequently, the limit function

$$M(t,a) = \lim_{n\to\infty} M_n(t,a)$$

exists. Then by using monotone convergence theorem, limit function $M(t,a)$ will be a solution to geometric equation (3.2.3). Moreover, if $F(x)$ is continuous, then the solution $M(t,a)$ is unique in any interval $[0, d]$ subject to $F(d) < 1$. To prove this result, assume that $0 < a \leq 1$. Then let

$C[0, d]$ be the Banach space of all continuous functions on $[0, d]$. Define a linear operator L on $C[0, d]$ such that for $G \in C[0, d], H = L(G)$ with

$$L: \quad H(t) = F(t) + \int_0^t G(a(t - u))dF(u), \quad \text{for } 0 \leq t \leq d. \quad (3.4.2)$$

It is easy to check that L is a contraction operator. In fact, for $G_1, G_2 \in C[0, d]$, let $H_1 = L(G_1)$ and $H_2 = L(G_2)$, then

$$\|H_1 - H_2\| \leq F(d)\|G_1 - G_2\|. \quad (3.4.3)$$

Then L is a contraction operator, since $F(d) < 1$. Therefore, by the fixed point theorem, there exists a unique fixed point in $C[0, d]$. As d is any positive number satisfying $F(d) < 1$, the solution to (3.2.3) is also unique.

On the other hand, given a GP $\{X_n, n = 1, 2, \ldots\}$ with ratio a, let the density function of X_1 be $f(x)$ and denote the Laplace transform of $f(x)$ by

$$f^*(s) = \int_0^\infty e^{-st} f(x)dx,$$

and the Laplace transform of $M(t, a)$ by

$$M^*(s, a) = \int_0^\infty e^{-st} M(t, a)dt.$$

Then taking the Laplace transform on the both sides of (3.2.3) gives

$$M^*(s, a) = \frac{f^*(s)}{s} + \frac{1}{a}M^*(\frac{s}{a}, a)f^*(s). \quad (3.4.4)$$

Now if $0 < a \leq 1$, Theorem 3.3.1 implies that the geometric function $M(t, a)$ is always finite, and (3.4.4) can be solved iteratively for $M^*(s, a)$. To do this, starting with

$$M_0^*(s, a) = \frac{f^*(s)}{s},$$

then by iteration, we have

$$M_n^*(s, a) = \frac{f^*(s)}{s} + \frac{1}{a}M_{n-1}^*(\frac{s}{a}, a)f^*(s).$$

By induction, it is easy to show that

$$M_n^*(s, a) = \frac{1}{s} \sum_{i=0}^n \{\prod_{j=0}^i f^*(\frac{s}{a^j})\}. \quad (3.4.5)$$

Let $n \to \infty$, by using ratio test, it is trivial that the series on the right hand side of (3.4.5) is convergent, since $0 < a \le 1$. Thus, the solution to (3.4.4) is given by

$$M^*(s,a) = \frac{1}{s} \sum_{i=0}^{\infty} \left\{ \prod_{j=0}^{i} f^*(\frac{s}{a^j}) \right\}. \tag{3.4.6}$$

Consequently, if $0 < a \le 1$, the geometric function $M(t,a)$ could be determined by inversion of $M^*(s,a)$.

However, if $a > 1$, series (3.4.6) is divergent. In fact, rewrite $M^*(s,a) = \sum_{i=0}^{\infty} a_i(s)$ with

$$a_i(s) = \frac{1}{s} \prod_{j=0}^{i} f^*(\frac{s}{a^j}).$$

Then the Raabe test gives

$$\lim_{i \to \infty} i \left\{ \frac{a_i(s)}{a_{i+1}(s)} - 1 \right\} = \lim_{i \to \infty} \frac{i\{1 - f^*(\frac{s}{a^{i+1}})\}}{f^*(\frac{s}{a^{i+1}})} = 0, \tag{3.4.7}$$

where (3.4.7) is due to (1.3.13). As a result, series (3.4.6) is divergent for $a > 1$, there is no solution to (3.4.4). This result agrees with the conclusion of Theorem 3.3.2.

In this section, two general methods for the solution of geometric equation are suggested. In practice, the implementation of (3.4.1) is a tedious job, the inversion of (3.4.6) is even an extravagant hope. Therefore, in Sections 3.5-3.7, we shall introduce some powerful methods for the solution of geometric equation.

3.5   * Analytic Solution to Geometric Equation

At first, consider a subset of $L^2[0,T]$, the space of all square integrable functions on $[0,T]$. Assume that $F(T) < 1$ and the density of X_1 is f. Then let

$$W[0,T] = \{u \mid u \text{ is continuous on } [0,T],\ u(0) = 0,\ u,\ u' \in L^2[0,T]\}.$$

The inner product is defined for $u, v \in W[0,T]$ is defined by

$$(u,v)_W = \int_0^T (u(x)v(x) + u'(x)v'(x))dx. \tag{3.5.1}$$

Then the norm of $u \in W[0,T]$ is given by

$$\|u\|_W = \sqrt{(u,u)_W}.$$

Clearly, $W[0,T]$ is a separable Hilbert space. Moreover, $W[0,T] \subset C[0,T]$, the space of all continuous functions on $[0,T]$. Note that in this section, we shall use $(\cdot,\cdot)_W$ and $(\cdot,\cdot)_L$ to denote the inner products in spaces $W[0,T]$ and $L^2[0,T]$, and use $\|\cdot\|_W$, $\|\cdot\|_L$ and $\|\cdot\|_C$ to denote the norms in spaces $W[0,T]$, $L^2[0,T]$ and $C[0,T]$, respectively. Then, let

$$R(x,y) = \frac{ch(x+y-T) + ch(|x-y|-T)}{2sh(T)}, \qquad (3.5.2)$$

where

$$ch(z) = \frac{e^z + e^{-z}}{2} \quad \text{and} \quad sh(z) = \frac{e^z - e^{-z}}{2}.$$

It is easy to check that for any function $u \in W[0,T]$, we have

$$(u(y), R(x,y))_W = u(x). \qquad (3.5.3)$$

Thus $R(x,y)$ is a reproducing kernel function (see e.g. Aronszajn (1950)). Because $R(x,y)$ is symmetric, (3.5.3) is equivalent to

$$(u(y), R(y,x))_W = u(x). \qquad (3.5.4)$$

Now, define an operator A on $W[0,T]$:

$$Ah(x): \quad h(x) \mapsto h(x) - \int_0^x h(a(x-y))f(y)dy. \qquad (3.5.5)$$

Thus, equation (3.2.7) is equivalent to

$$Ah = F. \qquad (3.5.6)$$

Now, let

$$G(x) = \int_0^x h(a(x-y))f(y)dy. \qquad (3.5.7)$$

Then, we have the following lemmas.

Lemma 3.5.1. Assume that $0 < a \leq 1$, $f \in L^2[0,T]$ and $h \in W[0,T]$. Then, we have

$$\|G\|_L^2 \leq \frac{T}{a}\|f\|_L^2\|h\|_L^2 \qquad (3.5.8)$$

and

$$\|G'\|_L^2 \leq aT\|f\|_L^2\|h'\|_L^2. \qquad (3.5.9)$$

Proof.

By using the Cauchy-Schwarz inequality, we have

$$\|G\|_L^2 = \int_0^T \{\int_0^x h(a(x-y))f(y)dy\}^2 dx$$

$$\leq \int_0^T \{\int_0^x h^2(a(x-y))dy \int_0^x f^2(y)dy\}dx$$

$$= \int_0^T \{\frac{1}{a}\int_0^{ax} h^2(z)dz \int_0^x f^2(y)dy\}dx$$

$$\leq \frac{T}{a}\|f\|_L^2\|h\|_L^2,$$

since $0 < a \leq 1$. Furthermore, because $h(0) = 0$, then

$$G'(x) = a \int_0^x h'(a(x-y))f(y)dy.$$

Thus, we have

$$\|G'\|_L^2 = \int_0^T [a \int_0^x h'(a(x-y))f(y)dy]^2 dx$$

$$\leq \int_0^T \{a^2 \int_0^x h'^2(a(x-y))dy \int_0^x f^2(y)dy\}dx$$

$$= \int_0^T \{a \int_0^{ax} h'^2(z)dz \int_0^x f^2(y)dy\}dx$$

$$\leq aT\|f\|_L^2\|h'\|_L^2.$$

This completes the proof of Lemma 3.5.1.

Lemma 3.5.2. Assume that $0 < a \leq 1$, $f \in L^2[0,T]$ and $h \in W[0,T]$. Then $Ah \in W[0,T]$.

Proof.

For any $h \in W[0,T]$, in order to prove $Ah \in W[0,T]$, we need to show that $Ah(x)$ is a continuous function on $[0,T]$, $Ah(0) = 0$, $Ah(x) \in L^2[0,T]$ and $(Ah(x))' \in L^2[0,T]$. The first two conditions are trivial. To show the third one, $Ah(x) \in L^2[0,T]$, we note that $h \in W[0,T]$ is continuous, then there exists $M > 0$ such that for all $x \in [0,T]$, $|h(x)| \leq M$. Consequently, we have

$$\|Ah\|_L^2 = (Ah, Ah)_L$$
$$= (h(x) - \int_0^x h(a(x-y))f(y)dy, h(x) - \int_0^x h(a(x-y))f(y)dy)_L$$
$$= \|h\|_L^2 - 2(h(x), \int_0^x h(a(x-y))f(y)dy)_L$$
$$+ \left\| \int_0^x h(a(x-y))f(y)dy \right\|_L^2. \tag{3.5.10}$$

It is easy to see that three terms in right hand side of (3.5.10) are all dominated by M^2T. As an example, from the definition of the inner product in $L^2[0,T]$, the second term becomes

$$\left| (h(x), \int_0^x h(a(x-y))f(y)dy)_L \right|$$
$$= \left| \int_0^T h(x) \int_0^x h(a(x-y))f(y)dydx \right|$$
$$\leq M^2 \int_0^T (\int_0^x f(y)dy)dx$$
$$\leq M^2T.$$

Thus, (3.5.10) yields

$$\|Ah\|_L^2 \leq \|h\|_L^2 + 2 \left| (h(x), \int_0^x h(a(x-y))f(y)dy)_L \right|$$
$$+ \left\| \int_0^x h(a(x-y))f(y)dy \right\|_L^2$$
$$\leq M^2T + 2M^2T + M^2T = 4M^2T.$$

Consequently, $Ah(x) \in L^2[0,T]$. Afterward, we shall show that $(Ah(x))' \in L^2[0,T]$.

$$\|(Ah(x))'\|_L^2 = (h'(x) - G'(x), h'(x) - G'(x))_L$$
$$= \|h'\|_L^2 - 2(h'(x), G'(x))_L + \|G'\|_L^2. \tag{3.5.11}$$

Once again, by using the Cauchy-Schwarz inequality and (3.5.9), we have

$$|(h'(x), G'(x))_L| = |\int_0^T h'(x) \cdot G'(x)dx|$$
$$\leq \|h'\|_L \|G'\|_L$$
$$\leq (aT)^{1/2}\|f\|_L\|h'\|_L^2.$$

Thus, (3.5.11) with the help of (3.5.9) yields

$$\|(Ah'(x))\|_L^2 \le \{1 + 2(aT)^{1/2}\|f\|_L + aT\|f\|_L^2\}\|h'\|_L^2. \qquad (3.5.12)$$

This implies that $(Ah(x))' \in L^2[0, T]$, and Lemma 3.5.2 follows.

Lemma 3.5.3. Assume that $0 < a \le 1$ and $f \in L^2[0, T]$. Then A is a bounded linear operator from $W[0, T]$ to $W[0, T]$.

Proof.

It follows from Lemma 3.5.2 that A is an operator from $W[0, T]$ to $W[0, T]$. Furthermore, A is obviously a linear operator. Now we shall show that A is a bounded operator. To this end, we note that

$$\begin{aligned}
\|Ah\|_W^2 &= (h(x) - G(x), h(x) - G(x))_W \\
&= \|h\|_W^2 - 2(h(x), G(x))_W + \|G\|_W^2 \\
&\le \|h\|_W^2 + 2\|h\|_W\|G\|_W + \|G\|_W^2. \qquad (3.5.13)
\end{aligned}$$

Because $G \in W[0, T]$, hence (3.5.1) and Lemma 3.5.1 imply that

$$\begin{aligned}
\|G\|_W^2 &= \|G\|_L^2 + \|G'\|_L^2 \\
&\le \frac{T}{a}\|f\|_L^2\|h\|_L^2 + aT\|f\|_L^2\|h'\|_L^2 \\
&\le \frac{T}{a}\|f\|_L^2\|h\|_W^2.
\end{aligned}$$

Therefore, (3.5.13) gives

$$\begin{aligned}
\|Ah\|_W^2 &\le \|h\|_W^2 + 2\{\frac{T}{a}\}^{1/2}\|f\|_L\|h\|_W^2 + \frac{T}{a}\|f\|_L^2\|h\|_W^2 \\
&= \{1 + (\frac{T}{a})^{1/2}\|f\|_L\}^2\|h\|_W^2.
\end{aligned}$$

Consequently

$$\|Ah\|_W \le \{1 + (\frac{T}{a})^{1/2}\|f\|_L\}\|h\|_W.$$

Thus, A is a bounded linear operator from $W[0, T]$ to $W[0, T]$ with norm

$$\|A\|_W \le 1 + (\frac{T}{a})^{1/2}\|f\|_L. \qquad (3.5.14)$$

Note that as $W[0, T]$ is a Hilbert space, the conjugate space of $W[0, T]$, i.e. the space of all bounded linear operator on $W[0, T]$, is itself in sense of isomorphism. Therefore the same W-norm could be used here.

Now, we can derive an analytic solution to equation (3.5.6). To do this, noting that $W[0, T]$ is a separable Hilbert space and $F \ne 0$, there exists a

68   *Geometric Process and Its Applications*

complete orthonormal basis $\{\alpha_n\}_{n=0}^{\infty}$ in $W[0,T]$ with $\alpha_0 = F/\|F\|_W$.

Then let A^* be the conjugate operator of A. Denote

$$V_n = \text{span}\{A^*\alpha_1, A^*\alpha_2, \ldots, A^*\alpha_n\},$$

and

$$V = \overline{\bigcup_{n=1}^{\infty} V_n},$$

where $\overline{\bigcup_{n=1}^{\infty} V_n}$ is the closure of $\bigcup_{n=1}^{\infty} V_n$. Moreover, let P be the projection operator from $A^*W[0,T]$ onto V. Denote $h_0 = A^*\alpha_0$, then we have the following result.

Theorem 3.5.4. The solution to equation (3.2.7) can be expressed by

$$h = \frac{\|F\|_W (h_0 - Ph_0)}{(h_0, h_0 - Ph_0)_W} \in \overline{A^*W[0,T]}, \tag{3.5.15}$$

where $\overline{A^*W[0,T]}$ is the closure of $A^*W[0,T]$.

Proof.

At first, we shall prove that equation (3.5.6) is equivalent to the following equations

$$\begin{cases} (h, A^*\alpha_0)_W = \|F\|_W, \\ (h, A^*\alpha_n)_W = 0 \quad n = 1, 2, \ldots. \end{cases} \tag{3.5.16}$$

To show that $(3.5.6) \Rightarrow (3.5.16)$, since $Ah = F$ and $\{\alpha_n\}_{n=0}^{\infty}$ is a complete orthonormal basis of $W[0,T]$, we have

$$(h, A^*\alpha_0)_W = (Ah, \alpha_0)_W = (F, \alpha_0)_W$$
$$= (\|F\|_W \cdot \alpha_0, \alpha_0)_W = \|F\|_W(\alpha_0, \alpha_0)_W = \|F\|_W.$$

Moreover,

$$(h, A^*\alpha_n)_W = (Ah, \alpha_n)_W = (F, \alpha_n)_W$$
$$= (\|F\|_W \cdot \alpha_0, \alpha_n)_W = \|F\|_W(\alpha_0, \alpha_n)_W = 0.$$

To show that $(3.5.16) \Rightarrow (3.5.6)$, by the Bessel equality and (3.5.16), we have

$$Ah = \sum_{n=0}^{\infty} (Ah, \alpha_n)_W \alpha_n = \sum_{n=0}^{\infty} (h, A^*\alpha_n)_W \alpha_n$$
$$= (h, A^*\alpha_0)_W \alpha_0 = \|F\|_W \cdot \alpha_0 = F.$$

Second, we shall prove that if equation (3.5.6) has a solution h, then $h_0 = A^*\alpha_0 \notin V$. Otherwise, if $h_0 \in V$, then there exists $\{v_n,\ n = 1, 2, \ldots\}$ such that $v_n \in V_n$ and

$$v_n = \sum_{i=1}^{n} a_{ni} A^* \alpha_i \to h_0 \quad \text{as} \quad n \to \infty,$$

where $a_{ni},\ i = 1, \ldots, n$ are some constants. Thus (3.5.16) gives

$$(h, v_n)_W = (h, \sum_{i=1}^{n} a_{ni} A^* \alpha_i)_W = \sum_{i=1}^{n} a_{ni}(h, A^* \alpha_i)_W = 0.$$

Therefore

$$(h, h_0)_W = \lim_{n \to \infty} (h, v_n)_W = 0.$$

That is, $(h, A^*\alpha_0)_W = 0$, which contradicts the first equation in (3.5.16) since h is the solution of (3.5.6) and $\|F\|_W \neq 0$. Consequently, $h_0 \notin V$ and hence $h_0 - Ph_0 \neq 0$.

Now let

$$h = \frac{\|F\|_W (h_0 - Ph_0)}{(h_0, h_0 - Ph_0)_W}.$$

It is easy to see that h satisfies (3.5.16). In fact,

$$(h, A^*\alpha_0)_W = (h, h_0)_W = (\frac{\|F\|_W (h_0 - Ph_0)}{(h_0, h_0 - Ph_0)_W}, h_0)_W$$

$$= \frac{\|F\|_W}{(h_0, h_0 - Ph_0)_W}(h_0 - Ph_0, h_0)_W = \|F\|_W,$$

$$(h, A^*\alpha_n)_W = (\frac{\|F\|_W (h_0 - Ph_0)}{(h_0, h_0 - Ph_0)_W}, A^*\alpha_n)_W$$

$$= \frac{\|F\|_W}{(h_0, h_0 - Ph_0)_W}(h_0 - Ph_0, A^*\alpha_n)_W = 0,$$

since P is the projection operator from $A^*W[0, T]$ onto V, and $h_0 - Ph_0 \perp V$. Therefore, (3.5.15) is an analytic solution to equation (3.5.6) or (3.2.7). This completes the proof of Theorem 3.5.4.

In application of Theorem 3.5.4, a complete orthonormal basis in $W[0, T]$ is needed. To do so, let $D = \{T_i, i = 1, 2, \ldots\}$ be a dense subset in $[0, T]$. Let $\{\varphi_i(x) = R(x, T_i), i = 1, 2, \ldots\}$ be a family of univariate functions obtained from the reproducing kernel function $R(x, y)$ by substituting $y = T_i$. Then $\{\varphi_i(x), i = 1, 2, \ldots\}$ are linear independent in $W[0, T]$.

To prove this, assume that there are some constants $\{c_i, i = 1, \ldots, n\}$ such that

$$\sum_{i=1}^{n} c_i \varphi_i(x) = 0.$$

Thus for any $u \in W[0, T]$, we have $(u, \sum_{i=1}^{n} c_i \varphi_i)_W = 0$. In particular, consider a sequence of functions $u_j \in W[0, T], j = 1, 2, \ldots$ such that $u_j(T_i) = \delta_{ij}$, where δ_{ij} is the Kronecker delta defined by

$$\delta_{ij} = \begin{cases} 1 & \text{if} \quad i = j, \\ 0 & \text{if} \quad i \neq j. \end{cases}$$

It follows from (3.5.4) that

$$u_j(T_i) = (u_j(x), R(x, T_i))_W = (u_j(x), \varphi_i(x))_W.$$

Then for any $j \leq n$, we have

$$c_j = \sum_{i=1}^{n} c_i u_j(T_i) = \sum_{i=1}^{n} c_i (u_j(x), \varphi_i(x))_W = (u_j(x), \sum_{i=1}^{n} c_i \varphi_i(x))_W = 0,$$

and so $\{\varphi_i(x), i = 1, 2, \ldots\}$ are linear independent. Because D is dense in $[0, T]$, then $\{\varphi_i(x), i = 1, 2, \ldots\}$ form a basis of $W[0, T]$.

Furthermore, define

$$\psi_i(x) = A^* \varphi_i(x). \tag{3.5.17}$$

Then $\{\psi_i(x), i = 1, 2, \cdots\}$ are linear independent in space $A^* W[0, T]$. To show this, assume that there are some constants $\{d_i, i = 1, \ldots, n\}$ such that

$$\sum_{i=1}^{n} d_i \psi_i(x) = 0.$$

Then for any $u \in W[0, T]$, we have

$$0 = (u(x), \sum_{i=1}^{n} d_i \psi_i(x))_W = (u(x), \sum_{i=1}^{n} d_i A^* \varphi_i(x))_W$$

$$= \sum_{i=1}^{n} d_i (u(x), A^* \varphi_i(x))_W$$

$$= \sum_{i=1}^{n} d_i (Au(x), \varphi_i(x))_W$$

$$= \sum_{i=1}^{n} d_i Au(T_i). \tag{3.5.18}$$

Because $\{T_i, i = 1, \cdots, n\}$ is a subset of D. We can arrange them in order so that $0 < T_1 < \cdots < T_n$. Then for each $j = 1, \ldots, n$, define a function $u_j, i = 1, \ldots, n$, on $[0, T]$ such that

$$u_j(x) = \begin{cases} 1 + \frac{x - T_j}{a(T_j - T_{j-1})} & T_j - a(T_j - T_{j-1}) < x \leq T_j, \\ 1 + \frac{T_j - x}{a(T_j - T_{j-1})} & T_j < x \leq T_j + a(T_j - T_{j-1}), \\ 0 & \text{elsewhere.} \end{cases} \quad (3.5.19)$$

By substituting $u = u_n$ into (3.5.18), it is easy to see that $Au_n(T_i) = 0$ for $i < n$. In fact,

$$Au_n(T_i) = u_n(T_i) - \int_0^{T_i} u_n(a(T_i - y))f(y)dy$$

$$= u_n(T_i) - \frac{1}{a}\int_0^{aT_i} u_n(z)f(T_i - \frac{z}{a})dz = 0,$$

since $0 < a \leq 1$ and $aT_i \leq aT_{n-1} \leq T_n - a(T_n - T_{n-1})$. On the other hand,

$$Au_n(T_n) = u_n(T_n) - \int_0^{T_n} u_n(a(T_n - y))f(y)dy$$

$$= 1 - \frac{1}{a}\int_0^{aT_n} u_n(z)f(T_n - \frac{z}{a})dz.$$

Now we shall prove that $Au_n(T_n) \neq 0$. In fact, if $aT_n \leq T_n - a(T_n - T_{n-1})$, then from (3.5.19) we have

$$Au_n(T_n) = 1 - \frac{1}{a}\int_0^{aT_n} u_n(z)f(T_n - \frac{z}{a})dy = 1.$$

If $aT_n > T_n - a(T_n - T_{n-1})$, again from (3.5.19)

$$Au_n(T_n) = 1 - \frac{1}{a}\int_0^{aT_n} u_n(z)f(T_n - \frac{z}{a})dy$$

$$= 1 - \frac{1}{a}\int_{T_n - a(T_n - T_{n-1})}^{aT_n} u_n(z)f(T_n - \frac{z}{a})dy$$

$$\geq 1 - \int_0^{\frac{2a-1}{a}T_n - T_{n-1}} f(x)dx \geq 1 - F(\frac{(2a-1)T}{a}) \geq 1 - F(T) > 0,$$

since

$$0 < \frac{2a - 1}{a}T_n - T_{n-1} < \frac{2a - 1}{a}T \leq T,$$

due to $F(T) < 1$. Thus $Au_n(T_n) \neq 0$. Then applying (3.5.18) to u_n gives

$$0 = \sum_{i=1}^{n} d_i Au_n(T_i) = d_n Au_n(T_n).$$

This implies that $d_n = 0$. Then we can substitute $u = u_{n-1}$ into (3.5.18) to obtain $d_{n-1} = 0$. In general, it follows by induction that $d_i = 0$ for $i = 1, \ldots, n$. Thus $\{\psi_i(x), i = 1, 2, \cdots\}$ are linear independent.

Thereafter, by the Schmidt method, a complete orthonormal basis $\{\tilde{\psi}_i(x), i = 1, 2, \ldots\}$ of $A^*W[0,T]$ can be constructed on the basis of $\{\psi_i(x), i = 1, 2, \ldots\}$, such that

$$\tilde{\psi}_i(x) = \sum_{k=1}^{i} \beta_{ki} A^* \varphi_k(x) \quad i = 1, 2, \ldots. \tag{3.5.20}$$

Now, let

$$Y_n = \text{span}\{\tilde{\psi}_1, \tilde{\psi}_2, \ldots, \tilde{\psi}_n\},$$

then

$$\overline{A^*W[0,T]} = \overline{\bigcup_{n=1}^{\infty} Y_n}.$$

Now, let P_n be the projection operator from $\overline{A^*W[0,T]}$ to Y_n.

Theorem 3.5.5. If $\{T_i, i = 1, 2, \ldots\}$ is a dense subset in $[0,T]$, then a series representation of the solution to equation (3.2.7) is given by

$$h(x) = \sum_{i=1}^{\infty} \left[\sum_{k=1}^{i} \beta_{ki} F(T_k) \right] \tilde{\psi}_i(x). \tag{3.5.21}$$

Proof.
Let

$$P_n h = \sum_{i=1}^{n} (h, \tilde{\psi}_i)_W \tilde{\psi}_i,$$

then

$$\|P_n h - h\|_W \to 0 \quad \text{as} \quad n \to \infty. \tag{3.5.22}$$

To prove (3.5.22), note that $h \in \overline{A^*W[0,T]} = \overline{\bigcup_{n=1}^{\infty} Y_n}$, there exists $g_m \in Y_m$ such that

$$\|g_m - h\|_W \to 0 \quad \text{as} \quad m \to \infty.$$

In virtue of the fact $\|P_n\|_W \le 1$, we have for $n \ge m$

$$
\begin{aligned}
\|P_n h - h\|_W &= \|P_n h - g_m + g_m - h\|_W \\
&\le \|P_n h - g_m\|_W + \|g_m - h\|_W \\
&= \|P_n(h - g_m)\|_W + \|g_m - h\|_W \\
&\le \|P_n\|_W \|h - g_m\|_W + \|g_m - h\|_W \\
&= (\|P_n\|_W + 1)\|g_m - h\|_W \\
&\le 2\|g_m - h\|_W \to 0, \quad \text{as} \ \ n \to \infty.
\end{aligned}
$$

Thus (3.5.22) follows. Then by the embedding theorem (see, e.g., Gilbarg and Trudinger (1977)), $\|h\|_C \le C\|h\|_W$ for some constant $C > 0$. Therefore, we have $P_n h \to h$ uniformly. Furthermore, by using the Bessel equality and the reproducing property of $R(x, y)$, it follows that

$$
\begin{aligned}
h(x) &= \sum_{i=1}^{\infty}(h, \tilde{\psi}_i)_W \tilde{\psi}_i(x) = \sum_{i=1}^{\infty}(h, \sum_{k=1}^{i} \beta_{ki} A^* \varphi_k)_W \tilde{\psi}_i(x) \\
&= \sum_{i=1}^{\infty}\sum_{k=1}^{i} \beta_{ki}(Ah, \varphi_k)_W \tilde{\psi}_i(x) = \sum_{i=1}^{\infty}\sum_{k=1}^{i} \beta_{ki}(F(x), R(T_k, x))_W \tilde{\psi}_i(x) \\
&= \sum_{i=1}^{\infty}\sum_{k=1}^{i} \beta_{ki} F(T_k) \tilde{\psi}_i(x).
\end{aligned}
$$

This completes the proof of Theorem 3.5.5.

Now, consider a truncated series of (3.5.22)

$$
h_n(x) = (P_n h)(x) = \sum_{i=1}^{n}[\sum_{k=1}^{i} \beta_{ki} F(T_k)]\tilde{\psi}_i(x), \quad n = 1, 2, \ldots \tag{3.5.23}
$$

Clearly, $\{h_n(x), n = 1, 2, \ldots\}$ is a sequence of approximate solutions to equation (3.2.7). The error of the approximate solution $h_n(x)$ is given by

$$
E_n(x) = h(x) - h_n(x).
$$

Then, we have the following result.

Theorem 3.5.6.

$$
\|E_{n+1}\|_W \le \|E_n\|_W, \quad n = 1, 2, \ldots \tag{3.5.24}
$$

Proof.

By the definition of $E_n(x)$, we have

$$E_{n+1}(x) = h(x) - h_{n+1}(x) = h(x) - h_n(x) - \sum_{k=1}^{n+1} \beta_{k,n+1} F(T_k) \tilde{\psi}_{n+1}(x)$$

$$= E_n(x) - \sum_{k=1}^{n+1} \beta_{k,n+1} F(T_k) \tilde{\psi}_{n+1}(x).$$

Consequently,

$$\|E_{n+1}\|_W^2 = (E_{n+1}(x), E_{n+1}(x))_W$$

$$= (E_n(x) - \sum_{j=1}^{n+1} \beta_{j,n+1} F(T_j) \tilde{\psi}_{n+1}(x), E_n(x) - \sum_{k=1}^{n+1} \beta_{k,n+1} F(T_k) \tilde{\psi}_{n+1}(x))_W$$

$$= \|E_n\|_W^2 - 2 \sum_{k=1}^{n+1} \beta_{k,n+1} F(T_k)(E_n(x), \tilde{\psi}_{n+1}(x))_W$$

$$+ \{\sum_{j=1}^{n+1} \beta_{j,n+1} F(T_j)\}\{\sum_{k=1}^{n+1} \beta_{k,n+1} F(T_k)\}(\tilde{\psi}_{n+1}(x), \tilde{\psi}_{n+1}(x))_W. \qquad (3.5.25)$$

Because of the orthogonality of sequence $\{\tilde{\psi}_i(x)\}$, we have $(h_n(x), \tilde{\psi}_{n+1}(x))_W = 0$. Then from (3.5.4), we have

$$(E_n(x), \tilde{\psi}_{n+1}(x))_W = (h(x) - h_n(x), \tilde{\psi}_{n+1}(x))_W$$

$$= (h(x), \tilde{\psi}_{n+1}(x))_W = (h(x), \sum_{j=1}^{n+1} \beta_{j,n+1} A^* \varphi_j(x))_W$$

$$= \sum_{j=1}^{n+1} \beta_{j,n+1} (h(x), A^* \varphi_j(x))_W = \sum_{j=1}^{n+1} \beta_{j,n+1} (Ah(x), \varphi_j(x))_W$$

$$= \sum_{j=1}^{n+1} \beta_{j,n+1} (F(x), R(x, T_j))_W = \sum_{j=1}^{n+1} \beta_{j,n+1} F(T_j).$$

Moreover, the orthonormality of sequence $\{\tilde{\psi}_i(x)\}$ gives

$$(\tilde{\psi}_{n+1}(x), \tilde{\psi}_{n+1}(x))_W = 1.$$

As a result, (3.5.25) becomes

$$\|E_{n+1}\|_W^2$$

$$= \|E_n\|_W^2 - 2\{\sum_{j=1}^{n+1} \beta_{j,n+1} F(T_j)\}\{\sum_{k=1}^{n+1} \beta_{k,n+1} F(T_k)\} + \{\sum_{k=1}^{n+1} \beta_{k,n+1} F(T_k)\}^2$$

$$= \|E_n\|_W^2 - \{\sum_{k=1}^{n+1} \beta_{k,n+1} F(T_k)\}^2 \leq \|E_n\|_W^2.$$

Thus Theorem 3.5.6 follows.

The following theorem gives an upper bound for error $E_n(x)$.

Theorem 3.5.7.

$$|E_n(x)| \leq \|h\|_W [R(x,x) - \sum_{i=1}^{n} \tilde{\psi}_i^2(x)]^{1/2}. \tag{3.5.26}$$

Proof.

By using reproducing property (3.5.3) or (3.5.4), we have

$$|E_n(x)|^2 = |h(x) - h_n(x)|^2 = |h(x) - \sum_{i=1}^{n} \left[\sum_{k=1}^{i} \beta_{ki} F(T_k) \right] \tilde{\psi}_i(x)|^2$$

$$= |(h(y), R(x,y))_W - \sum_{i=1}^{n} \left[\sum_{k=1}^{i} \beta_{ki}(F(y), R(y,T_k))_W \right] \tilde{\psi}_i(x)|^2$$

$$= |(h(y), R(x,y))_W - \sum_{i=1}^{n} \sum_{k=1}^{i} \beta_{ki}(Ah(y), \varphi_k(y))_W \tilde{\psi}_i(x)|^2$$

$$= |(h(y), R(x,y))_W - \sum_{i=1}^{n} \sum_{k=1}^{i} \beta_{ki}(h(y), A^*\varphi_k(y))_W \tilde{\psi}_i(x)|^2$$

$$= |(h(y), R(x,y))_W - (h(y), \sum_{i=1}^{n} \sum_{k=1}^{i} \beta_{ki} A^*\varphi_k(y)\tilde{\psi}_i(x))_W|^2$$

$$= |(h(y), R(x,y))_W - (h(y), \sum_{i=1}^{n} \tilde{\psi}_i(y)\tilde{\psi}_i(x))_W|^2 \tag{3.5.27}$$

$$= |(h(y), R(x,y) - \sum_{i=1}^{n} \tilde{\psi}_i(y)\tilde{\psi}_i(x))_W|^2$$

$$\leq \|h\|_W^2 \|R(x,y) - \sum_{i=1}^{n} \tilde{\psi}_i(y)\tilde{\psi}_i(x)\|_W^2$$

$$= \|h\|_W^2 \|R(x,y) - \sum_{i=1}^{n} (\tilde{\psi}_i(t), R(x,t))_W \tilde{\psi}_i(y)\|_W^2$$

$$= \|h\|_W^2 (R(x,y) - \sum_{i=1}^{n} (\tilde{\psi}_i(t), R(x,t))_W \tilde{\psi}_i(y),$$

$$R(x,y) - \sum_{j=1}^{n} (\tilde{\psi}_j(t), R(x,t))_W \tilde{\psi}_j(y))_W$$

$$\leq \|h\|_W^2 \left\{ \|R(x,y)\|_W^2 - 2\sum_{i=1}^{n}(\tilde{\psi}_i(t), R(x,t))_W (R(x,y), \tilde{\psi}_i(y))_W \right.$$

$$\left. + \sum_{i=1}^{n}\sum_{j=1}^{n}(\tilde{\psi}_i(t), R(x,t))_W (\tilde{\psi}_j(t), R(x,t))_W (\tilde{\psi}_i(y), \tilde{\psi}_j(y))_W \right\}$$

$$= \|h\|_W^2 \left\{ \|R(x,y)\|_W^2 - \sum_{i=1}^{n}(\tilde{\psi}_i(t), R(x,t))_W^2 \right\}, \tag{3.5.28}$$

where (3.5.27) is due to (3.5.20) and (3.5.28) is from the reproducing property. Now, because

$$\|R(x,y)\|_W^2 = (R(x,y), R(x,y))_W = R(x,x).$$

we have

$$|E_n(x)|^2 \leq \|h\|_W^2 \left\{ R(x,x) - \sum_{i=1}^{n} \tilde{\psi}_i^2(x) \right\}. \tag{3.5.29}$$

Hence Theorem 3.5.7 follows.

In practice, the determination of normalization coefficients $\{\beta_{ki}, k \leq i\}$ is not easy. A computational method is suggested as follows: first express the $\tilde{\psi}$'s in terms of ψ's; then rewrite the approximate solution as

$$h_n(x) = \sum_{i=1}^{n} c_i \psi_i(x). \tag{3.5.30}$$

To determine the coefficients $\{c_i, i = 1, \ldots, n\}$, we can apply operator A to both sides of (3.5.30) and let $x = T_j, j = 1, \ldots, n$. It follows that

$$(Ah_n)(T_j) = \sum_{i=1}^{n} c_i(A\psi_i)(T_j) \quad j = 1, \ldots, n. \tag{3.5.31}$$

Since the projection operator P_n is a self-conjugate operator, $P_n = P_n^*$, and $A^*\varphi_j(x) = \psi_j(x) \in Y_n$, $P_n\psi_j(x) = \psi_j(x), j = 1, 2, \ldots, n$. By using the property of conjugate operator, the left hand side of (3.5.31) becomes

$$
\begin{aligned}
(Ah_n)(T_j) &= (AP_n h)(T_j) = ((AP_n h)(y), R(y, T_j))_W = ((AP_n h)(y), \varphi_j(y))_W \\
&= ((P_n h)(y), A^* \varphi_j(y))_W = ((P_n h)(y), \psi_j(y))_W \\
&= (h(y), P_n \psi_j(y))_W = (h(y), \psi_j(y))_W = (h(y), A^* \varphi_j(y))_W \\
&= (Ah(y), \varphi_j(y))_W = (F(y), R(T_j, y))_W = F(T_j).
\end{aligned}
$$

The right hand side of (3.5.31) gives

$$
\sum_{i=1}^{n} c_i (A\psi_i(y), \varphi_j(y)) = \sum_{i=1}^{n} c_i (\psi_i(y), A^* \varphi_j(y))_W = \sum_{i=1}^{n} c_i (\psi_i(y), \psi_j(y))_W.
$$

Consequently, equation (3.5.31) becomes

$$
\sum_{i=1}^{n} (\psi_i, \psi_j)_W c_i = F(T_j), \quad j = 1, \cdots, n. \tag{3.5.32}
$$

As the coefficient matrix of linear equations (3.5.32) is symmetric and positive definite, (3.5.32) will have a unique solution c_i, $i = 1, \ldots, n$. Therefore, (3.5.30) could be applied for finding an approximate solution to (3.2.7).

In order to find the inner product of ψ_i and ψ_j, we need to consider the representation of operator A and its conjugate operator A^*. To derive a representation of operator A, by the reproducing property and the linearity of the inner product, for any continuous function $g(s)$, the following equation holds

$$
g(s) u(s) = (u(t), g(s) R(s, t))_W.
$$

Furthermore, for any $u \in W[0, T]$, we have

$$
\begin{aligned}
Au(s) &= A(u(t), R(s, t))_W \\
&= (u(t), R(s, t))_W - \int_0^s (u(t), R(a(s-y), t))_W f(y) dy \\
&= (u(t), R(s, t))_W - (u(t), \int_0^s R(a(s-y), t) f(y) dy)_W \\
&= (u(t), R(s, t) - \int_0^s R(a(s-y), t) f(y) dy)_W. \tag{3.5.33}
\end{aligned}
$$

Therefore, for any $u, v \in W[0, T]$, (3.5.1) gives

$$(u, A^*v)_W = (Au, v)_W = \int_0^T [(Au(x))v(x) + (Au(x))'v'(x)]dx$$

$$= \int_0^T \{(u(t), R(x, t) - \int_0^x R(a(x - y), t)f(y)dy)_W \cdot v(x)$$

$$+ (u(t), R(x, t) - \int_0^x R(a(x - y), t)f(y)dy)'_W \cdot v'(x)\}dx$$

$$= \int_0^T \{(u(t), v(x)[R(x, t) - \int_0^x R(a(x - y), t)f(y)dy])_W$$

$$+ (u(t), v'(x)\frac{\partial[R(x, t) - \int_0^x R(a(x - y), t)f(y)dy]}{\partial x})_W\}dx$$

$$= \int_0^T \{(u(t), v(x)[R(x, t) - \int_0^x R(a(x - y), t)f(y)dy]$$

$$+ v'(x)\frac{\partial[R(x, t) - \int_0^x R(a(x - y), t)f(y)dy]}{\partial x})_W\}dx\,.$$

Consequently, we have

$$(u, A^*v)_W = (u(t), \int_0^T \{v(x)[R(x, t) - \int_0^x R(a(x - y), t)f(y)dy]$$

$$+ v'(x)\frac{\partial[R(x, t) - \int_0^x R(a(x - y), t)f(y)dy]}{\partial x}\}dx)_W$$

$$= (u(t), (v(x), R(x, t) - \int_0^x R(a(x - y), t)f(y)dy)_W)_W\,.$$

Then, we have

$$A^*v(t) = (v(x), R(x, t) - \int_0^x R(a(x - y), t)f(y)dy)_W\,. \qquad (3.5.34)$$

Thus from (3.5.17) and using the reproducing property, a representation of $\psi_i(t)$ is given by

$$\psi_i(t) = (\varphi_i(x), R(x, t) - \int_0^x R(a(x - y), t)f(y)dy)_W$$

$$= (R(x, T_i), R(x, t) - \int_0^x R(a(x - y), t)f(y)dy)_W$$

$$= R(T_i, t) - \int_0^{T_i} R(a(T_i - y), t)f(y)dy \qquad (3.5.35)$$

$$= \varphi_i(t) - \int_0^{T_i} R(a(T_i - y), t)f(y)dy\,. \qquad (3.5.36)$$

Moreover, we have

$$(\psi_i, \psi_j)_W = (\varphi_i(t) - \int_0^{T_i} R(a(T_i - y), t) f(y) dy, \psi_j(t))_W$$

$$= (\varphi_i(t), \psi_j(t))_W - (\int_0^{T_i} R(a(T_i - y), t) f(y) dy, \psi_j(t))_W$$

$$= (R(t, T_i), \psi_j(t))_W - \int_0^{T_i} (R(a(T_i - y), t), \psi_j(t))_W f(y) dy,$$

$$= \psi_j(T_i) - \int_0^{T_i} \psi_j(a(T_i - y)) f(y) dy, \tag{3.5.37}$$

or by the symmetry of the inner product

$$(\psi_i, \psi_j)_W = \psi_i(T_j) - \int_0^{T_j} \psi_i(a(T_j - y)) f(y) dy. \tag{3.5.38}$$

With the help of (3.5.32) and (3.5.37) or (3.5.38), we can easily find an approximate solution $h_n(x)$ by (3.5.30).

3.6 Numerical Solution to Geometric Equation

In this section, for $0 < a \leq 1$, a numerical solution to equation (3.2.7) for the geometric function is studied by using a trapezoidal integration rule. First of all, we introduce the following lemma that is useful for estimation of the error in the numerical method.

Lemma 3.6.1. Given a nonnegative sequence $\{y_n, n = 0, \ldots, N\}$. Assume that

$$(1) \; y_0 = 0,$$

$$(2) \; y_n \leq A + Bh \sum_{j=0}^{n-1} y_j, \quad 1 \leq n \leq N,$$

where $h = 1/N$, A and B are two positive constants independent of h. Then

$$\max_{0 \leq i \leq N} y_i \leq A e^B. \tag{3.6.1}$$

Proof.

First of all, we prove an inequality

$$y_n \leq A(1 + Bh)^{n-1}. \tag{3.6.2}$$

In fact, by condition (2), (3.6.2) is trivial for $n = 1$. Now, assume that (3.6.2) holds for $j = 1, \ldots, n-1$. Then

$$y_n \leq A + Bh \sum_{j=1}^{n-1} y_j$$

$$\leq A + Bh \sum_{j=1}^{n-1} A(1 + Bh)^{j-1}$$

$$= A(1 + Bh)^{n-1}.$$

By induction, (3.6.2) holds for any integer n. Because $h = 1/N$ we have

$$y_n \leq A(1 + \frac{B}{N})^{n-1}$$

$$\leq A(1 + \frac{B}{N})^N \leq Ae^B.$$

Now rewrite $M(t, a)$ as $\Lambda(t)$ for convenience. Then, by taking a transformation

$$s = a(t - y),$$

equation (3.2.7) will become

$$\Lambda(t) = F(t) + \frac{1}{a} \int_0^{at} \Lambda(s)f(t - \frac{s}{a})ds. \tag{3.6.3}$$

Without loss of generality, suppose that $t \in [0, T]$ and $f(0) = 0$ for simplicity. Then, partition interval $[0, T]$ by points $T_i = ih$, $i = 0, 1, \ldots, N$ with step width $h = T/N$. Afterward, let

$$\Lambda(T_i) = F(T_i) + \frac{1}{a} \int_0^{aT_i} \Lambda(s)f(T_i - \frac{s}{a})ds$$

$$= F(T_i) + \frac{1}{a} \int_0^{T_{[ai]}} \Lambda(s)f(T_i - \frac{s}{a})ds + \frac{1}{a} \int_{T_{[ai]}}^{aT_i} \Lambda(s)f(T_i - \frac{s}{a})ds$$

$$= F(T_i) + I_1 + I_2, \tag{3.6.4}$$

where $[x]$ denotes the integer part of a real number x with

$$I_1 = \frac{1}{a} \int_0^{T_{[ai]}} \Lambda(s)f(T_i - \frac{s}{a})ds \tag{3.6.5}$$

and

$$I_2 = \frac{1}{a} \int_{T_{[ai]}}^{aT_i} \Lambda(s)f(T_i - \frac{s}{a})ds. \tag{3.6.6}$$

Now write

$$g(s) = \frac{1}{a}\Lambda(s)f(T_i - \frac{s}{a}).$$

Then a trapezoidal integration rule with partition points $\{T_i,\ i = 0, 1, \ldots, [ai]\}$ on interval $(T_0, T_{[ai]}) = (0, T_{[ai]})$ can be applied to I_1. Thus

$$T_1(g) = \frac{h}{2}g(T_0) + h\sum_{k=1}^{[ai]-1} g(T_k) + \frac{h}{2}g(T_{[ai]}).$$

Because $\Lambda(T_0) = \Lambda(0) = 0$, it follows that

$$
\begin{aligned}
I_1 &= \int_0^{T_{[ai]}} g(s)ds = T_1(g) + E_1(g) \\
&= \frac{h}{2a}\Lambda(T_0)f(T_i - \frac{T_0}{a}) + \frac{h}{a}\sum_{k=1}^{[ai]-1}\Lambda(T_k)f(T_i - \frac{T_k}{a}) \\
&\quad + \frac{h}{2a}\Lambda(T_{[ai]})f(T_i - \frac{T_{[ai]}}{a}) + E_1(g) \\
&= \frac{h}{a}\sum_{k=1}^{[ai]-1}\Lambda(T_k)f(T_i - \frac{T_k}{a}) + \frac{h}{2a}\Lambda(T_{[ai]})f(T_i - \frac{T_{[ai]}}{a}) \\
&\quad + E_1(g),
\end{aligned}
\tag{3.6.7}
$$

where $E_1(g) = I_1 - T_1(g)$ is the error of $T_1(g)$. It is well known that if $g \in C^2[0, T]$, the space of all functions with continuous second derivative on $[0, T]$, then the error of using a trapezoidal integration rule is of order 2, i.e.

$$E_1(g) = O(h^2). \tag{3.6.8}$$

Similarly, a trapezoidal integration rule with 2 partition points $T_{[ai]}, aT_i$ on interval $[T_{[ai]}, aT_i]$ is applied to I_2. Then

$$
\begin{aligned}
T_2(g) &= \frac{aT_i - T_{[ai]}}{2a}\left\{\Lambda(T_{[ai]})f(T_i - \frac{T_{[ai]}}{a}) + \Lambda(aT_i)f(T_i - \frac{aT_i}{a})\right\} \\
&= \frac{aT_i - T_{[ai]}}{2a}\Lambda(T_{[ai]})f(T_i - \frac{T_{[ai]}}{a}),
\end{aligned}
$$

since $f(0) = 0$. Therefore I_2 becomes

$$
\begin{aligned}
I_2 &= \int_{T_{[ai]}}^{aT_i} g(s)ds = T_2(g) + E_2(g) \\
&= \frac{aT_i - T_{[ai]}}{2a}\Lambda(T_{[ai]})f(T_i - \frac{T_{[ai]}}{a}) + E_2(g),
\end{aligned}
\tag{3.6.9}
$$

where $E_2(g) = I_2 - T_2(g)$ is the error of $T_2(g)$. Similar to (3.6.8), we have

$$E_2(g) = O((aT_i - T_{[ai]})^2) = O((aih - [ai]h)^2) = O(h^2), \quad (3.6.10)$$

since $\mid ai - [ai] \mid \leq 1$.

By using (3.6.7) and (3.6.9), (3.6.4) becomes

$$\Lambda(0) = 0, \qquad (3.6.11)$$

and

$$\Lambda(T_i) = F(T_i) + \frac{h}{a} \sum_{k=1}^{[ai]-1} \Lambda(T_k) f(T_i - \frac{T_k}{a})$$
$$+ \frac{h}{2a} \Lambda(T_{[ai]}) f(T_i - \frac{T_{[ai]}}{a}) + \frac{aT_i - T_{[ai]}}{2a} \Lambda(T_{[ai]}) f(T_i - \frac{T_{[ai]}}{a})$$
$$+ (E_1(g) + E_2(g)) \quad i = 1, \dots, N. \qquad (3.6.12)$$

To obtain a numerical solution, in view of the fact $\Lambda(0) = 0$, we can start with

$$\Lambda_0 = 0. \qquad (3.6.13)$$

In general, an approximate solution Λ_i of $\Lambda(T_i)$ could be obtained from (3.6.12) by neglecting the sum of errors $(E_1(g)+E_2(g))$. In other words, an approximate solution Λ_i can be determined recursively from the following equation

$$\Lambda_i = F(T_i) + \frac{h}{a} \sum_{k=1}^{[ai]-1} \Lambda_k f(T_i - \frac{t_k}{a}) + \frac{h}{2a} \Lambda_{[ai]} f(T_i - \frac{T_{[ai]}}{a})$$
$$+ \frac{aT_i - T_{[ai]}}{2a} \Lambda_{[ai]} f(T_i - \frac{T_{[ai]}}{a}) \quad i = 1, \dots, N. \qquad (3.6.14)$$

Denote the error of Λ_i by $e_i = \Lambda(T_i) - \Lambda_i$. Then

$$e_0 = 0. \qquad (3.6.15)$$

In general, by subtracting (3.6.14) from (3.6.12), we have

$$e_i = \frac{h}{a} \sum_{k=1}^{[ai]-1} e_k f(T_i - \frac{T_k}{a}) + \left\{ \frac{h}{2a} + \frac{aT_i - T_{[ai]}}{2a} \right\} e_{[ai]} f(T_i - \frac{T_{[ai]}}{a})$$
$$+ (E_1(g) + E_2(g)) \quad i = 1, \dots, N. \qquad (3.6.16)$$

Then, write

$$A = \max_{1 \leq i \leq N} |(E_1(g) + E_2(g))|,$$

and

$$
B_{ik} = \begin{cases}
\frac{1}{a} f(T_i - \frac{T_k}{a}) & \text{for } 1 \le k \le [ai] - 1, \\
(\frac{1}{2a} + \frac{aT_i - T_{[ai]}}{2ah}) f(T_i - \frac{T_{[ai]}}{a}) & \text{for } k = [ai], \\
0 & \text{for } [ai] < k \le i.
\end{cases}
$$

Moreover, let

$$
B = \max_{1 \le k \le i \le N} \{B_{ik}\} < \infty.
$$

Then an upper bound for the error e_i can be derived from (3.6.16). In fact, we have

$$
\begin{aligned}
|e_i| \\
\le h \sum_{k=1}^{[ai]-1} |e_k| \frac{1}{a} f(T_i - \frac{T_k}{a}) + h \left(\frac{1}{2a} + \frac{aT_i - T_{[ai]}}{2ah} \right) |e_{[ai]}| f(T_i - \frac{T_{[ai]}}{a}) + A \\
\le A + Bh \sum_{k=1}^{i-1} |e_k|, \quad i = 1, \ldots, N,
\end{aligned}
$$

$$(3.6.17)$$

with

$$
|e_0| = 0.
$$

Now Lemma 3.6.1 implies that

$$
\max_{1 \le i \le N} |e_i| \le A e^B.
\tag{3.6.18}
$$

Note that by (3.6.8) and (3.6.10), $A = O(h^2)$. As a result, (3.6.18) yields that

$$
\max_{1 \le i \le N} |e_i| \le ch^2
\tag{3.6.19}
$$

for some constant c which is independent of h. In conclusion, the error of Λ_i is of order h^2.

3.7 Approximate Solution to Geometric Equation

Given a GP $\{X_n, n = 1, 2, \ldots\}$ with ratio a, let the density function of X_1 be $f(x)$ with $E(X_1) = \lambda$, $\text{Var}(X_1) = \sigma^2$ and $\mu_k = E(X_1^k)$, $k = 1, 2, \ldots$. Another possible approach to determination of geometric function $M(t, a)$ is to find its Laplace transform $M^*(s, a)$ and then obtain $M(t, a)$ by inversion. If $a = 1$, the GP reduces to a renewal process. Then the geometric function

$M(t, 1)$ becomes the renewal function of the renewal process. It follows from (1.3.11) that

$$M^*(s, 1) = \frac{f^*(s)}{s(1 - f^*(s))}, \qquad (3.7.1)$$

where $M^*(s, 1)$ is the Laplace transform of $M(t, 1)$, and $f^*(s)$ is the Laplace transform of $f(x)$. In general, the inversion of (3.7.1) is not easy except for some special cases. Of course, if $a \neq 1$, the problem will be more troublesome, since one can see from (3.4.4) or (3.4.6) that there even exists no simple expression of $M^*(s, a)$. Alternatively, one may try to find an approximate expression for $M(t, a)$. In practice, an approximate formula of $M(t, a)$ might be good enough for application. For example, if $a = 1$, according to Theorem 1.3.4, the renewal function $M(t, 1)$ is given by

$$M(t, 1) = \frac{t}{\lambda} + \frac{\sigma^2 - \lambda^2}{2\lambda^2} + o(1). \qquad (3.7.2)$$

To derive an approximate expression for $M(t, a)$, we can expand $M^*(s, a)$ as a Taylor series with respect to a at $a = 1$ in the following way.

$$M^*(s, a) = M^*(s, 1) + \frac{\partial M^*(s, a)}{\partial a}\Big|_{a=1}(a - 1)$$
$$+ \frac{1}{2}\frac{\partial^2 M^*(s, a)}{\partial a^2}\Big|_{a=1}(a - 1)^2 + o\{(a - 1)^2\}. \qquad (3.7.3)$$

To do this, first of all, (3.7.1) yields that

$$\frac{\partial M^*(s, 1)}{\partial s} = \frac{sf^{*\prime}(s) - f^*(s)(1 - f^*(s))}{s^2(1 - f^*(s))^2}, \qquad (3.7.4)$$

and

$$\frac{\partial^2 M^*(s, 1)}{\partial s^2} = \frac{1}{s^3[1 - f^*(s)]^3}$$
$$\times \left\{ 2f^*(s)[1 - f^*(s)]^2 + [s^2 f^{*\prime\prime}(s) - 2sf^{*\prime}(s)][1 - f^*(s)] + 2s^2[f^{*\prime}(s)]^2 \right\}.$$
$$(3.7.5)$$

By substituting $u = s/a$ and differentiating both sides of (3.4.4) with respect to a, it follows that

$$\frac{\partial M^*(s, a)}{\partial a}$$
$$= f^*(s) \left\{ -\frac{1}{a^2} M^*(u, a) - \frac{s}{a^3}\frac{\partial M^*(u, a)}{\partial u} + \frac{1}{a}\frac{\partial M^*(u, a)}{\partial a} \right\}. \qquad (3.7.6)$$

By letting $a = 1$, (3.7.6) with the help of (3.7.1) and (3.7.4) gives

$$\frac{\partial M^*(s,a)}{\partial a}\Big|_{a=1} = -\frac{f^*(s)}{1 - f^*(s)}\left\{M^*(s,1) + s\frac{\partial M^*(s,1)}{\partial s}\right\}$$
$$= -\frac{f^*(s)f^{*\prime}(s)}{[1 - f^*(s)]^3}. \tag{3.7.7}$$

Thus from (3.7.7), we have

$$\frac{\partial^2 M^*(s,a)}{\partial a \partial s}\Big|_{a=1} = -\frac{1}{[1 - f^*(s)]^4}$$
$$\times \left\{f^*(s)f^{*\prime\prime}(s)[1 - f^*(s)] + [f^{*\prime}(s)]^2 + 2f^*(s)[f^{*\prime}(s)]^2\right\}. \tag{3.7.8}$$

Again by differentiating the both sides of (3.7.6) with respect to a, we have

$$\frac{\partial^2 M^*(s,a)}{\partial a^2}$$
$$= f^*(s)\left\{\frac{2}{a^3}M^*(u,a) + \frac{4s}{a^4}\frac{\partial M^*(u,a)}{\partial u} - \frac{2}{a^2}\frac{\partial M^*(u,a)}{\partial a}\right.$$
$$\left. + \frac{s^2}{a^5}\frac{\partial^2 M^*(u,a)}{\partial u^2} - \frac{2s}{a^3}\frac{\partial^2 M^*(u,a)}{\partial a \partial u} + \frac{1}{a}\frac{\partial^2 M^*(u,a)}{\partial a^2}\right\}. \tag{3.7.9}$$

Now, substituting $a = 1$ into (3.7.9) yields that

$$\frac{\partial^2 M^*(s,a)}{\partial a^2}\Big|_{a=1}$$
$$= \frac{f^*(s)}{1 - f^*(s)}\left\{2M^*(s,1) - 2\frac{\partial M^*(s,a)}{\partial a}\Big|_{a=1} + 4s\frac{\partial M^*(s,1)}{\partial s}\right.$$
$$\left. -2s\frac{\partial^2 M^*(s,a)}{\partial a \partial s}\Big|_{a=1} + s^2\frac{\partial^2 M^*(s,1)}{\partial s^2}\right\}$$
$$= \frac{f^*(s)}{[1 - f^*(s)]^5}\left\{[sf^{*\prime\prime}(s) + 2f^{*\prime}(s)][1 - f^*(s)]^2\right.$$
$$+ 2\{s[f^*(s)f^{*\prime\prime}(s) + (f^{*\prime}(s))^2] + f^*(s)f^{*\prime}(s)\}[1 - f^*(s)]$$
$$\left. + 2s(f^{*\prime}(s))^2(1 + 2f^*(s))\right\}. \tag{3.7.10}$$

On the other hand, we have the following approximate expansions

$$(1)\quad f^*(s) = 1 - \lambda s + \frac{1}{2}(\lambda^2 + \sigma^2)s^2 - \frac{1}{6}\mu_3 s^3 + \frac{1}{24}\mu_4 s^4 + o(s^4), \tag{3.7.11}$$

$$(2)\quad f^{*\prime}(s) = -\lambda + (\lambda^2 + \sigma^2)s - \frac{1}{2}\mu_3 s^2 + \frac{1}{6}\mu_4 s^3 + o(s^3), \tag{3.7.12}$$

$$(3)\quad f^{*\prime\prime}(s) = (\lambda^2 + \sigma^2) - \mu_3 s + \frac{1}{2}\mu_4 s^2 + o(s^2). \tag{3.7.13}$$

By substitution of (3.7.11)-(3.7.13) into (3.7.1), (3.7.7) and (3.7.10) respectively, we have

$$M^*(s,1) = \frac{1}{\lambda s^2} + \frac{\sigma^2 - \lambda^2}{2\lambda^2 s} + O(1), \qquad (3.7.14)$$

$$\frac{\partial M^*(s,a)}{\partial a}\Big|_{a=1} = \frac{1}{\lambda^2 s^3} + \frac{\sigma^2 - \lambda^2}{2\lambda^3 s^2} + O(1), \qquad (3.7.15)$$

and

$$\begin{aligned}
\frac{\partial^2 M^*(s,a)}{\partial a^2}\Big|_{a=1} &= \frac{4}{\lambda^3 s^4} + \frac{3(\sigma^2 - \lambda^2)}{\lambda^4 s^3} + \frac{1}{6\lambda^5 s^2}[9(\sigma^2 + \lambda^2)^2 - 12\lambda^2\sigma^2 - 4\lambda\mu_3] \\
&+ \frac{1}{12\lambda^6 s}[(9\lambda^2 + 15\sigma^2)(\lambda^2 + \sigma^2)^2 - 4\lambda\mu_3(3\lambda^2 + 4\sigma^2) + 3\lambda^2\mu_4] \\
&+ O(1). \qquad (3.7.16)
\end{aligned}$$

Therefore, it follows from (3.7.3) that

$$\begin{aligned}
M^*&(s,a) \\
&= \frac{1}{\lambda s^2} + \frac{\sigma^2 - \lambda^2}{2\lambda^2 s} + \Big\{\frac{1}{\lambda^2 s^3} + \frac{\sigma^2 - \lambda^2}{2\lambda^3 s^2}\Big\}(a-1) \\
&+ \Big\{\frac{2}{\lambda^3 s^4} + \frac{3(\sigma^2 - \lambda^2)}{2\lambda^4 s^3} + \frac{1}{12\lambda^5 s^2}[9(\sigma^2 + \lambda^2)^2 - 12\lambda^2\sigma^2 - 4\lambda\mu_3] \\
&+ \frac{1}{24\lambda^6 s}[(9\lambda^2 + 15\sigma^2)(\lambda^2 + \sigma^2)^2 - 4\lambda\mu_3(3\lambda^2 + 4\sigma^2) + 3\lambda^2\mu_4]\Big\}(a-1)^2 \\
&+ O(1). \qquad (3.7.17)
\end{aligned}$$

Consequently, by taking an inversion of (3.7.17), we can obtain an approximate expression for $M(t,a)$ with the help of the Tauberian theorem below

$$\lim_{t\to+\infty} M(t,a) = \lim_{s\to 0} sM^*(s,a).$$

Theorem 3.7.1. For $0 < a \le 1$, we have

$$\begin{aligned}
M&(t,a) \\
&= \frac{t}{\lambda} + \frac{\sigma^2 - \lambda^2}{2\lambda^2} + \Big\{\frac{t^2}{2\lambda^2} + \frac{(\sigma^2 - \lambda^2)t}{2\lambda^3}\Big\}(a-1) \\
&+ \Big\{\frac{t^3}{3\lambda^3} + \frac{3(\sigma^2 - \lambda^2)t^2}{4\lambda^4} + \frac{t}{12\lambda^5}[9(\lambda^2 + \sigma^2)^2 - 12\lambda^2\sigma^2 - 4\lambda\mu_3] \\
&+ \frac{1}{24\lambda^6}[(9\lambda^2 + 15\sigma^2)(\lambda^2 + \sigma^2)^2 - 4\lambda\mu_3(3\lambda^2 + 4\sigma^2) + 3\lambda^2\mu_4]\Big\}(a-1)^2 \\
&+ o(1). \qquad (3.7.18)
\end{aligned}$$

Clearly, if $a = 1$, (3.7.18) reduces to (3.7.2). In other words, Theorem 3.7.1 is a generalization of Theorem 1.3.4.

Now we consider some special distributions that are very popular in life testing and reliability.

(1) Exponential distribution

Suppose that X_1 has an exponential distribution $Exp(1/\lambda)$ with density function

$$f(x) = \begin{cases} \frac{1}{\lambda}\exp(-\frac{x}{\lambda}) & x > 0, \\ 0 & x \le 0. \end{cases}$$

Then

$$E[X_1] = \lambda, \quad \text{Var}[X_1] = \lambda^2$$

and

$$\mu_k = E[X_1^k] = \lambda^k \Gamma(k+1), k = 1, 2, \ldots$$

Consequently, from Theorem 3.7.1 we have

Corollary 3.7.2. If $0 < a \le 1$ and X_1 has an exponential distribution $Exp(1/\lambda)$, then

$$M(t, a) = \frac{t}{\lambda} + \frac{t^2}{2\lambda^2}(a - 1) + \frac{t^3}{3\lambda^3}(a - 1)^2 + o(1). \qquad (3.7.19)$$

(2) Gamma distribution

Suppose that X_1 has a gamma distribution $\Gamma(\alpha, \beta)$ with density function

$$f(x) = \begin{cases} \frac{\beta^\alpha}{\Gamma(\alpha)} x^{\alpha-1}\exp(-\beta x) & x > 0, \\ 0 & x \le 0. \end{cases}$$

Then

$$E[X_1] = \frac{\alpha}{\beta}, \quad \text{Var}[X_1] = \frac{\alpha}{\beta^2}$$

and

$$\mu_k = E[X_1^k] = \frac{\Gamma(k + \alpha)}{\beta^k \Gamma(\alpha)}, k = 1, 2, \ldots$$

Consequently, from Theorem 3.7.1 we have the following result.

Corollary 3.7.3. If $0 < a \le 1$ and X_1 has a gamma distribution $\Gamma(\alpha, \beta)$, then

$$M(t,a) = \frac{\beta t}{\alpha} + \frac{1-\alpha}{2\alpha} + \frac{\beta}{2\alpha^2}[\beta t^2 + (1-\alpha)t](a-1)$$

$$+ \frac{1}{24\alpha^3}\left\{8\beta^3 t^3 + 18(1-\alpha)\beta^2 t^2 + 2(1-\alpha)(1-5\alpha)\beta t\right.$$

$$\left. + (1-\alpha^2)\right\}(a-1)^2 + o(1). \tag{3.7.20}$$

(3) Weibull distribution

Suppose that X_1 has a Weibull distribution $W(\alpha, \beta)$ with density function

$$f(x) = \begin{cases} \alpha\beta x^{\alpha-1}\exp(-\beta x^\alpha) & x > 0, \\ 0 & x \le 0. \end{cases}$$

Then

$$E[X_1] = \frac{\Gamma(1+\frac{1}{\alpha})}{\beta^{1/\alpha}}, \quad \mathrm{Var}[X_1] = \frac{\Gamma(1+\frac{2}{\alpha}) - [\Gamma(1+\frac{1}{\alpha})]^2}{\beta^{2/\alpha}}$$

and

$$\mu_k = E[X_1^k] = \frac{\Gamma(1+\frac{k}{\alpha})}{\beta^{k/\alpha}}, k = 1, 2, \ldots$$

Consequently, from Theorem 3.7.1 we have the following result.

Corollary 3.7.4. If $0 < a \le 1$ and X_1 has a Weibull distribution $W(\alpha, \beta)$, then

$$M(t,a) = \frac{\beta^{1/\alpha}t}{\Gamma(1+\frac{1}{\alpha})} + \frac{\Gamma(1+\frac{2}{\alpha})}{2[\Gamma(1+\frac{1}{\alpha})]^2} - 1$$

$$+ \frac{1}{2[\Gamma(1+\frac{1}{\alpha})]^2}\left\{\beta^{2/\alpha}t^2 + \frac{\{\Gamma(1+\frac{2}{\alpha}) - 2[\Gamma(1+\frac{1}{\alpha})]^2\}\beta^{1/\alpha}t}{\Gamma(1+\frac{1}{\alpha})}\right\}(a-1)$$

$$+ \frac{1}{24[\Gamma(1+\frac{1}{\alpha})]^3}\left\{8\beta^{3/\alpha}t^3 + \frac{18\{\Gamma(1+\frac{2}{\alpha}) - 2[\Gamma(1+\frac{1}{\alpha})]^2\}\beta^{2/\alpha}t^2}{\Gamma(1+\frac{1}{\alpha})}\right.$$

$$+ \frac{2\beta^{1/\alpha}t}{[\Gamma(1+\frac{1}{\alpha})]^2}\{9[\Gamma(1+\frac{2}{\alpha})]^2 - 12[\Gamma(1+\frac{1}{\alpha})]^2[\Gamma(1+\frac{2}{\alpha}) - (\Gamma(1+\frac{1}{\alpha}))^2]$$

$$- 4\Gamma(1+\frac{1}{\alpha})\Gamma(1+\frac{3}{\alpha})\} + \frac{[\Gamma(1+\frac{2}{\alpha})]^2}{[\Gamma(1+\frac{1}{\alpha})]^3}\{15\Gamma(1+\frac{2}{\alpha}) - 6(\Gamma(1+\frac{1}{\alpha}))^2]$$

$$- 4\Gamma(1+\frac{1}{\alpha})\Gamma(1+\frac{3}{\alpha})[4\Gamma(1+\frac{2}{\alpha}) - (\Gamma(1+\frac{1}{\alpha}))^2]$$

$$\left. + 3(\Gamma(1+\frac{1}{\alpha}))^2\Gamma(1+\frac{4}{\alpha})\}\right\}(a-1)^2 + o(1). \tag{3.7.21}$$

(4) Lognormal distribution

Suppose that X_1 has a lognormal distribution $LN(\mu, \tau^2)$ with density function

$$f(x) = \begin{cases} \frac{1}{\sqrt{2\pi}\tau x}\exp[-\frac{1}{2\tau^2}(\ell n x - \mu)^2] & x > 0, \\ 0 & \text{elsewhere.} \end{cases}$$

Then

$$\lambda = E[X_1] = \exp(\mu + \frac{1}{2}\tau^2), \quad \sigma^2 = \text{Var}[X_1] = \lambda^2[\exp(\tau^2) - 1].$$

and

$$\mu_k = E[X_1^k] = \lambda^k \exp\{\frac{1}{2}k(k-1)\tau^2\}, k = 1, 2, \ldots$$

Consequently, from Theorem 3.7.1 we have the following result.

Corollary 3.7.5. If $0 < a \leq 1$ and X_1 has a lognormal distribution $LN(\mu, \tau^2)$, let $\delta = \exp(\tau^2)$, then

$$\begin{aligned}
M(t,a) = {} & \frac{t}{\lambda} + \frac{\delta - 2}{2} + \left\{\frac{t^2}{2\lambda^2} + \frac{(\delta - 2)t}{2\lambda}\right\}(a - 1) \\
& + \left\{\frac{t^3}{3\lambda^3} + \frac{3(\delta - 2)t^2}{4\lambda^2} - \frac{t}{12\lambda}(4\delta^3 - 9\delta^2 + 12\delta - 12) \right. \\
& \left. + \frac{1}{24}(3\delta^6 - 16\delta^4 + 19\delta^3 - 6\delta^2)\right\}(a - 1)^2 + o(1). \quad (3.7.22)
\end{aligned}$$

3.8 Comparison with Simulation Solution to Geometric Equation

We have studied the analytic, numerical and approximate methods for the solution of equation (3.2.7). In practice, a simulation method is also applicable. To demonstrate and compare these four methods, we shall consider four numerical examples each with exponential distribution, gamma distribution, Weibull distribution and lognormal distribution respectively. For each example, the solutions obtained by these four methods will be compared.

To do this, assume that $\{X_i, i = 1, 2, \ldots\}$ is a GP with ratio $0 < a \leq 1$, and the distribution and density of X_1 are F and f respectively.

At first, from (3.5.30), we shall evaluate an analytic solution on $[0, T]$ where T is determined by the distribution or the requirement in practical problem. The procedure for the analytic solution is as follows.

Step A1: Partition interval $[0, T]$ into N subintervals with equal length $h = T/N$. Then evaluate respectively the values $F(T_i)$ and $f(T_i)$ of F and f at nodes $T_i = ih, i = 0, \ldots, N$.

Step A2: Evaluate the values $R(T_i, T_j)$, $R(aT_i, T_j)$, $R(T_i, aT_j)$ and $R(aT_i, aT_j)$ of reproducing kernel function respectively.

Step A3: Calculate the values of $\psi_i(T_j)$ for $j = 0, \ldots, N$. To do so, from (3.5.35), the trapezoidal integration rule yields,

$$\psi_0(T_j) = R(0, T_j),$$

and for $i \geq 1$,

$$\psi_i(T_j) = R(T_i, T_j) - \int_0^{T_i} R(a(T_i - y), T_j) f(y) dy$$

$$\doteq R(T_i, T_j)$$

$$- \frac{h}{2} \left(R(aT_i, T_j) f(0) + 2 \sum_{k=1}^{i-1} R(a(T_i - T_k), T_j) f(T_k) + R(0, T_j) f(T_i) \right).$$

Then calculate the values of $\psi_i(aT_j)$ for $j = 1, 2, \ldots, N$ in a similar way. In fact, (3.5.35) gives

$$\psi_0(aT_j) = R(0, aT_j),$$

and for $i \geq 1$,

$$\psi_i(aT_j) = R(T_i, aT_j) - \int_0^{T_i} R(a(T_i - y), aT_j) f(y) dy$$

$$\doteq R(T_i, aT_j) - \frac{h}{2} \left(R(aT_i, aT_j) f(0) \right.$$

$$\left. + 2 \sum_{k=1}^{i-1} R(a(T_i - T_k), aT_j) f(T_k) + R(0, aT_j) f(T_i) \right).$$

Step A4: Determine the values of inner products $A(i, j) = (\psi_i, \psi_j)_W$. To do this, from (3.5.37), for $i \geq 1$, using trapezoidal integration gives

$$A(i, j) \doteq \psi_j(T_i)$$

$$- \frac{h}{2} \left(\psi_j(aT_i) f(0) + 2 \sum_{k=1}^{i-1} \psi_j(a(T_i - T_k)) f(T_k) + \psi_j(0) f(T_i) \right).$$

Step A5: Solve linear equations (3.5.32) that are now given by

$$\sum_{i=1}^{N} A(i,j)c_i = F(T_j), \quad j = 1,\ldots,N, \qquad (3.8.1)$$

for c_i, $i = 1,\ldots,N$.

Step A6: Determine the values of an approximate analytic solution $\Lambda_N(t)$. From (3.5.30), it is given by

$$\Lambda_N(T_j) = \sum_{i=1}^{N} c_i \psi_i(T_j) \qquad (3.8.2)$$

with $\Lambda_N(0) = 0$.

Second, we shall evaluate an numerical solution from (3.6.14) on $[0, T]$. The procedure for the numerical solution is much simpler.

Step N1: Partition interval $[0, T]$ into N subintervals with equal length $h = T/N$. Then calculate the values $F(T_j)$ and $f(T_j)$ of F and f at nodes $T_j = jh, j = 1,\ldots,N$.

Step N2: Let $\Lambda_0 = 0$ and evaluate Λ_j, $j = 1,\ldots,N$ from (3.6.14) recursively. The values of Λ_j, $j = 0, 1,\ldots,N$, form a numerical solution to (3.2.7).

Third, according to (3.7.18) or (3.7.19)-(3.7.22), an approximate solution could be obtained.

Finally, we could obtain a simulation solution in the following way.

Step S1: For each $k = 1,\ldots,n$, generate a sequence of i.i.d. random numbers $Y_1^{(k)}, Y_2^{(k)},\ldots$, each having distribution F, using a subroutine of a software, MATLAB say.

Step S2: By taking transformation $X_i^{(k)} = Y_i^{(k)}/a^{i-1}$, $\{X_i^{(k)}, i = 1, 2,\ldots\}$ form a realization of the GP with ratio a.

Step S3: Calculate the partial sums $S_i^{(k)} = \sum_{j=1}^{i} X_j^{(k)}, i = 1, 2,\ldots$.

Step S4: Count $N^{(k)}(T_j)$, the number of events occurred by time $T_j = jh$ with $h = T/N$, for $j = 1,\ldots,N$.

Step S5: After n, 2000 for example, times simulation, take

$$\bar{\Lambda}(T_j) = \frac{1}{n} \sum_{k=1}^{n} N^{(k)}(T_j)$$

as a simulation solution of $\bar{\Lambda}(t)$ at T_j, $j = 1, \ldots, N$.

After evaluation, the analytic solution, numerical solution, approximate solution and simulation solution are plotted together in the same figure for comparison. For the analytic solution, we shall use a dash line in the figure; for the numerical solution, we shall apply a solid line; for the approximate solution, we shall use a dash-dotted line; while for the simulation solution, we shall apply a dotted line. For easy comparison, in the same figure, we shall also plot the 95% and 105% of the values of simulation solution as the lower and upper bounds of the geometric function $\Lambda(T_j)$ by using dotted lines.

Because real data analysis that will be conducted in Chapter 4 shows that the ratio a of the fitted GP model will satisfy the condition

$$0.95 \leq a \leq 1.05,$$

the ratios in all four examples are taken as $a = 0.95$.

Example 3.8.1. The Exponential Distribution

In this example, assume that X_1 has an exponential distribution $Exp(2)$ with density function

$$f(x) = \begin{cases} 2e^{-2x} & x > 0, \\ 0 & x \leq 0. \end{cases} \tag{3.8.3}$$

The distribution function is $F(x) = 1 - e^{-2x}$ for $x > 0$ and 0 otherwise. The ratio is $a = 0.95$. The value of T is taken to be $10E(X_1) = 5$. We first divide interval $[0, T] = [0, 5]$ into $N = 1000 = 200T$ subintervals with equal length 0.005.

Afterward, we can follow Steps A1-A6 to obtain an analytic solution $\Lambda_N(T_j)$, $j = 1, \ldots, N$. Then, Steps N1 and N2 are used to obtain a numerical solution Λ_i, $i = 1, \ldots, N$. Thereafter, from (3.7.19), an approximate solution is given by

$$\tilde{\Lambda}(t) = 2t - 0.1t^2 + \frac{0.02}{3}t^3. \tag{3.8.4}$$

Finally, the subroutine *exprnd* in MATLAB is applied to generate a sequence of i.i.d. random numbers each having exponential distribution $Exp(2)$. For this purpose, Steps S1-S5 are applied to evaluate a simulation solution $\bar{\Lambda}(T_j)$. Then, the 95% and 105% of the values of simulation solution are taken as the lower and upper bounds of the solution $\Lambda(T_j)$.

The analytic, numerical, approximate and simulation solutions with the lower and upper bounds of the solution are plotted together in Figure 3.8.1.

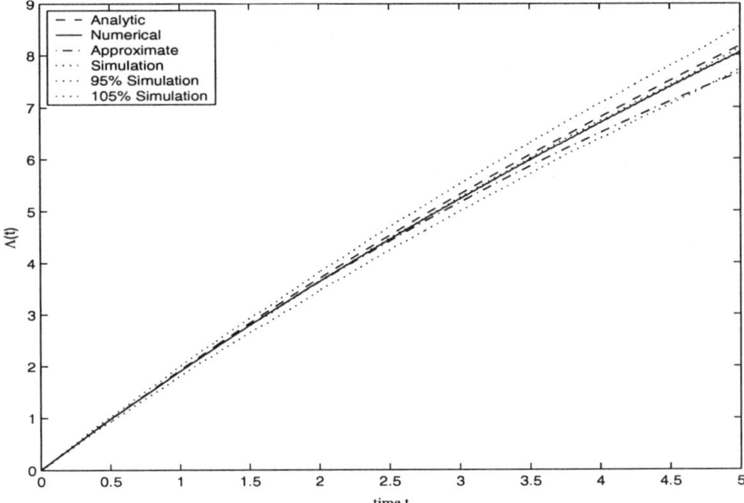

Figure 3.8.1. Exponential distribution $Exp(2)$

Figure 3.8.1 shows that the results obtained by four methods are very close. The values of $\Lambda_N(T_j)$, Λ_j, $\tilde{\Lambda}(T_j)$ and $\bar{\Lambda}(T_j)$ all lie inside the lower and upper bounds of the solution. The relative errors of four methods are all smaller than 5%.

Example 3.8.2. The Gamma Distribution

In this example, assume that X_1 has a gamma distribution $\Gamma(2,1)$ with density function

$$f(x) = \begin{cases} xe^{-x} & x > 0, \\ 0 & x \leq 0. \end{cases} \tag{3.8.5}$$

The distribution function is $F(x) = 1-(1+x)e^{-x}$ for $x > 0$ and 0 otherwise. The ratio a is still 0.95. Now choose $T = 10E(X_1) = 20$. Then interval $[0,T] = [0,20]$ is divided into $N = 100T = 2000$ subintervals with equal length 0.01.

An analytic solution and a numerical solution can be obtained according to Steps A1-A6 and Steps N1-N2 respectively. On the other hand, from (3.7.20), an approximate solution is given by

$$\tilde{\Lambda}(t) = \frac{1}{192}\{0.02t^3 - 1.245t^2 + 97.245t - 48.0075\}. \tag{3.8.6}$$

Then, the subroutine *gamrnd* of the MATLAB is applied to generate a sequence of i.i.d. random numbers each having gamma distribution $\Gamma(2,1)$. Afterward, a simulation solution will be obtained by following Steps S1-S5. Again, the 95% and 105% of the values of the simulation solution are taken as the lower and upper bounds of the solution $\Lambda(T_j)$. The numerical results obtained by four methods are plotted together in Figure 3.8.2 for comparison.

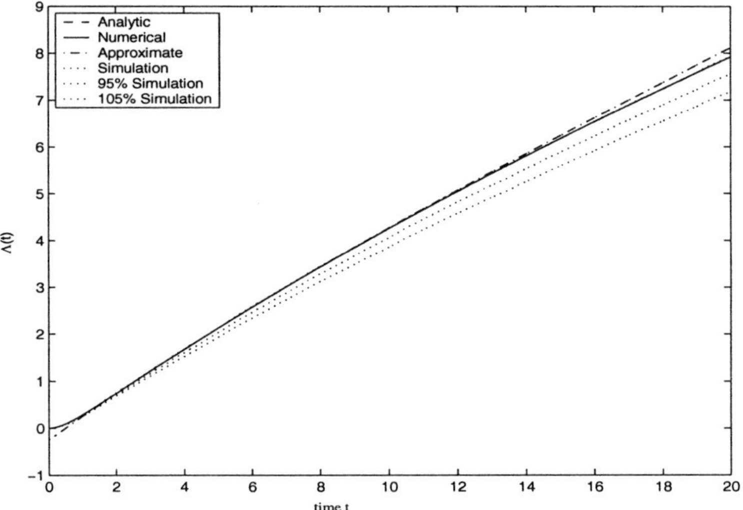

Figure 3.8.2. Gamma distribution $\Gamma(2,1)$

Figure 3.8.2 shows that the results obtained by the analytic, numerical and simulation methods are all very close. The values of $\Lambda_N(T_j)$ and Λ_j and $\bar{\Lambda}(T_j)$ all lie inside the lower and upper bounds of the solution. The relative errors of these three methods are all smaller than 5%. However, it is not the case for the values of approximate solution $\tilde{\Lambda}(T_j)$. We can see that the approximate solution still lies inside the lower and upper bounds in an intermediate interval [1,14]. However, the deviation of the approximate solution in two end intervals [0,1] and [14, 20] is larger and the error is more than 5% but still less than 10%.

Example 3.8.3. The Weibull Distribution

In this example, assume that X_1 has a Weibull distribution $W(2,1)$ with density function

$$f(x) = \begin{cases} 2x\exp(-x^2) & x > 0, \\ 0 & x \le 0. \end{cases} \tag{3.8.7}$$

Now, the distribution function is $F(x) = 1 - e^{-x^2}$ for $x > 0$, and 0 otherwise. The ratio is still $a = 0.95$. We choose $T = 10 > 10E(X_1) = 5\sqrt{\pi}$. Then, interval $[0, T] = [0, 10]$ is partitioned into $N = 100T = 1000$ subintervals with equal length 0.01.

According to (3.7.21), an approximate solution is given by

$$\begin{aligned}
\tilde{\Lambda}(t) = {} & \frac{2t}{\sqrt{\pi}} + \frac{2-\pi}{\pi} - \frac{0.1}{\pi}\{t^2 + \frac{(2-\pi)t}{\sqrt{\pi}}\} \\
& + \frac{0.0025}{3\pi^{3/2}}\left\{8t^3 + \frac{18(2-\pi)}{\sqrt{\pi}}t^2 + \frac{8t}{\pi}(\frac{3\pi^2}{4} - \frac{9\pi}{2} + 9)\right. \\
& \left. + \frac{8}{\pi^{3/2}}(\frac{3\pi^2}{8} - 6\pi + 15)\right\}.
\end{aligned} \tag{3.8.8}$$

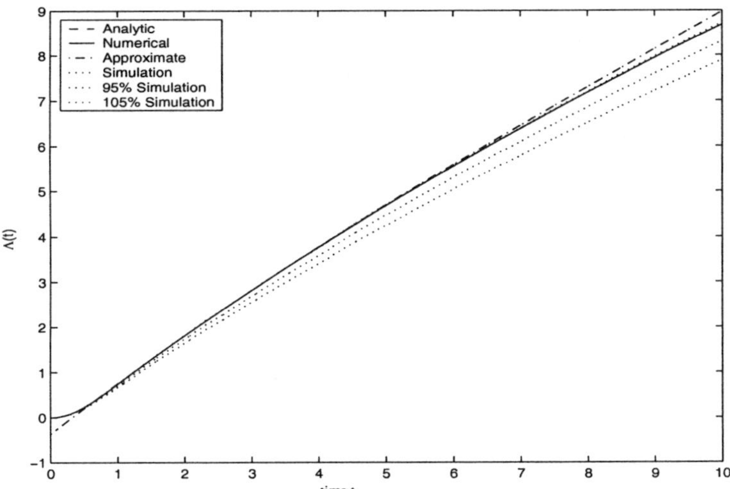

Figure 3.8.3. Weibull distribution $W(1,2)$

Again, the analytic, numerical and approximate solutions are compared with a simulation solution by plotting them together in Figure 3.8.3. Note that the subroutine *weibrnd* in MATLAB is now used for obtaining a sequence of i.i.d. random numbers from $W(2,1)$ distribution. Moreover, the 95% and 105% of the values of the simulation solution are taken as the lower and upper bounds of the solution $\Lambda(T_j)$.

We can see from Figure 3.8.3 that the differences among the values obtained by analytic, numerical and simulation methods are all very small. The values $\Lambda_N(T_j)$, Λ_j and $\bar{\Lambda}(T_j)$ all lie inside the lower and upper bounds of the solution. This means that the relative errors of these three methods are all smaller than 5% . However for the values of approximate solution $\tilde{\Lambda}(T_j)$, although in an intermediate interval $[0.5, 6.5]$, the values $\tilde{\Lambda}(T_j)$ still lie inside the lower and upper bounds, the deviation of the approximate solution from the simulation solution in two end intervals $[0, 0.5]$ and $[6.5, 10]$, is larger and the error is more than 5% but still less than 10%.

Example 3.8.4. The Lognormal Distribution

In this example, assume that X_1 has a lognormal distribution $LN(0,1)$ with density function

$$f(x) = \begin{cases} \frac{1}{\sqrt{2\pi}x}\exp\{-\tfrac{1}{2}(\ell nx)^2\} & x > 0, \\ 0 & x \leq 0. \end{cases} \tag{3.8.9}$$

Now $F(x) = \Phi(\ell nx)$ for $x > 0$, and 0 otherwise, where $\Phi(x)$ is the standard normal distribution function. The ratio is still $a = 0.95$. We take $T = 18 > 10E(X_1) = 10e^{1/2}$. Then, interval $[0,T] = [0,18]$ is divided into $N = 100T = 1800$ subintervals with equal length 0.01.

Now, we can also evaluate the analytic and numerical solution following Steps A1-A6 and Steps N1-N2 respectively. Afterward, from (3.7.22), an approximate solution can be obtained by

$$\begin{aligned}
\tilde{\Lambda}(t) &= \frac{t}{e^{1/2}} + \frac{e-2}{2} - 0.05\{\frac{t^2}{2e} + \frac{(e-2)t}{2e^{1/2}}\} \\
&\quad + 0.0025\left\{\frac{t^3}{3e^{3/2}} + \frac{3(e-2)t^2}{4e} - \frac{t}{12e^{1/2}}(4e^3 - 9e^2 + 12e - 12)\right. \\
&\quad \left. + \frac{1}{24}(3e^6 - 16e^4 + 19e^3 - 6e^2)\right\}.
\end{aligned} \tag{3.8.10}$$

Finally, a simulation solution is obtained following Steps S1-S5 as we did in previous examples, except the subroutine *logrnd* in MATLAB now is applied for generating a sequence of i.i.d. random numbers from $LN(0,1)$ distribution. Moreover, the 95% and 105% of the values of the simulation solution are taken as the lower and upper bounds of the solution. Similarly, the results obtained by four methods are plotted together in Figure 3.8.4 for comparison.

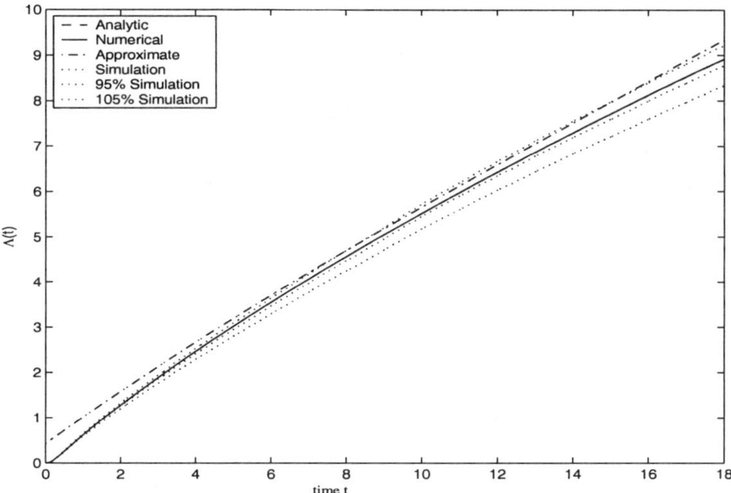

Figure 3.8.4. Lognormal distribution $LN(0,1)$

Again, we can see from Figure 3.8.4 that the differences among the results obtained by analytic, numerical and simulation methods are also very small. The values $\Lambda_N(T_j)$, Λ_i and $\bar{\Lambda}(t)$ all lie inside the lower and upper bounds of the solution. The relative errors of these three methods are all smaller than 5% . However, for the values of approximate solution $\tilde{\Lambda}(T_j)$, although in an intermediate interval [7, 16], the approximate solution still lies inside the lower and upper bounds, its deviation from the simulation solution in two end intervals [0, 7] and [16, 18] is larger so that the error is more than 5% but still less than 10%.

From the study of four numerical examples, we can make the following comments.

(1) The analytic, numerical and simulation methods are all powerful in the determination of the geometric function $M(t,a)$.

(2) The error of the approximate solution is less than 10% on a reasonably large interval, larger than $[0, 10E(X_1)]$ say. In an intermediate interval, the error is even less than 5%. For the exponential distribution case, the

approximate solution is accurate on $[0, 10E(X_1)]$. Overall, the approximate solutions given by (3.7.19)-(3.7.22) are good approximate solutions.

(3) In theoretical research, an approximate solution for $M(t, a)$ is useful. Formula (3.7.18) or (3.7.19)-(3.7.22) could be an appropriate choice. For exponential distribution, (3.7.19) is an accurate approximation. However, for the other three distributions, formulas (3.7.20)-(3.7.22) are accurate in an intermediate interval. Therefore, in using an approximate solution of (3.7.18), it is suggested to compare the values between the approximate solution and a simulation solution to determine an appropriate interval so that the approximate solution can be applied more accurately.

3.9 Exponential Distribution Case

As a particular case, suppose that $0 < a \leq 1$ and X_1 has an exponential distribution $Exp(1/\lambda)$ with density function

$$f(x) = \begin{cases} \frac{1}{\lambda}e^{-x/\lambda} & x > 0, \\ 0 & x \leq 0. \end{cases}$$

If in addition $a = 1$, then $\{N(t), t \geq 0\}$ is a Poisson process with rate $1/\lambda$. It is well known that

$$M(t, 1) = E[N(t)] = \frac{t}{\lambda}. \tag{3.9.1}$$

In general, Braun et al. (2005) derived an upper and lower bounds for the geometric function $M(t, a)$.

Theorem 3.9.1. If $0 < a \leq 1$, and X_1 has an exponential distribution $Exp(1/\lambda)$, then

$$\frac{a}{1-a}\ell n\left\{\frac{(1-a)t}{a\lambda} + 1\right\} \leq M(t, a) \leq \frac{1}{1-a}\ell n\left\{\frac{(1-a)t}{\lambda} + 1\right\}. \tag{3.9.2}$$

Proof.

Let $p_i(t) = P(N(t) = i)$. Then from (2.5.4) and (2.5.5), it is straightforward to derive the following equations:

$$\frac{d}{dt}M(t, a) = \frac{1}{\lambda}E[a^{N(t)}], \tag{3.9.3}$$

$$\frac{d}{dt}E[a^{N(t)}] = -\frac{1-a}{\lambda}E[a^{2N(t)}]. \tag{3.9.4}$$

Because function x^2 is convex, then by using Jensen's inequality, (3.9.4) yields

$$\frac{d}{dt}E[a^{N(t)}] \le -\frac{1-a}{\lambda}\{E[a^{N(t)}]\}^2.\tag{3.9.5}$$

Note that $N(0) = 0$, then from (3.9.5) we have

$$E[a^{N(t)}] \le \frac{1}{\frac{(1-a)t}{\lambda}+1}.\tag{3.9.6}$$

Thus (3.9.3) and (3.9.6) yield that

$$\frac{d}{dt}M(t,a) \le \frac{1}{(1-a)t+\lambda}.\tag{3.9.7}$$

Then by noting that $M(0,a) = 0$, the right hand side of the inequality (3.9.2) follows. To show the left hand side of (3.9.2), we note that $1/x$ is a convex function for $x > 0$, then Jensen's inequality implies that

$$E[a^{N(t)}] \ge \frac{1}{E[a^{-N(t)}]}.\tag{3.9.8}$$

Therefore, the combination of (2.5.2), (3.9.3) and (3.9.8) yields that

$$\frac{d}{dt}M(t,a) \ge \frac{1}{\frac{(1-a)t}{a}+\lambda}.\tag{3.9.9}$$

As a result, the left hand side of (3.9.2) follows. This completes the proof of Theorem 3.9.1.

In particular, if $a = 1$, the GP reduces to a Poisson process, then (3.9.2) reduces to the following equality

$$M(t,1) = \frac{t}{\lambda}.\tag{3.9.10}$$

This agrees with the well known result (3.9.1). Thus, Theorem 3.9.1 is a generalization of the well known result in Poisson process.

Furthermore, by expanding the logarithm function as the Taylor series at $a = 1$, we have the following result.

Corollary 3.9.2. If $0 < a \le 1$, and X_1 has an exponential distribution $Exp(1/\lambda)$, then

$$\frac{t}{\lambda} + \frac{t^2}{2a\lambda^2}(a-1) + \frac{t^3}{3a^2\lambda^3}(a-1)^2 + o\{(a-1)^2\}$$

$$\le M(t,a) \le \frac{t}{\lambda} + \frac{t^2}{2\lambda^2}(a-1) + \frac{t^3}{3\lambda^3}(a-1)^2 + o\{(a-1)^2\}.\tag{3.9.11}$$

In comparison with Corollary 3.7.2 and Corollary 3.9.2, we can see that in the exponential distribution case, the approximate solution to the geometric function $M(t, a)$ is in fact its upper bound. This is the reason why the approximate solution for exponential distribution case is as accurate as the other three solutions we discussed in Section 3.8.

3.10   Notes and References

The geometric function was first introduced by Lam (1988a, b). Sections 3.2 - 3.4 are based on Lam (1988b). However, Theorem 3.3.2 is new. It is due to Lam (2005b) that is a generalization and improvement of a result in Braun et al. (2005). By using a reproducing kernel technique, in Section 3.5, a series expansion of the geometric function is obtained. That is based on Lam and Tang's paper (2007). Section 3.6 is based on Tang and Lam (2007) in which a numerical solution to the geometric function is studied by using a trapezoidal integration rule, see Stoer and Bulirsch (1980) for the reference. The results in Section 3.7 are originally due to Lam (2005b). By expanding the Laplace transform of $M(t, a)$ in the Taylor series, an approximate expression of $M(t, a)$ is obtained. Theorem 3.7.1 gives a simple approximate formula for $M(t, a)$, it is a new result. The material in Section 3.8 is also new in which the analytic solution, numerical solution and approximate solution of $M(t, a)$ are compared with the simulation solution through four numerical examples each with exponential distribution, gamma distribution, Weibull distribution and lognormal distribution respectively. Section 3.9 studies an exponential distribution case. That is a particular and important case. Theorem 3.9.1 is due to Braun et al. (2005) that gives a lower bound and an upper bound of the geometric function for the exponential distribution case.

Chapter 4

Statistical Inference of Geometric Process

4.1 Introduction

Suppose we want to apply a model for analysis of data, three questions will arise. First, how can we justify if the data agree with the model? Or how do we test whether the data are consistent with the model? Second, if the data agree with the model, how do we estimate the parameter in the model? Third, after fitting the model to the data, how well is the fitting? What is the distribution or the limiting distribution of the parameter estimator? In using a GP model for analysing a data set $\{X_i, i = 1, 2, \ldots, n\}$, we are also faced with these three questions. In this chapter, we shall answer these questions through a nonparametric as well as a parametric approach.

Given a stochastic process $\{X_i, i = 1, 2, \ldots\}$, in Section 4.2 we shall introduce some statistics for testing if $\{X_i, i = 1, 2, \ldots\}$ is a GP. In Section 4.3, under the assumption that a data set comes from a GP, a least squares estimator is suggested for the ratio a of the GP. Moreover, the moment estimators are introduced for the mean λ and variance σ^2 of X_1. Then Section 4.4 studies the asymptotic distributions of the above estimators. The methods used in Sections 4.2 to 4.4 are nonparametric. In Section 4.5, we study the parametric inference problem of the GP by making an additional assumption that X_1 follows a lognormal distribution.

4.2 Hypothesis Testing for Geometric Process

To answer the first question, suppose we are given a stochastic process $\{X_i, i = 1, 2, \ldots\}$, we need to test if $\{X_i, i = 1, 2, \ldots\}$ agrees with a GP. For

101

this purpose, we can form the following two sequences of random variables.

$$U_i = X_{2i}/X_{2i-1}, i = 1, 2, \ldots, \tag{4.2.1}$$

and

$$U_i' = X_{2i+1}/X_{2i}, i = 1, 2, \ldots. \tag{4.2.2}$$

Moreover, for a fixed integer m, we can also form two more sequences of random variables.

$$V_i = X_i X_{2m+1-i}, i = 1, 2, \ldots, m, \tag{4.2.3}$$

and

$$V_i' = X_{i+1} X_{2m+2-i}, i = 1, 2, \ldots, m. \tag{4.2.4}$$

The following two theorems are due to Lam (1992b). They are essential for testing whether the stochastic process $\{X_i, i = 1, 2, \ldots\}$ is a GP.

Theorem 4.2.1. If $\{X_i, i = 1, 2, \ldots\}$ is a GP, then $\{U_i, i = 1, 2, \ldots\}$ and $\{U_i', i = 1, 2, \ldots\}$ are respectively two sequences of i.i.d. random variables.
Proof.

Assume that $\{X_i, i = 1, 2, \ldots\}$ is a GP. Clearly, $\{U_i, i = 1, 2, \ldots\}$ are independent. Now, we shall show that the distributions of $U_i, i = 1, 2, \ldots$ are identical. To do this, let the probability density function of X_i be f_i. Then it follows from Definition 2.2.1 that

$$f_i(x) = a^{i-1} f(a^{i-1} x)$$

where a is the ratio of GP $\{X_i, i = 1, 2, \ldots\}$ and f is the density function of X_1. Then the density function g_i of U_i is given by

$$
\begin{aligned}
g_i(u) &= \int_0^\infty t f_{2i}(ut) f_{2i-1}(t) dt \\
&= \int_0^\infty a^{4i-3} t f(a^{2i-1} ut) f(a^{2i-2} t) dt \\
&= \int_0^\infty a y f(auy) f(y) dy,
\end{aligned}
$$

that does not depend on i. This implies that $\{U_i, i = 1, 2, \ldots\}$ is a sequence of i.i.d. random variables. A similar argument shows that $\{U_i', i = 1, 2, \ldots\}$ is also a sequence of i.i.d. random variables. This completes the proof of

Theorem 4.2.1.

Theorem 4.2.2. If $\{X_i, i = 1, 2, \ldots\}$ is a GP, then for any fixed integer m, $\{V_i, i = 1, 2, \ldots, m\}$ and $\{V_i', i = 1, 2, \ldots, m\}$ are respectively two sequences of i.i.d. random variables.

Proof.

Assume that $\{X_i, i = 1, 2, \ldots\}$ is a GP. Then, $\{V_i, i = 1, 2, \ldots, m\}$ are clearly independent. To show that the distributions of $V_i, i = 1, 2, \ldots, m$ are identical, let the density function of X_i be f_i. Then the density function h_i of V_i is given by

$$
\begin{aligned}
h_i(v) &= \int_0^\infty \frac{1}{t} f_i(\frac{v}{t}) f_{2m+1-i}(t) dt \\
&= \int_0^\infty a^{2m-1} \frac{1}{t} f(a^{i-1}\frac{v}{t}) f(a^{2m-i}t) dt \\
&= \int_0^\infty a^{2m-1} \frac{1}{y} f(\frac{v}{ay}) f(a^{2m}y) dy.
\end{aligned}
$$

Again, it does not depend on i. This implies that $\{V_i, i = 1, 2, \ldots, m\}$ is a sequence of i.i.d. random variables. By a similar argument, $\{V_i', i = 1, 2, \ldots, m\}$ is also a sequence of i.i.d. random variables. This completes the proof of Theorem 4.2.2.

In practice, in order to apply Theorems 4.2.1 and 4.2.2 to a data set $\{X_i, i = 1, 2, \ldots, n\}$, we should use all the information involved in the data set. For this purpose, the following auxiliary sequences are constructed.

(1) If $n = 2m$ is even, form

$$\{U_i, i = 1, 2, \ldots, m\} \quad \text{and} \quad \{V_i, i = 1, 2, \ldots, m\}. \tag{4.2.5}$$

(2) If $n = 2m + 1$ is odd, form

$$\{U_i', i = 1, 2, \ldots, m\} \quad \text{and} \quad \{V_i, i = 1, 2, \ldots, m\}; \tag{4.2.6}$$

or

$$\{U_i, i = 1, 2, \ldots, m\} \quad \text{and} \quad \{V_i', i = 1, 2, \ldots, m\}. \tag{4.2.7}$$

Then according to the parity of n, we can test whether the data set $\{X_i, i = 1, 2, \ldots, n\}$ comes from a GP by testing whether the random variables in sequences (4.2.5), (4.2.6) or (4.2.7) are respectively i.i.d. or not. To do this, let I_A be the indicator of an event A. For testing whether random variables $\{W_i, i = 1, 2, \ldots, n\}$ are i.i.d. or not, the following tests can be applied.

(1) The turning point test.
Define

$$T_W = \sum_{i=2}^{m-1} I_{[(W_i - W_{i-1})(W_{i+1} - W_i) < 0]}.$$

If $\{W_i, i = 1, 2, \ldots, n\}$ are i.i.d., then asymptotically

$$T(W) = \left[T_W - \frac{2(m-2)}{3}\right] / \left[\frac{16m - 29}{90}\right]^{1/2} \sim N(0, 1). \qquad (4.2.8)$$

(2) The difference-sign test.
Let

$$D_W = \sum_{i=2}^{m} I_{[W_i > W_{i-1}]}.$$

If $\{W_i, i = 1, 2, \ldots, n\}$ are i.i.d., then asymptotically

$$D(W) = \left[D_W - \frac{m-1}{2}\right] / \left[\frac{m+1}{12}\right]^{1/2} \sim N(0, 1). \qquad (4.2.9)$$

See Ascher and Feingold (1984) for some other testing statistics.

4.3   Estimation of Parameters in Geometric Process

To answer the second question, assume that a data set $\{X_i, i = 1, 2, \ldots, n\}$ follows a GP. Let

$$Y_i = a^{i-1} X_i, \quad i = 1, 2, \ldots, n. \qquad (4.3.1)$$

Then $\{Y_i, i = 1, 2, \ldots, n\}$ is a sequence of i.i.d. random variables, so is $\{\ell n Y_i, i = 1, 2, \ldots, n\}$. Thus, we can denote its common mean and variance by $\mu = E[\ell n Y_i]$ and $\tau^2 = \text{Var}[\ell n Y_i]$ respectively. Taking logarithm on the both sides of (4.3.1) gives

$$\ell n Y_i = (i - 1)\ell n a + \ell n X_i, \quad i = 1, 2, \ldots, n. \qquad (4.3.2)$$

On the other hand, we can rewrite

$$\ell n Y_i = \mu + e_i, \quad i = 1, 2, \ldots, n, \qquad (4.3.3)$$

where $e's$ are i.i.d. random variables each having mean 0 and variance τ^2. Then, (4.3.2) becomes

$$\ell n X_i = \mu - (i-1)\ell n a + e_i, \quad i = 1, 2, \ldots, n. \tag{4.3.4}$$

Consequently, (4.3.4) is a simple regression equation. By using linear regression technique, Lam (1992b) obtained the following least squares estimators of $\mu, \beta = \ell n a$ and τ^2.

$$\hat{\mu} = \frac{2}{n(n+1)} \sum_{i=1}^{n} (2n - 3i + 2)\ell n X_i, \tag{4.3.5}$$

$$\hat{\beta} = \frac{6}{(n-1)n(n+1)} \sum_{i=1}^{n} (n - 2i + 1)\ell n X_i \tag{4.3.6}$$

and

$$\hat{\tau^2} = \frac{1}{n-2} \left\{ \sum_{i=1}^{n} (\ell n X_i)^2 - \frac{1}{n} (\sum_{i=1}^{n} \ell n X_i)^2 - \frac{\hat{\beta}}{2} \sum_{i=1}^{n} (n - 2i + 1)\ell n X_i \right\}. \tag{4.3.7}$$

As a result, an estimator of a is given by

$$\hat{a} = \exp(\hat{\beta}). \tag{4.3.8}$$

For convenience, write $\hat{Y}_i = \hat{a}^{i-1} X_i$, $\bar{\hat{Y}} = \sum_{i=1}^{n} \hat{Y}_i / n$ and $\bar{X} = \sum_{i=1}^{n} X_i / n$. Then the moment estimators for λ and σ^2 are given respectively by

$$\hat{\lambda} = \begin{cases} \bar{\hat{Y}}, & a \neq 1, \\ \bar{X}, & a = 1, \end{cases} \tag{4.3.9}$$

and

$$\hat{\sigma}^2 = \begin{cases} \frac{1}{n-1} \sum_{i=1}^{n} (\hat{Y}_i - \bar{\hat{Y}})^2 & a \neq 1, \\ \frac{1}{n-1} \sum_{i=1}^{n} (X_i - \bar{X})^2 & a = 1. \end{cases} \tag{4.3.10}$$

Clearly, estimators $\hat{a}$, $\hat{\lambda}$ and $\hat{\sigma}^2$ are nonparametric estimators. Moreover, the estimator $\hat{a}$ of a can be obtained by minimizing the sum squares of errors

$$Q = \sum_{i=1}^{n} [\ell n X_i - \mu + (i-1)\ell n a]^2. \tag{4.3.11}$$

Another nonparametric estimators for a and λ can be obtained by minimizing directly the sum square of errors

$$Q_D = \sum_{i=1}^{n}[X_i - a^{-(i-1)}\lambda]^2. \tag{4.3.12}$$

Then we can derive the following two equations

$$\left(\sum_{i=1}^{n}\frac{X_i}{a^{i-1}}\right)\left(\sum_{i=1}^{n}\frac{i-1}{a^{2i-1}}\right)$$
$$= \left(\sum_{i=1}^{n}\frac{(i-1)X_i}{a^i}\right)\left(\sum_{i=1}^{n}\frac{1}{a^{2(i-1)}}\right), \tag{4.3.13}$$

and

$$\lambda = \frac{\displaystyle\sum_{i=1}^{n} a^{-(i-1)}X_i}{\displaystyle\sum_{i=1}^{n} a^{-2(i-1)}}. \tag{4.3.14}$$

Let the solution to equations (4.3.13) and (4.3.14) be $\hat{a}_D$ and $\hat{\lambda}_D$. Then they are another least squares estimators of a and λ.

4.4　Asymptotic Distributions of the Estimators

In this section, we shall answer the third question by studying the limit distributions of the estimators $\hat{a}, \hat{\lambda}$ and $\hat{\sigma}^2$. The following four theorems are due to Lam et al. (2004).

Theorem 4.4.1. If $E[(\ell n Y)^2] < \infty$, then

$$n^{3/2}(\hat{\beta} - \beta) \xrightarrow{L} N(0, 12\tau^2). \tag{4.4.1}$$

Proof.
　　First of all, because $\beta = \ell n a$, $\sum_{i=1}^{n}(n - 2i + 1) = 0$ and

$$\sum_{i=1}^{n}(n - 2i + 1)i = -(n-1)n(n+1)/6,$$

then it follows from (4.3.6) that

$$\hat{\beta} - \beta = \frac{6}{(n-1)n(n+1)} \sum_{i=1}^{n} (n - 2i + 1)\ln X_i - \ln a$$

$$= \frac{6}{(n-1)n(n+1)} \left\{ \sum_{i=1}^{n} (n - 2i + 1)\ln X_i + \sum_{i=1}^{n} (n - 2i + 1)(i - 1)\ln a \right\}$$

$$= \frac{6}{(n-1)n(n+1)} \sum_{i=1}^{n} (n - 2i + 1)\ln Y_i$$

$$= \frac{6}{(n-1)n(n+1)} \sum_{i=1}^{n} (n - 2i + 1)(\ln Y_i - \mu).$$

Then, we have

$$[(n-1)n(n+1)]^{1/2}(\hat{\beta} - \beta) = \frac{1}{[(n-1)n(n+1)]^{1/2}} \sum_{i=1}^{n} \beta_{ni}, \quad (4.4.2)$$

where

$$\beta_{ni} = 6(n - 2i + 1)(\ln Y_i - \mu) \qquad i = 1, 2, \ldots, n$$

are independent. Let

$$B_n = \sum_{i=1}^{n} \text{Var}[\beta_{ni}] = 36\tau^2 \sum_{i=1}^{n} (n - 2i + 1)^2$$

$$= 12\tau^2 (n - 1)n(n + 1). \qquad (4.4.3)$$

Now, denote the distribution of β_{ni} by G_{ni} and the common distribution of $\ln Y_i - \mu$ by G. Then we check the Lindeberg condition

$$L_n = \frac{1}{B_n} \sum_{i=1}^{n} \int_{|z| \geq \epsilon B_n^{1/2}} z^2 dG_{ni}(z)$$

$$= \frac{1}{B_n} \sum_{i=1}^{n} \int_{6|(n-2i+1)w| \geq \epsilon B_n^{1/2}} 36(n - 2i + 1)^2 w^2 dG(w) \quad (4.4.4)$$

$$\leq \frac{36n(n - 1)^2}{B_n} \int_{|w| \geq \epsilon B_n^{1/2}/[6(n-1)]} w^2 dG(w)$$

$$\to 0 \quad \text{as} \quad n \to \infty, \qquad (4.4.5)$$

where (4.4.5) is due to (4.4.3) and the condition $E(\ln Y)^2 < \infty$. Thus, by the Lindeberg-Feller theorem,

$$[(n-1)n(n+1)]^{1/2}(\hat{\beta} - \beta) \xrightarrow{L} N(0, 12\tau^2),$$

and (4.4.1) follows.

Now from Theorem 4.4.1, using the Cramér δ theorem (see, e.g. Arnold (1990)) yields straightforwardly the following result.

Theorem 4.4.2. If $E[(\ell nY)^2] < \infty$, then

$$n^{3/2}(\hat{a} - a) \xrightarrow{L} N(0, 12a^2\tau^2). \tag{4.4.6}$$

As a result, on the basis of Theorem 4.4.1 or 4.4.2, one can test whether $a = 1$ or not. This is equivalent to test

$$H_0 : \beta = 0 \quad \text{against} \quad H_1 : \beta \neq 0.$$

In fact, from (4.4.1), the following testing statistic could be applied

$$R = n^{3/2}\hat{\beta}/(\sqrt{12}\hat{\tau}), \tag{4.4.7}$$

where $\hat{\tau}$ is computed from (4.3.7). Under H_0, $R \sim N(0, 1)$ approximately. Note that the problem of testing whether $a = 1$ is equivalent to the problem of testing whether the data set agrees with a renewal process or an homogeneous Poisson process. Therefore an alternative way is to apply the Laplace test, it is based on the statistic

$$L = \frac{[12(n - 1)]^{1/2}}{T_n}\Big(\frac{\sum_{i=1}^{n-1} T_i}{n - 1} - \frac{T_n}{2}\Big), \tag{4.4.8}$$

where $T_i = \sum_{j=1}^{i} X_j$ with $T_0 = 0$, and T_i is the occurrence time of the ith event. Under H_0, $L \sim N(0, 1)$ approximately (see Cox and Lewis (1966) for reference).

The following two theorems study the limit distributions of estimators $\hat{\lambda}$ and $\hat{\sigma}^2$ respectively.

Theorem 4.4.3.
(1) If $a \neq 1$, $E[Y^2] < \infty$ and $E[(\ell nY)^2] < \infty$, then

$$\sqrt{n}(\hat{\lambda} - \lambda) \xrightarrow{L} N(0, \sigma^2 + 3\lambda^2\tau^2). \tag{4.4.9}$$

(2) If $a = 1$ and $E[X^2] < \infty$, then

$$\sqrt{n}(\hat{\lambda} - \lambda) \xrightarrow{L} N(0, \sigma^2). \tag{4.4.10}$$

Proof.
To prove (4.4.9), at first, it follows from (4.3.9) that

$$\hat{\lambda} - \lambda = \frac{1}{n}\sum_{i=1}^{n}(\hat{Y}_i - Y_i) + \frac{1}{n}\sum_{i=1}^{n}(Y_i - \lambda). \tag{4.4.11}$$

Then Theorem 4.4.1 yields that

$$\hat{\beta} - \beta = O_p(n^{-3/2}).$$

Moreover, because

$$E\left[\frac{1}{n(n-1)}\sum_{i=1}^{n}(i-1)Y_i\right] = \frac{\lambda}{2}$$

and

$$\mathrm{Var}\left[\frac{1}{n(n-1)}\sum_{i=1}^{n}(i-1)Y_i\right] = \frac{2n-1}{6n(n-1)}\sigma^2,$$

we have

$$\frac{1}{n(n-1)}\sum_{i=1}^{n}(i-1)Y_i = \frac{\lambda}{2} + O_p(n^{-1/2}).$$

On the other hand, in virtue of

$$\begin{aligned}
\hat{Y}_i &= Y_i + \{(\hat{a}/a)^{i-1} - 1\}Y_i \\
&= Y_i + \{\exp[(i-1)(\hat{\beta} - \beta)] - 1\}Y_i \\
&= Y_i + (i-1)(\hat{\beta} - \beta)Y_i + O_p(n^{-1}),
\end{aligned} \tag{4.4.12}$$

hence

$$\begin{aligned}
\frac{1}{n}\sum_{i=1}^{n}(\hat{Y}_i - Y_i) &= \frac{1}{n}(\hat{\beta} - \beta)\sum_{i=1}^{n}(i-1)Y_i + O_p(n^{-1}) \\
&= \frac{\lambda}{2}(n-1)(\hat{\beta} - \beta) + O_p(n^{-1}).
\end{aligned} \tag{4.4.13}$$

Then with the help of (4.4.2), (4.4.11) becomes

$$\begin{aligned}
\sqrt{n}(\hat{\lambda} - \lambda) &= \frac{\lambda}{2}n^{1/2}(n-1)(\hat{\beta} - \beta) + n^{-1/2}\sum_{i=1}^{\infty}(Y_i - \lambda) + O_p(n^{-1/2}) \\
&= n^{-1/2}\sum_{i=1}^{n}\lambda_{ni} + O_p(n^{-1/2}),
\end{aligned} \tag{4.4.14}$$

where

$$\lambda_{ni} = Y_i - \lambda + 3\lambda(1 - \frac{2i}{n+1})(\ell nY_i - \mu), \quad i = 1, 2, \ldots, n$$

are independent. Now, let

$$B_n = \sum_{i=1}^{n}\mathrm{Var}[\lambda_{ni}] = n\sigma^2 + \frac{3n(n-1)}{n+1}\lambda^2\tau^2.$$

Denote respectively the distribution of λ_{ni} by H_{ni} and the common distribution of Y_i by F. Then, we check the Lindeberg condition

$$L_n = \frac{1}{B_n} \sum_{i=1}^{n} \int_{|z| \geq \epsilon B_n^{1/2}} z^2 dH_{ni}(z)$$

$$= \frac{1}{B_n} \sum_{i=1}^{n} \int_{D_{ni}} \left\{ y - \lambda + 3\lambda(1 - \frac{2i}{n+1})(\ell n y - \mu) \right\}^2 dF(y)$$

$$\leq \frac{n}{B_n} \int_{D_n} \{| y - \lambda | + 3\lambda | \ell n y - \mu |\}^2 dF(y) \to 0 \quad \text{as} \quad n \to \infty,$$

$$(4.4.15)$$

where

$$D_{ni} = \{y : | y - \lambda + 3\lambda(1 - \frac{2i}{n+1})(\ell n y - \mu) | \geq \epsilon B_n^{1/2}\},$$

$$D_n = \{y : | y - \lambda | + 3\lambda | \ell n y - \mu | \geq \epsilon B_n^{1/2}\}$$

and (4.4.15) holds since

$$E[(| Y - \lambda | + 3\lambda | \ell n Y - \mu |)^2] < \infty.$$

Then by using the Lindeberg-Feller theorem and Slutsky theorem, (4.4.9) follows directly from (4.4.15).

On the other hand, if $a = 1$, then from (4.3.9), (4.4.10) is trivial. This completes the proof of Theorem 4.4.3.

Theorem 4.4.4.
(1) If $a \neq 1, E[Y^4] < \infty$ and $E[(\ell n Y)^2] < \infty$, then

$$\sqrt{n}(\hat{\sigma^2} - \sigma^2) \xrightarrow{L} N(0, \omega^2 + 12\sigma^4 \tau^2), \qquad (4.4.16)$$

where $\omega^2 = \text{Var}(Y - \lambda)^2$.
(2) If $a = 1$ and $E[X^4] < \infty$ then

$$\sqrt{n}(\hat{\sigma^2} - \sigma^2) \xrightarrow{L} N(0, \omega^2), \qquad (4.4.17)$$

where $\omega^2 = \text{Var}(X - \lambda)^2$.
Proof.
To prove (4.4.16), first of all, we shall show that for $a \neq 1$

$$\hat{\sigma^2} = \frac{1}{n} \sum_{i=1}^{n} (Y_i - \lambda)^2 + n\sigma^2 \left(\hat{\beta} - \beta\right) + O_p\left(n^{-1}\right). \qquad (4.4.18)$$

It follows from (4.4.12) that

$$\bar{\hat{Y}} = \frac{1}{n} \sum_{i=1}^{n} \hat{Y}_i = \bar{Y} + \frac{1}{n}(\hat{\beta} - \beta) \sum_{i=1}^{n} (i - 1)Y_i + O_p(n^{-1}). \qquad (4.4.19)$$

Subtracting (4.4.19) from (4.4.12) yields

$$\hat{Y}_i - \bar{\hat{Y}} = Y_i - \bar{Y} + (\hat{\beta} - \beta)[(i-1)Y_i - \frac{1}{n}\sum_{j=1}^{n}(i-1)Y_j] + O_p(n^{-1}).$$

In view of Theorem 4.4.1, we have

$$\hat{\beta} - \beta = O_p(n^{-3/2}),$$

then

$$\hat{\sigma}^2 = \frac{1}{n-1}\sum_{i=1}^{n}(\hat{Y}_i - \bar{\hat{Y}})^2$$

$$= \frac{1}{n-1}\left\{\sum_{i=1}^{n}(Y_i - \bar{Y})^2 + 2(\hat{\beta} - \beta)\sum_{i=1}^{n}[(i-1)Y_i - \frac{1}{n}\sum_{j=1}^{n}(j-1)Y_j](Y_i - \bar{Y})\right.$$

$$\left. + (\hat{\beta} - \beta)^2 \sum_{i=1}^{n}[(i-1)Y_i - \frac{1}{n}\sum_{j=1}^{n}(j-1)Y_j]^2\right\} + O_p(n^{-1})$$

$$= \frac{1}{n-1}\sum_{i=1}^{n}(Y_i - \bar{Y})^2 + \frac{2}{n-1}(\hat{\beta} - \beta)[\sum_{i=1}^{n}(i-1)Y_i^2 - \bar{Y}\sum_{j=1}^{n}(j-1)Y_j]$$

$$+ O_p(n^{-1}). \tag{4.4.20}$$

As a result, by estimating the orders in probability in (4.4.20), we have

$$\hat{\sigma}^2 = \frac{1}{n-1}\left\{\sum_{i=1}^{n}(Y_i - \lambda)^2 - n(\bar{Y} - \lambda)^2\right\}$$

$$+ n(\hat{\beta} - \beta)\left\{\frac{2}{n(n-1)}\sum_{i=1}^{n}(i-1)[Y_i^2 - E(Y^2)]\right.$$

$$\left. + E(Y^2) - \frac{2}{n(n-1)}\bar{Y}\sum_{j=1}^{n}(j-1)(Y_j - \lambda) - \lambda(\bar{Y} - \lambda) - \lambda^2\right\} + O_p(n^{-1})$$

$$= \frac{1}{n}\sum_{i=1}^{n}(Y_i - \lambda)^2 + n(\hat{\beta} - \beta)[E(Y^2) - \lambda^2] + O_p(n^{-1}). \tag{4.4.21}$$

Consequently, (4.4.18) follows. Now we are able to prove (4.4.16). In fact, it follows from (4.4.18) and (4.4.2) that

$$\sqrt{n}(\hat{\sigma}^2 - \sigma^2) = \frac{1}{\sqrt{n}}\sum_{i=1}^{n}\sigma_{ni} + O_p(n^{-1/2}), \tag{4.4.22}$$

where
$$\sigma_{ni} = (Y_i - \lambda)^2 - \sigma^2 + \frac{6\sigma^2(n - 2i + 1)}{n}(\ell n Y_i - \mu), \qquad i = 1, 2, \ldots, n.$$
Obviously, σ_{ni} are i.i.d. random variables for $i = 1, 2, \ldots, n$ with $E(\sigma_{ni}) = 0$. Then by a simple algebra, we have
$$B_n = \sum_{i=1}^{n} \mathrm{Var}(\sigma_{ni}) = n\omega^2 + \frac{12(n-1)(n+1)}{n}\sigma^4\tau^2$$
subject to $E[Y^4] < \infty$ and $E[(\ell n Y)^2] < \infty$. Now, by a similar argument as we did in (4.4.15), it is straightforward to show that the Lindeberg condition holds. In fact, let the distribution of σ_{ni} be I_{ni} and the common distribution of Y_i be F respectively. Denote
$$E_{ni} = \{y \ : \mid (y - \lambda)^2 - \sigma^2 + 6\sigma^2(1 - \frac{2i-1}{n})(\ell n y - \mu) \mid \geq \epsilon B_n^{1/2}\}$$
and
$$E_n = \{y \ : \ (y - \lambda)^2 + \sigma^2 + 6\sigma^2 \mid \ell n y - \mu \mid \geq \epsilon B_n^{1/2}\},$$
then
$$L_n = \frac{1}{B_n} \sum_{i=1}^{n} \int_{|z| \geq \epsilon B_n^{1/2}} z^2 dI_{ni}(z)$$
$$= \frac{1}{B_n} \sum_{i=1}^{n} \int_{E_{ni}} \left\{(y - \lambda)^2 - \sigma^2 + 6\sigma^2(1 - \frac{2i-1}{n})(\ell n y - \mu)\right\}^2 dF(y)$$
$$\leq \frac{n}{B_n} \int_{E_n} [(y - \lambda)^2 + \sigma^2 + 6\sigma^2 \mid \ell n y - \mu \mid]^2 dF(y) \to 0$$
$$\text{as} \quad n \to \infty, \tag{4.4.23}$$
where (4.4.23) holds, since
$$E[((Y - \lambda)^2 + \sigma^2 + 6\sigma^2 \mid \ell n Y - \mu \mid)^2] < \infty.$$
Consequently, it follows from the Lindeberg-Feller theorem that
$$\frac{1}{B_n^{1/2}} \sum_{i=1}^{n} \sigma_{ni} \xrightarrow{L} N(0, 1), \tag{4.4.24}$$
Hence (4.4.16) follows from (4.4.24) by using the Slutsky theorem.

For $a = 1$, by central limit theorem, (4.4.17) is a well known result. This completes the proof of Theorem 4.4.4.

As a result, by using Theorems 4.4.1-4.4.4, the p-values of the estimates or an approximate confidence interval for parameters a, λ and σ^2 can be determined. In practice, the unknown parameter values can be replaced by their estimates. For example, from (4.4.6) an approximate 95% confidence interval for a is given by
$$(\hat{a} - 1.96\sqrt{12}\hat{a}\hat{\tau}n^{-3/2}, \ \hat{a} + 1.96\sqrt{12}\hat{a}\hat{\tau}n^{-3/2}). \tag{4.4.25}$$

4.5 Parametric Inference for Geometric Process

In this chapter, so far we have just applied a nonparametric approach to the statistical inference problem of a GP. In this section, we shall study the same problem by using a parametric approach. By making an additional condition that X_1 follows a specific life distribution, the maximum likelihood estimators (MLE) of a, λ and σ^2 of the GP and their asymptotic distributions will be studied. The results are then compared with that obtained by the nonparametric approach.

Now assume that $\{X_i, i = 1, 2, \cdots\}$ form a GP with ratio a and X_1 follows a lognormal distribution $LN(\mu, \tau^2)$ with probability density function

$$f(x) = \begin{cases} \frac{1}{\sqrt{2\pi}\tau x}\exp\{-\frac{1}{2\tau^2}(\ell n x - \mu)^2\} & x > 0, \\ 0 & \text{elsewhere.} \end{cases} \tag{4.5.1}$$

Let $Y_i = a^{i-1}X_i$, it is easy to show that $E[\ell n Y_i] = E[\ell n X_1] = \mu$ and $\text{Var}[\ell n Y_i] = \text{Var}[\ell n X_1] = \tau^2$. On the other hand, we have

$$\lambda = E[X_1] = \exp(\mu + \frac{1}{2}\tau^2) \tag{4.5.2}$$

and

$$\sigma^2 = \text{Var}[X_1] = \lambda^2[\exp(\tau^2) - 1]. \tag{4.5.3}$$

Now, it follows from (4.5.1) that the likelihood function is given by

$$L(a, \mu, \tau^2)$$
$$= (2\pi)^{-n/2}\tau^{-n}(\prod_{i=1}^{n} X_i)^{-1}\exp\left\{-\frac{1}{2\tau^2}\sum_{i=1}^{n}[\ell n(a^{i-1}X_i) - \mu]^2\right\}. \tag{4.5.4}$$

Therefore, the log-likelihood function is

$$\ell n L(a, \mu, \tau^2)$$
$$= -\frac{n}{2}\ell n(2\pi) - \frac{n}{2}\ell n(\tau^2) - \ell n\left\{\prod_{i=1}^{n} X_i\right\}$$
$$\quad -\frac{1}{2\tau^2}\sum_{i=1}^{n}[\ell n(a^{i-1}X_i) - \mu]^2. \tag{4.5.5}$$

Now, let the maximum likelihood estimators (MLE) of a, μ and τ^2 be, respectively, $\hat{a}_L, \hat{\mu}_L$ and $\hat{\tau}_L^2$. Then from (4.5.5), $\hat{a}_L, \hat{\mu}_L$ and $\hat{\tau}_L^2$ will be the solution of the following equations:

$$\sum_{i=1}^{n}(n - 2i + 1)\ell n(a^{i-1}X_i) = 0, \tag{4.5.6}$$

$$\mu = \frac{1}{n} \sum_{i=1}^{n} \ell n(a^{i-1} X_i) \tag{4.5.7}$$

and

$$\tau^2 = \frac{1}{n} \sum_{i=1}^{n} [\ell n(a^{i-1} X_i) - \mu]^2. \tag{4.5.8}$$

Consequently, (4.5.6) yields

$$\ell n \hat{a}_L = -\frac{\sum_{i=1}^{n}(n - 2i + 1)\ell n X_i}{\sum_{i=1}^{n}(i - 1)(n - 2i + 1)}$$

$$= \frac{6}{(n-1)n(n+1)} \sum_{i=1}^{n}(n - 2i + 1)\ell n X_i. \tag{4.5.9}$$

Substitution of (4.5.9) into (4.5.7) yields that

$$\hat{\mu}_L = \frac{1}{n} \sum_{i=1}^{n}[(i - 1)\ell n \hat{a}_L + \ell n X_i]$$

$$= \frac{1}{n} \left\{ \frac{3}{n+1} \sum_{i=1}^{n}(n - 2i + 1)\ell n X_i + \sum_{i=1}^{n} \ell n X_i \right\}$$

$$= \frac{2}{n(n+1)} \sum_{i=1}^{n}(2n - 3i + 2)\ell n X_i. \tag{4.5.10}$$

Furthermore, it follows from (4.5.7)-(4.5.9) that

$$\hat{\tau}_L^2 = \frac{1}{n} \sum_{i=1}^{n} \left\{ (i - 1)\ell n \hat{a}_L + \ell n X_i - \frac{1}{n} \sum_{j=1}^{n}[(j - 1)\ell n \hat{a}_L + \ell n X_j] \right\}^2$$

$$= \frac{1}{n} \sum_{i=1}^{n} \left\{ \ell n X_i - \frac{1}{n} \sum_{j=1}^{n} \ell n X_j - \frac{n - 2i + 1}{2}\ell n \hat{a}_L \right\}^2$$

$$= \frac{1}{n} \left\{ \sum_{i=1}^{n}(\ell n X_i - \frac{1}{n} \sum_{j=1}^{n} \ell n X_j)^2 \right.$$

$$-\ell n \hat{a}_L \sum_{i=1}^{n}(n - 2i + 1)(\ell n X_i - \frac{1}{n} \sum_{j=1}^{n} \ell n X_j)$$

$$\left. +\frac{1}{4}(\ell n \hat{a}_L)^2 \sum_{i=1}^{n}(n - 2i + 1)^2 \right\}$$

$$= \frac{1}{n} \left\{ \sum_{i=1}^{n}(\ell n X_i)^2 - \frac{1}{n}\left(\sum_{i=1}^{n} \ell n X_i \right)^2 - \frac{1}{2}\ell n \hat{a}_L \sum_{i=1}^{n}(n - 2i + 1)\ell n X_i \right\}.$$

$$\tag{4.5.11}$$

Therefore, in comparison with the least squares estimators of a, μ and τ^2 given by (4.3.8), (4.3.5) and (4.3.7), we have the following theorem.

Theorem 4.5.1.

(1) $\hat{a}_L = \hat{a}$;

(2) $\hat{\mu}_L = \hat{\mu}$;

(3) $\hat{\tau}_L^2 = \frac{n-2}{n}\hat{\tau}^2$.

Then by using the invariance property of the MLE, it follows from (4.5.2) and (4.5.3) that the MLE of λ and σ^2 are given respectively by

$$\hat{\lambda}_L = \exp(\hat{\mu}_L + \frac{1}{2}\hat{\tau}_L^2) \tag{4.5.12}$$

and

$$\hat{\sigma}_L^2 = \hat{\lambda}_L^2[\exp(\hat{\tau}_L^2) - 1]. \tag{4.5.13}$$

If $a = 1$, we can replace a by 1 in (4.5.7) and (4.5.8). Accordingly, the MLE of μ and τ^2 are given by

$$\hat{\mu}_L = \frac{1}{n}\sum_{i=1}^{n} \ln X_i \tag{4.5.14}$$

and

$$\hat{\tau}_L^2 = \frac{1}{n}\sum_{i=1}^{n}[\ln X_i - \hat{\mu}_L]^2$$

$$= \frac{1}{n}\left\{\sum_{i=1}^{n}(\ln X_i)^2 - \frac{1}{n}\left(\sum_{i=1}^{n}\ln X_i\right)^2\right\}. \tag{4.5.15}$$

Note that under lognormal distribution assumption, we can see from Theorem 4.5.1 that the MLE of μ, β and τ^2 are essentially the least squares estimators. Then, from linear regression theory, we have the following result.

Lemma 4.5.2.

(1)

$$\begin{pmatrix} \hat{\mu} \\ \hat{\beta} \end{pmatrix} \sim N\left(\begin{pmatrix} \mu \\ \beta \end{pmatrix}, \frac{2\tau^2}{n(n+1)}\begin{pmatrix} 2n-1 & 3 \\ 3 & \frac{6}{n-1} \end{pmatrix}\right) \tag{4.5.16}$$

(2)

$$\frac{(n-2)\hat{\tau}^2}{\tau^2} \sim \chi^2(n-2), \tag{4.5.17}$$

(3)

$$\begin{pmatrix} \hat{\mu} \\ \hat{\beta} \end{pmatrix} \text{ is independent of } \hat{\tau}^2. \qquad (4.5.18)$$

For the proof of Lemma 4.5.2, see for instance Seber (1977).

Lemma 4.5.3.

$$\sqrt{n}\begin{pmatrix} \hat{\mu}_L - \mu \\ \hat{\tau}_L^2 - \tau^2 \end{pmatrix} \xrightarrow{L} N\left(\begin{pmatrix} 0 \\ 0 \end{pmatrix}, \begin{pmatrix} 4\tau^2 & 0 \\ 0 & 2\tau^4 \end{pmatrix}\right). \qquad (4.5.19)$$

Proof.

It follows from (4.5.16) that

$$\sqrt{n}(\hat{\mu} - \mu) \sim N(0, \frac{2(2n-1)\tau^2}{n+1}).$$

Then by applying the Slutsky theorem and Theorem 4.5.1, we have

$$\sqrt{n}(\hat{\mu}_L - \mu) \xrightarrow{L} N(0, 4\tau^2).$$

On the other hand, From (4.5.17) and Theorem 4.5.1, we have

$$\frac{n\hat{\tau}_L^2}{\tau^2} = \frac{(n-2)\hat{\tau}^2}{\tau^2} \sim \chi^2(n-2).$$

This means that $n\hat{\tau}_L^2/\tau^2$ is the sum of $(n-2)$ i.i.d random variables each having a chi-square distribution $\chi^2(1)$. Therefore, the central limit theorem gives

$$\frac{n\hat{\tau}_L^2/\tau^2 - (n-2)}{\sqrt{2(n-2)}} = \frac{\sqrt{n-2}(n\hat{\tau}_L^2/(n-2) - \tau^2)}{\sqrt{2}\tau^2} \xrightarrow{L} N(0,1).$$

Consequently,

$$\sqrt{n}(\hat{\tau}_L^2 - \tau^2) \xrightarrow{L} N(0, 2\tau^4).$$

Finally, we note that from (4.5.18), $\hat{\mu}_L$ and $\hat{\tau}_L^2$ are independent. This completes the proof of Lemma 4.5.3.

Furthermore, we have the following theorem.

Theorem 4.5.4.

(1) $n^{3/2}(\hat{a}_L - a) \xrightarrow{L} N(0, 12a^2\tau^2)$,

(2) $n^{1/2}(\hat{\lambda}_L - \lambda) \xrightarrow{L} N(0, \lambda^2\tau^2(4 + \frac{1}{2}\tau^2))$,

(3) $n^{1/2}(\hat{\sigma}_L^2 - \sigma^2) \xrightarrow{L} N(0, 16\sigma^4\tau^2 + 2\tau^4(\lambda^2 + 2\sigma^2)^2)$.

Proof.

Because $\hat{a}_L = \hat{a}$, part (1) of Theorem 4.5.4 follows directly from Theorem 4.4.2. To prove part (2), first of all, define

$$g(\mu, \tau^2) = \exp(\mu + \frac{1}{2}\tau^2),$$

then by using Lemma 4.5.3 and the Cramér δ theorem, (4.5.2) yields that

$$n^{1/2}(\hat{\lambda}_L - \lambda) \xrightarrow{L} N(0, \sigma_\lambda^2), \tag{4.5.20}$$

where

$$\sigma_\lambda^2 = (\partial g/\partial \mu \ \ \partial g/\partial \tau^2) \begin{pmatrix} 4\tau^2 & 0 \\ 0 & 2\tau^4 \end{pmatrix} \begin{pmatrix} \partial g/\partial \mu \\ \partial g/\partial \tau^2 \end{pmatrix}$$

$$= (\lambda \ \lambda/2) \begin{pmatrix} 4\tau^2 & 0 \\ 0 & 2\tau^4 \end{pmatrix} \begin{pmatrix} \lambda \\ \lambda/2 \end{pmatrix}$$

$$= \lambda^2\tau^2 \left(4 + \frac{1}{2}\tau^2\right) \tag{4.5.21}$$

(see Lehmann (1983) for reference). This completes the proof of part (2). By using (4.5.3), the proof of part (3) is similar.

Now, we shall compare the estimators of λ and σ^2 obtained by the maximum likelihood method and nonparametric method. To do this, assume that $a \neq 1$, then from Theorems 4.4.3 and 4.5.4, the asymptotic relative efficiency of $\hat{\lambda}$ to $\hat{\lambda}_L$ (see Lehmann (1983) for reference) is given by

$$e(\hat{\lambda}, \hat{\lambda}_L) = \frac{\text{asy. Var}[\hat{\lambda}_L]}{\text{asy. Var}[\hat{\lambda}]} = \frac{\lambda^2\tau^2(4 + \frac{1}{2}\tau^2)}{\sigma^2 + 3\lambda^2\tau^2}$$

$$= \frac{\tau^2(4 + \frac{1}{2}\tau^2)}{\exp(\tau^2) - 1 + 3\tau^2} < 1, \tag{4.5.22}$$

(4.5.22) is due to (4.5.3) and the following inequality

$$e^x > 1 + x + \frac{1}{2}x^2 \quad \text{for} \quad x > 0.$$

To study the asymptotic relative efficiency of $\hat{\sigma}^2$ to σ_L^2, we need the following lemma that can be proved by induction.

Lemma 4.5.5.

$$h(n) = 6^n - 4 \times 3^n - 2^n(2n^2 + 1) + 8(n^2 + 1) > 0 \quad \text{for} \quad n = 0, 1, \ldots$$

Now, because $X_1 \sim LN(\mu, \tau^2)$, the rth moment of X_1 is then given by

$$E(X_1^r) = \exp(r\mu + \frac{1}{2}r^2\tau^2). \tag{4.5.23}$$

Thus (4.5.3) gives

$$\omega^2 = \text{Var}(X_1 - \lambda)^2 = E[\{(X_1 - \lambda)^2 - \sigma^2\}^2]$$
$$= \lambda^4\{\exp(6\tau^2) - 4\exp(3\tau^2) - \exp(2\tau^2) + 8\exp(\tau^2) - 4\}.$$

Furthermore, we have

$$\omega^2 + 12\sigma^4\tau^2 - \{16\sigma^4\tau^2 + 2\tau^4(\lambda^2 + 2\sigma^2)^2\}$$
$$= \lambda^4\{\exp(6\tau^2) - 4\exp(3\tau^2) - \exp(2\tau^2) + 8\exp(\tau^2) - 4$$
$$-4\tau^2[\exp(\tau^2) - 1]^2 - 2\tau^4[2\exp(\tau^2) - 1]^2\}$$
$$= \lambda^4 \sum_{n=3}^{\infty} h(n)\tau^{2n}/n! > 0, \tag{4.5.24}$$

where (4.5.24) is derived by the Taylor expansion. Therefore the asymptotic relative efficiency of $\hat{\sigma}^2$ to $\hat{\sigma}_L^2$ is given by

$$e(\hat{\sigma}^2, \hat{\sigma}_L^2) = \frac{\text{asy. Var}[\hat{\sigma}_L^2]}{\text{asy. Var}[\hat{\sigma}^2]} = \frac{16\sigma^4\tau^2 + 2\tau^4(\lambda^2 + 2\sigma^2)^2}{\omega^2 + 12\sigma^4\tau^2}$$
$$< 1. \tag{4.5.25}$$

In conclusion, the combination of (4.5.22) and (4.5.25) yields the following result.

Theorem 4.5.6.
(1) $e(\hat{\lambda}, \hat{\lambda}_L) < 1$;
(2) $e(\hat{\sigma}^2, \hat{\sigma}_L^2) < 1$.

  Therefore asymptotically, for $a \neq 1$, $\hat{\lambda}_L$ is more efficient than $\hat{\lambda}$ and hence better than $\hat{\lambda}$. Similarly, $\hat{\sigma}_L^2$ is more efficient than $\hat{\sigma}^2$ and hence better than $\hat{\sigma}^2$.

4.6   Notes and References

In this chapter, we study the problems for the statistical inference of a GP, including the hypothesis testing and estimation of parameters a, λ and σ^2 in the GP. Sections 4.2 and 4.3 are based on Lam (1992b), but the estimators $\hat{a}_D$ and $\hat{\lambda}_D$ were studied by Chan et al. (2006). Section 4.4 is due to Lam et al. (2004). The results in Sections 4.2-4.4 are nonparametric. However, a parametric approach can be applied. Section 4.5 is due to Lam and Chan (1998), it studies the parametric estimation of the three parameters a, λ

and σ^2 of a GP under an additional assumption that X_1 has a lognormal distribution. Lam and Chan (1998) also did some simulation study for justification of the parametric method. Thereafter, Chan et al. (2004) also considered the statistical inference problem for a GP with a gamma distribution by assuming that X_1 has a $\Gamma(\alpha, \beta)$ distribution with density function (1.4.17). The MLE for parameters a, α and β are then derived. Consequently, the MLE for λ and σ^2 are obtained accordingly. See Chan et al. (2004) for more details. Refer to Leung (2005) for some related work.

Chapter 5

Application of Geometric Process to Data Analysis

5.1 Introduction

In Chapter 4, we consider the problem of testing if a data set, from a sequence of events with trend for example, comes from a GP. Then we study the problem of estimation of the three important parameters a, λ and σ^2 in the GP. In this chapter, we shall apply the theory developed in Chapter 4 to analysis of data. To do this, in Section 5.2, the methodology of using a GP model for data analysis is discussed. First of all, we shall test if a data set $\{X_i, i = 1, 2, \ldots, n\}$ is consistent with a GP, i.e. if a data set can be modelled by a GP. Then three parameters a, λ and σ^2 of the GP are estimated. Consequently, the fitted values of the data set can be simply evaluated by

$$\hat{X}_i = \hat{\lambda}/\hat{a}^{n-1} \quad i = 1, 2, \ldots, n, \tag{5.1.1}$$

where $\hat{a}$ and $\hat{\lambda}$ are the estimates of a and λ respectively. Then, Section 5.3 briefly studies the methodology of data analysis by using two nonhomogeneous Poisson process models including the Cox-Lewis model and the Weibull process (WP) model. Afterward in Section 5.4, 10 real data sets are fitted by the GP model, the renewal process (RP) or the homogeneous Poisson process (HPP) model, the Cox-Lewis model and the WP model. The numerical results obtained by the four models are compared. The results show that on average the GP model is the best model among these four models. Thereafter in Section 5.5, a threshold GP model is introduced to analyse the data with multiple trends. As an example, four real SARS data sets are then studied.

121

5.2   Data Analysis by Geometric Process Model

Now, we shall introduce the methodology for applying a GP model to data analysis. To do this, we assume that a data set $\{X_i, i = 1, 2, \ldots, n\}$ is given. We may interpret the data set as a sequence of the interarrival times from a series of events. Then the arrival time $T_i, i = 1, 2, \ldots, n$, of the ith event is determined by

$$T_i = \sum_{j=1}^{i} X_j$$

with $T_0 = 0$. Moreover, the fitted value of X_j is denoted by $\hat{X}_j, j = 1, 2, \ldots, n$. Then the fitted value of T_i will be given by

$$\hat{T}_i = \sum_{j=1}^{i} \hat{X}_i$$

with $\hat{T}_0 = 0$.

To test if a data set fits a GP model, we can form two sequences of random variables and test whether the random variables in these sequences are i.i.d according to Theorems 4.2.1 and 4.2.2. If the size n is even, two sequences in (4.2.5) are constructed, and if the size n is odd, two sequences in (4.2.6) or (4.2.7) are formed. Let $T(U)$ and $T(V)$ (or $T(U^{'})$ and $T(V^{'})$) be the values calculated from (4.2.8) by replacing W_i with U_i and V_i (or U_i' and V_i') respectively. Similarly, let $D(U)$ and $D(V)$ (or $D(U')$ and $D(V')$) be the values evaluated from (4.2.9) by replacing W_i with U_i and V_i (or U_i' and V_i') respectively. Then we can compute the following p-values:

$$P_T^U = \begin{cases} P(\mid Z \mid \geq T(U)) & \text{if } W_i \text{ is replaced by } U_i, \\ P(\mid Z \mid \geq T(U')) & \text{if } W_i \text{ is replaced by } U_i', \end{cases}$$

$$P_D^U = \begin{cases} P(\mid Z \mid \geq D(U)) & \text{if } W_i \text{ is replaced by } U_i, \\ P(\mid Z \mid \geq D(U')) & \text{if } W_i \text{ is replaced by } U_i'. \end{cases}$$

If a p-value is small, less than 0.05 say, we shall reject the null hypothesis that the data set fits a GP. Similary, we can also evaluate the p-values

$$P_T^V = \begin{cases} P(\mid Z \mid \geq T(V)) & \text{if } W_i \text{ is replaced by } V_i, \\ P(\mid Z \mid \geq T(V')) & \text{if } W_i \text{ is replaced by } V_i', \end{cases}$$

$$P_D^V = \begin{cases} P(\mid Z \mid \geq D(V)) & \text{if } W_i \text{ is replaced by } V_i, \\ P(\mid Z \mid \geq D(V')) & \text{if } W_i \text{ is replaced by } V_i'. \end{cases}$$

To test whether the data set agrees with a RP (or HPP), we first calculate the values of R and L from (4.4.7) and (4.4.8). Then we can evaluate $P_R = P(|Z| \geq R)$ and $P_L = P(|Z| \geq L)$. If a p-value is small, less than 0.05 say, we shall reject the null hypothesis that the data set fits a RP (or HPP).

If a data set comes from a GP, the parameters a, λ and σ^2 will be estimated respectively by (4.3.8)-(4.3.10). As a result, the fitted value of X_i can be simply evaluated by (5.1.1).

5.3 Data Analysis by Poisson Process Models

Suppose that we want to analyse a data set $\{X_i, i = 1, 2, \ldots, n\}$ where X_i is the interarrival time between the $(i-1)$th and ith events by using a Poisson process model. If a renewal process (RP) model with mean λ or a homogeneous Poisson process (HPP) model with rate λ is applied, the estimate of λ will be given by

$$\hat{\lambda} = \bar{X} = \frac{1}{n} \sum_{i=1}^{n} X_i. \tag{5.3.1}$$

Then for all $i = 1, \ldots, n$, the fitted value of X_i is simply given by

$$\hat{X}_i = \hat{\lambda} = \bar{X}. \tag{5.3.2}$$

Now suppose a nonhomogeneous Poisson process model with intensity function $\lambda(t)$ is applied. We shall discuss two popular models here, namely the Cox-Lewis model and the Weibull process model.

(1) The Cox-Lewis model

By (1.2.21), the intensity function is

$$\lambda(x) = \exp(\alpha_0 + \alpha_1 x), \qquad -\infty < \alpha_0, \alpha_1 < \infty, \quad x > 0. \tag{5.3.3}$$

Clearly, if $\alpha_1 > 0$, the Cox-Lewis model is applicable to the case with decreasing successive interarrival times. If $\alpha_1 < 0$, it can be applied to the case with increasing successive interarrival times. If $\alpha_1 = 0$, the Cox-Lewis model becomes a HPP model. Therefore, we can assume that $\alpha_1 \neq 0$. Now denote $w = \exp(\alpha_0)/\alpha_1$. The conditional density of T_i given $T_1, \ldots, T_{i-1}$

is given by

$$f[t \mid T_1, \ldots, T_{i-1}] = \lambda(t) \exp\{-\int_{T_{i-1}}^{t} \lambda(x)dx\}$$

$$= \exp(\alpha_0 + \alpha_1 t) \exp\{-w(e^{\alpha_1 t} - e^{\alpha_1 T_{i-1}})\}, \quad t > T_{i-1}.$$

Then, it is straightforward to show that

$$E[T_i \mid T_1, \ldots, T_{i-1}]$$

$$= \int_{T_{i-1}}^{\infty} t \exp(\alpha_0 + \alpha_1 t) \exp\{-w(e^{\alpha_1 t} - e^{\alpha_1 T_{i-1}})\}dt \qquad (5.3.4)$$

$$= \exp(\alpha_0 + we^{\alpha_1 T_{i-1}}) \int_{T_{i-1}}^{\infty} t \exp(\alpha_1 t - we^{\alpha_1 t})dt. \qquad (5.3.5)$$

It is easy to show that the MLE $\hat{\alpha}_0$ and $\hat{\alpha}_1$ of α_0 and α_1 are the solution to the following equations.

$$\exp(\alpha_0) = \frac{n\alpha_1}{\exp(\alpha_1 T_n) - 1}, \qquad (5.3.6)$$

$$\sum_{i=1}^{n} T_i + \frac{n}{\alpha_1} - \frac{nT_n \exp(\alpha_1 T_n)}{\exp(\alpha_1 T_n) - 1} = 0. \qquad (5.3.7)$$

Thus, $\hat{w} = \exp(\hat{\alpha}_0)/\hat{\alpha}_1$. Moreover, starting with $\widehat{T}_0 = 0$, the fitted value of T_i using (5.3.5) will be given iteratively by

$$\widehat{T}_i = \exp(\hat{\alpha}_0 + \hat{w}e^{\hat{\alpha}_1 \widehat{T}_{i-1}}) \int_{\widehat{T}_{i-1}}^{\infty} t \exp(\hat{\alpha}_1 t - \hat{w}e^{\hat{\alpha}_1 t})dt.$$

Consequently, the fitted value of X_i is given by

$$\hat{X}_i = \widehat{T}_i - \widehat{T}_{i-1}. \qquad (5.3.8)$$

(2) The Weibull process model

Now, suppose that a data set $\{X_i, i = 1, 2, \ldots, n\}$ follows a Weibull process (WP) model. From (1.2.22), the intensity function is

$$\lambda(x) = \alpha\theta x^{\theta-1}, \qquad \alpha, \theta > 0, x > 0. \qquad (5.3.9)$$

Obviously, if $\theta > 1$, the WP model is applicable to the case that the successive interarrival times are decreasing, and if $0 < \theta < 1$, it is suitable to the case that the successive interarrival times are increasing. If $\theta = 1$, the WP model reduces to a HPP model. Therefore, we can assume that $\theta \neq 1$. The conditional density of T_i given $T_1, \ldots, T_{i-1}$ is given by

$$f[t \mid T_1, \ldots, T_{i-1}] = \lambda(t) \exp\{-\int_{T_{i-1}}^{t} \lambda(x)dx\}$$

$$= \alpha\theta t^{\theta-1} \exp\{-\alpha(t^{\theta} - T_{i-1}^{\theta})\}, \quad t > T_{i-1}.$$

Then, it is straightforward to show that

$$E[T_i \mid T_1, \ldots, T_{i-1}] = \frac{\exp(u_{i-1})}{\alpha^{1/\theta}} \int_{u_{i-1}}^{\infty} t^{1/\theta} e^{-t} dt, \qquad (5.3.10)$$

where $u_{i-1} = \alpha T_{i-1}^{\theta}$ with $u_0 = 0$. Moreover, the MLE $\hat{\alpha}$ and $\hat{\theta}$ of α and θ are given respectively by

$$\hat{\alpha} = \frac{n}{T_n^{\hat{\theta}}}, \qquad (5.3.11)$$

and

$$\hat{\theta} = \frac{n}{\left[\sum_{i=1}^{n} \ell n(T_n/T_i) \right]}. \qquad (5.3.12)$$

Based on (5.3.11) and (5.3.12), we can fit T_i by (5.3.10) iteratively as in the Cox-Lewis model. The fitted value is given by

$$\widehat{T}_i = \frac{\exp(\hat{u}_{i-1})}{\hat{\alpha}^{1/\hat{\theta}}} \int_{\hat{u}_{i-1}}^{\infty} t^{1/\hat{\theta}} e^{-t} dt, \qquad (5.3.13)$$

where $\hat{u}_{i-1} = \hat{\alpha} \widehat{T}_{i-1}^{\hat{\theta}}$ with $\hat{u}_0 = 0$. Again, the fitted value of X_i can be evaluated by (5.3.8).

5.4 Real Data Analysis and Comparison

In this section, ten real data sets will be analyzed by the GP model, the RP or HPP model, the Cox-Lewis model and the WP model. We shall test if a data set can be modelled by a GP model and if the ratio of the GP equals 1. Then for each data set, the parameters in four models are estimated respectively. Afterward, the data are fitted by these four models.

To compare the fitted results obtained by different models, we define the mean squared error (MSE) by

$$\text{MSE} = \frac{1}{n} \sum_{i=1}^{n} (X_i - \hat{X}_i)^2,$$

and the maximum percentage error (MPE) by

$$\text{MPE} = \max_{1 \leq i \leq n} \{ \mid T_i - \widehat{T}_i \mid /T_i \}.$$

These two quantities are used as the criteria for measuring the goodness-of-fit of a model and for the comparison of different models. Roughly speaking, the MSE measures the overall fitting while the MPE concerns more on an individual fitting of a data set. Then the values of MSE and MPE are evaluated for model comparison.

In the 10 examples below, if the RP (HPP) model is rejected at 0.05 level of significance, the values P_R and/or P_L are marked by '*'. If $a = 1$, i.e. the RP (HPP) model is not rejected, we can fitted two GP models, one is the GP model with ratio $\hat{a}$, and the other one is the RP (HPP) model, then choose a better model with a smaller MSE or MPE as the modified GP model (MGP). If $a \neq 1$, i.e. the RP (HPP) is rejected, the GP model is also the modified GP model. Accordingly, the values of MSE or MPE corresponding to the modified GP model will also be marked by '*'.

Example 5.4.1. The coal-mining disaster data.

This data set was originally studied by Maguire, Pearson and Wynn (1952), and was corrected and extended by Jarrett (1979). Also see Andrews and Herzberg (1985) for its source. The data set contains one zero because there were two accidents on the same day. This zero is replaced by 0.5 since two accidents occuring on the same day are usually not at exactly the same time. Hence the interarrival time could be approximated by 0.5 day. Then the size of the adjusted data is 190.

(1) Testing if the data agree with a GP.

H_0: it is a GP				H_0: it is a RP (or HPP)	
P_T^U	P_D^U	P_T^V	P_D^V	P_R	P_L
0.46108	0.47950	0.14045	0.47950	*8.76585×10^{-7}	*1.84297×10^{-7}

The p-values P_T^U, P_D^U, P_T^V and P_D^V are all insignificant and hence we conclude that the data set could be modelled by a GP model. Moreover the p-values P_R and P_L reveal a strong evidence that the data follow a GP with $a \neq 1$.

(2) Estimation and comparison.

		GP		RP (HPP)		Cox-Lewis		WP
MSE		*8.17888×10^4		9.77941×10^4		7.88506×10^4		8.77217×10^4
MPE		*0.50313		2.45817		0.51646		0.86325
	a	0.99091	a	1	α_0	-4.52029	α	0.15404
Est.	λ	78.00898	λ	213.41842	α_1	-4.96387×10^{-5}	θ	0.67082
	σ^2	8700.84032	σ^2	98311.51844				

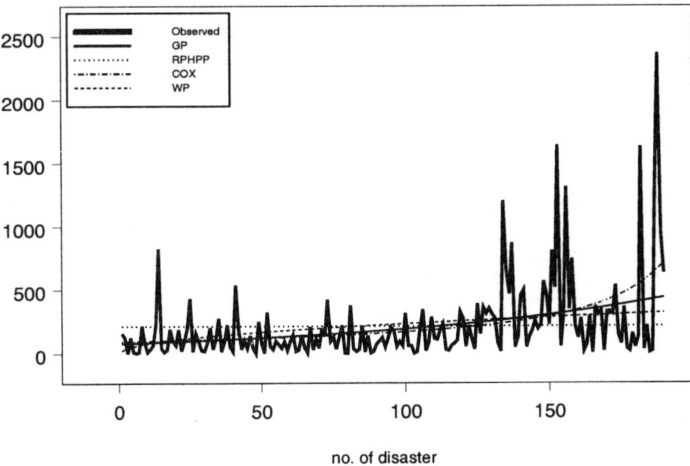

Figure 5.4.1. Observed and fitted X_i in Example 5.4.1.

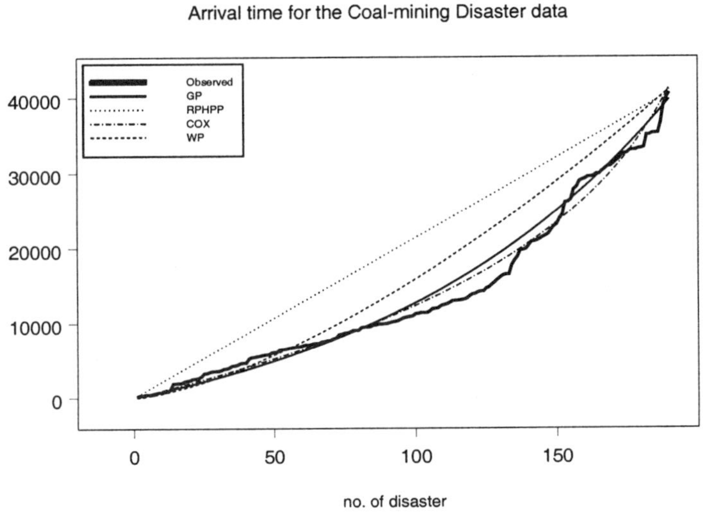

Figure 5.4.2. Observed and fitted T_i in Example 5.4.1.

Example 5.4.2. The aircraft 3 data.

Thirteen data sets of the time intervals between successive failures of air conditioning equipment in 13 Boeing 720 aircraft were studied by Proschan

(1963). Later on, the Cox-Lewis model was applied to the analysis of these data sets (see Cox and Lewis (1966)). Totally, there are 13 data sets and we shall analyze the largest three of them, namely the data sets of aircraft 3, 6 and 7. The aircraft 3 data set is of size 29 which has the second largest size among the 13 data sets.

(1) Testing if the data agree with a GP.

H_0: it is a GP				H_0: it is a RP (or HPP)	
P_T^U	P_D^U	P_T^V	P_D^V	P_R	P_L
0.49691	0.65472	0.17423	0.65472	*0.04935	0.13877

The results show that a GP model is appropriate for the aircraft 3 data. Moreover, since $P_R = 0.04935$, the hypothesis of a RP (or HPP) is rejected at 0.05 significance level.

(2) Estimation and comparison.

		GP		RP (HPP)		Cox-Lewis		WP	
MSE		*4.26817×10^3		4.84059×10^3		4.27368×10^3		4.64721×10^3	
MPE		*0.45221		0.67034		0.33067		0.36280	
	a	0.96528	a	1	α_0	-4.07980	α	2.59523×10^{-2}	
Est.	λ	49.30115	λ	83.51724	α_1	-3.03625×10^{-4}	θ	0.90073	
	σ^2	1630.72965	σ^2	5013.47291					

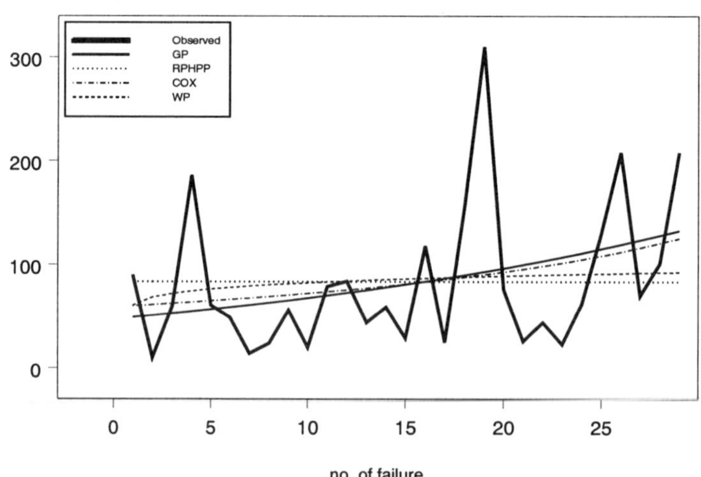

Figure 5.4.3. Observed and fitted X_i in Example 5.4.2.

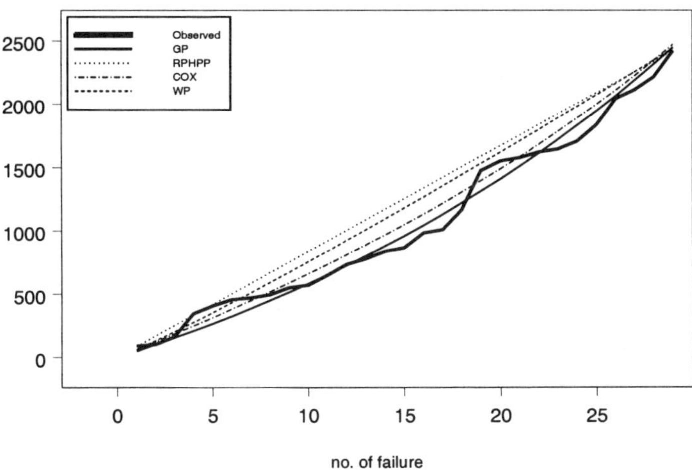

Figure 5.4.4. Observed and fitted T_i in Example 5.4.2.

Example 5.4.3. The aircraft 6 data.

The aircraft 6 data set is of size 30 which has the largest size among the 13 data sets.

(1) Testing if the data agree with a GP.

H_0: it is a GP				H_0: it is a RP (or HPP)	
P_T^U	P_D^U	P_T^V	P_D^V	P_R	P_L
0.66327	0.38648	0.82766	1.00000	0.07306	*0.02735

The results show that the data set can be modelled by a GP but not a RP (or HPP) at 0.05 significance level.

(2) Estimation and comparison.

		GP	RP (HPP)		Cox-Lewis		WP	
MSE		*4.38519×10^3	4.99517×10^3		4.45453×10^3		4.88807×10^3	
MPE		*3.90365	1.59130		4.84261		6.33027	
	a	1.05009	a	1	α_0	-5.01813	α	3.79604×10^{-4}
Est.	λ	112.78403	λ	59.60000	α_1	9.17952×10^{-4}	θ	1.50592
	σ^2	14966.79465	σ^2	5167.42069				

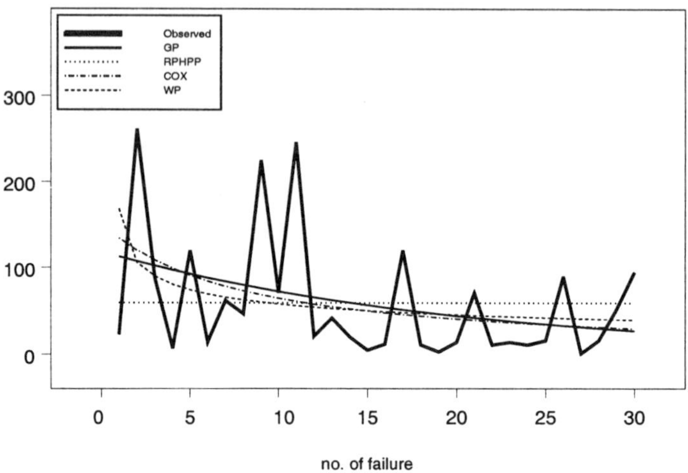

Figure 5.4.5. Observed and fitted X_i in Example 5.4.3.

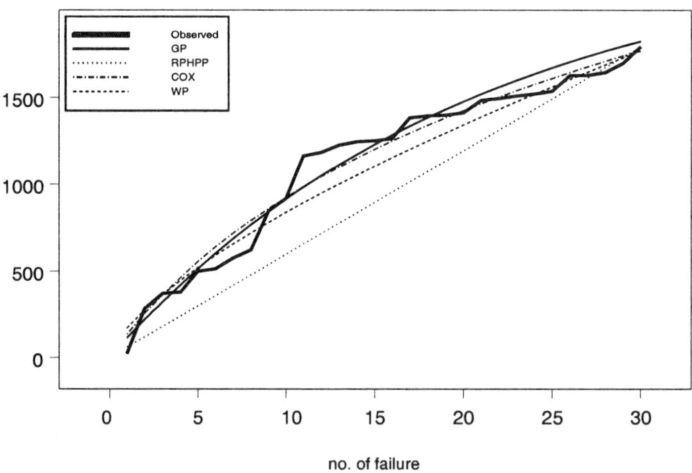

Figure 5.4.6. Observed and fitted T_i in Example 5.4.3.

Example 5.4.4. The aircraft 7 data.

The aircraft 7 data set has 27 interarrival times and has the third largest size among the 13 data sets.

(1) Testing if the data agree with a GP.

H_0: it is a GP				H_0: it is a RP (or HPP)	
P_T^U	P_D^U	P_T^V	P_D^V	P_R	P_L
0.63641	0.35454	0.23729	1.00000	0.44498	0.58593

Results show that a GP model is a reasonable model. Moreover, since $P_R = 0.44498$ and $P_L = 0.58593$, a RP (or HPP) model is also acceptable.

(2) Estimation and comparison.

		GP	RP (HPP)		Cox-Lewis		WP	
MSE		$^*3.90667 \times 10^3$	3.90897×10^3		3.87200×10^3		3.89746×10^3	
MPE		$^*0.42138$	0.88503		0.78491		0.84606	
	a	0.97639	a	1	α_0	-4.27505	α	1.42946×10^{-2}
Est.	λ	56.12595	λ	76.81481	α_1	-6.47003×10^{-5}	θ	0.98775
	σ^2	2080.77753	σ^2	4059.31054				

Inter-arrival time for the Aircraft 7 data

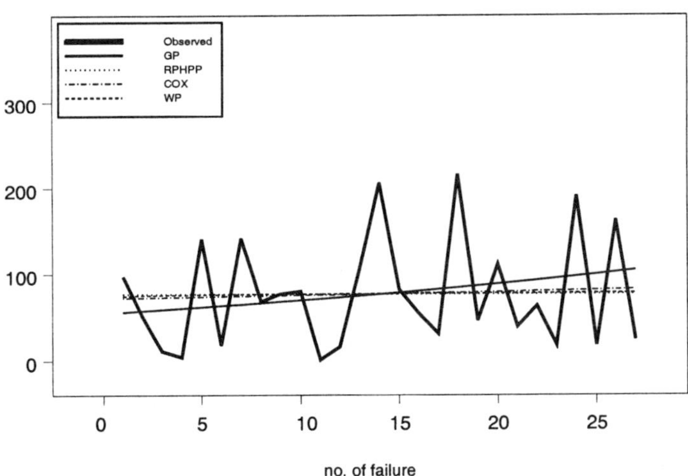

Figure 5.4.7. Observed and fitted X_i in Example 5.4.4.

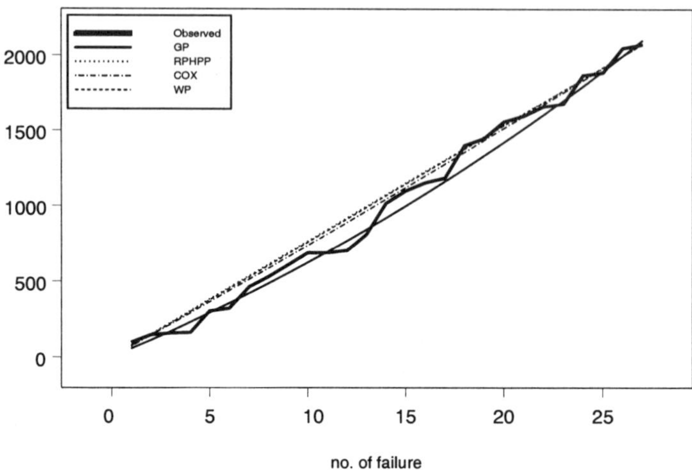

Figure 5.4.8. Observed and fitted T_i in Example 5.4.4.

Example 5.4.5. The computer data.

The computer data recorded the failure times of an electronic computer in unspecified units. The size of data set is 257 (see Cox and Lewis (1966) for reference).

(1) Testing if the data agree with a GP.

H_0: it is a GP				H_0: it is a RP (or HPP)	
P_T^U	P_D^U	P_T^V	P_D^V	P_R	P_L
0.83278	0.64731	0.67283	0.16991	0.62773	0.40020

From the results, the data set could be modelled by either a GP model or a RP (or HPP) model.

(2) Estimation and comparison.

		GP		RP (HPP)		Cox		WP
MSE		*2.63970×10^5		2.64204×10^5		2.73778×10^5		2.64607×10^5
MPE		1.02546		*0.95401		1.14872		0.76350
Est.	a	1.00031	a	1	α_0	-5.99797	α	4.34914×10^{-3}
	λ	377.24250	λ	362.85992	α_1	2.19202×10^{-6}	θ	0.96013
	σ^2	284228.18562	σ^2	265235.87093				

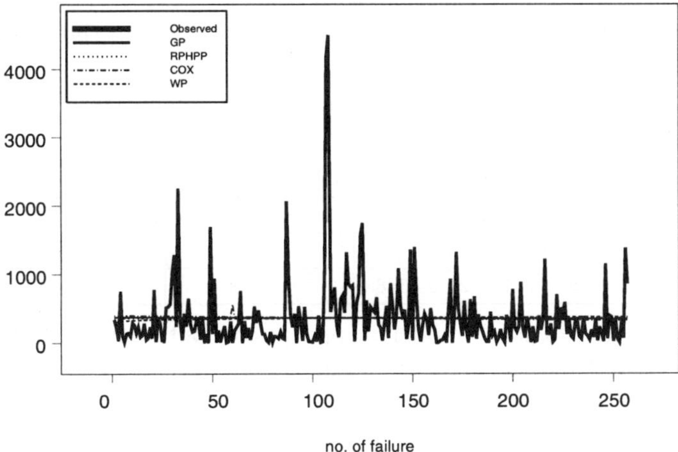

Figure 5.4.9. Observed and fitted X_i in Example 5.4.5.

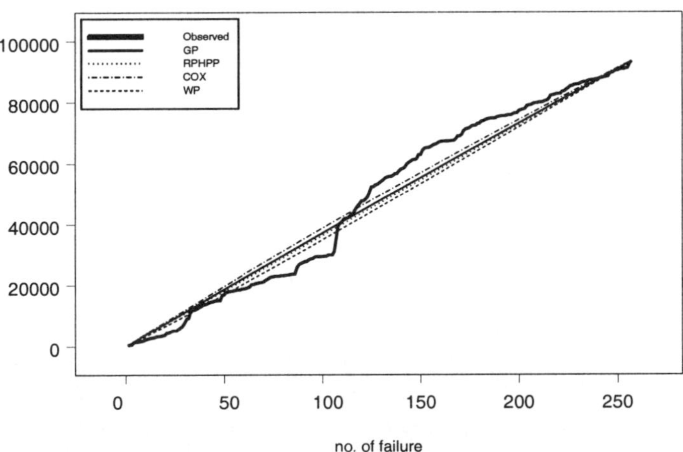

Figure 5.4.10. Observed and fitted T_i in Example 5.4.5.

Example 5.4.6. The patient data.

The patient data set recorded the arrival time of the patients at an intensive care unit of a hospital. The data set contains some zeros because 2 patients arrived in one unit time interval (5 minutes). As in Example

5.4.1, these zeros are replaced by 0.5. Moreover the arrival time of the 247th patient is earlier than that of the 246th patient. This is clearly a wrong record. Hence the data set is adjusted by deleting the arrival times of the 247th and all later patients. After adjustment, the adjusted data set will have 245 interarrival times coming from 246 patients (see Cox and Lewis (1966) for reference).

(1) Testing if the data agree with a GP.

H_0: it is a GP				H_0: it is a RP (or HPP)	
P_T^U	P_D^U	P_T^V	P_D^V	P_R	P_L
0.66525	0.08581	0.82872	0.43488	0.22009	*0.00961

From the results, a GP model is appropriate, but a RP (or HPP) model is rejected at 0.01 significance level.

(2) Estimation and comparison.

		GP		RP (HPP)		Cox		WP
MSE		*1.17648×10^3		1.21071×10^3		1.17018×10^3		1.18411×10^3
MPE		*6.59249		5.33424		7.55167		12.21500
Est.	a	1.00158	a	1	α_0	-3.94854	α	5.68196×10^{-3}
	λ	45.55491	λ	38.00751	α_1	6.24328×10^{-5}	θ	1.16764
	σ^2	1671.27418	σ^2	1215.37058				

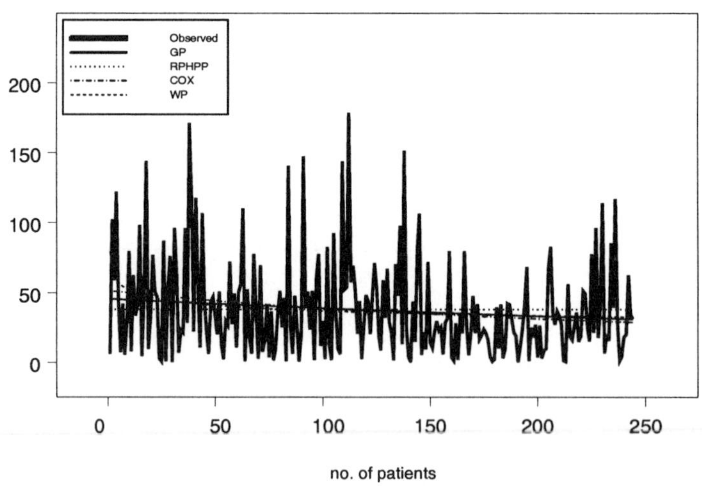

Figure 5.4.11. Observed and fitted X_i in Example 5.4.6.

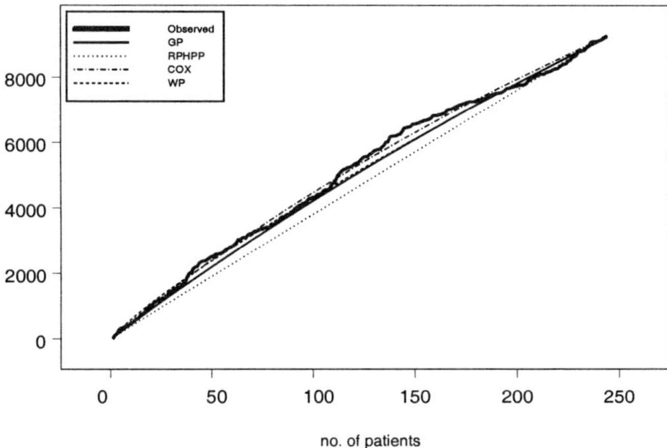

Figure 5.4.12. Observed and fitted T_i in Example 5.4.6.

Example 5.4.7. The No.3 data.

The No.3 data set contains the arrival times to unscheduled maintenance actions for the U.S.S Halfbeak No.3 main propulsion diesel engine. The No.3 data were studied by Ascher and Feingold (1969 and 1984). Since we are interested only in the arrival times of failures which cause the unscheduled maintenance actions, the arrival times to scheduled engine overhauls are then discarded. As a result, the adjusted data set contains 71 interarrival times.

(1) Testing if the data agree with a GP.

H_0: it is a GP				H_0: it is a RP (or HPP)	
P_T^U	P_D^U	P_T^V	P_D^V	P_R	P_L
0.41029	0.56370	0.68056	1.00000	*3.74283×10^{-5}	*9.8658×10^{-14}

The results show a strong evidence that the data set could be modelled by a GP with $a \neq 1$.

(2) Estimation and comparison.

	GP		RP (HPP)		Cox-Lewis		WP	
MSE	*1.96179×10^5		3.32153×10^5		2.19096×10^5		3.76349×10^5	
MPE	*0.40626		0.78942		1.06518		2.50796	
Est.	a	1.04165	a	1	α_0	-8.33532	α	4.8626×10^{-11}
	λ	1.07621×10^3	λ	3.59408×10^2	α_1	1.49352×10^{-4}	θ	2.76034
	σ^2	2.11031×10^6	σ^2	3.36898×10^5				

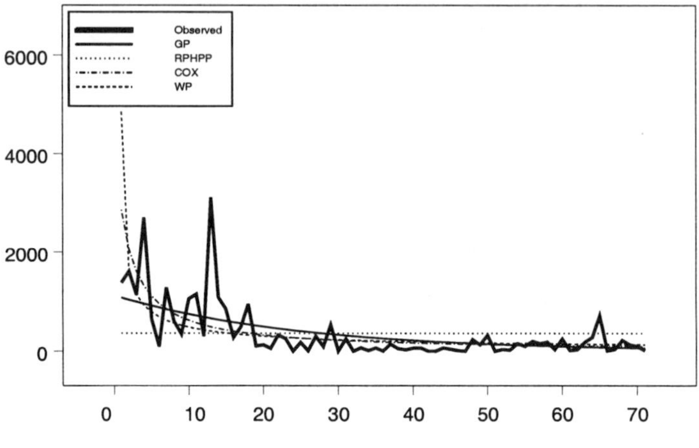

Figure 5.4.13. Observed and fitted X_i in Example 5.4.7.

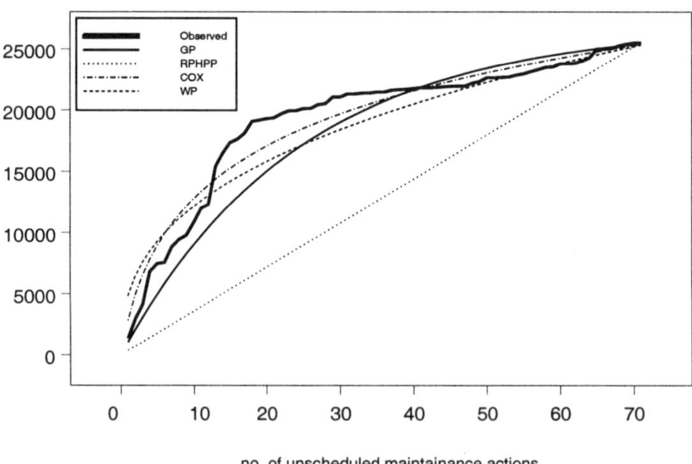

Figure 5.4.14. Observed and fitted T_i in Example 5.4.7.

Example 5.4.8. The No.4 data.

The No.4 data set contains the arrival times to unscheduled maintenance actions for the U.S.S. Grampus No.4 main propulsion diesel engine.
The data set was studied by Lee (1980a and b). The largest interarrival

time of 6930 is extremely outlying because "the person who recorded failures went on leave and nobody took his place until his return" (see Ascher and Feingold (1984)). Thus the datum 6930 and its successor 575 are scrapped. Moreover, as in Example 5.4.7, the arrival times of the scheduled engine overhauls are also discarded. Consequently, the adjusted data set contains 56 interarrival times with a zero interarrival time replaced by 0.5 as before.

(1) Testing if the data agree with a GP.

H_0: it is a GP				H_0: it is a RP (or HPP)	
P_T^U	P_D^U	P_T^V	P_D^V	P_R	P_L
0.75734	0.33459	0.08925	0.74773	0.12270	0.31765

From the results, although a GP model is still applicable, a RP or an HPP model is also acceptable.

(2) Estimation and comparison.

		GP		RP (HPP)		Cox-Lewis		WP
MSE		6.99051×10^4		$^*6.86290 \times 10^4$		6.76273×10^4		6.57068×10^4
MPE		$^*0.48788$		0.68707		0.58417		0.39750
Est.	a	1.01809	a	1	α_0	-5.89286	α	4.56957×10^{-4}
	λ	440.42242	λ	269.11607	α_1	3.77321×10^{-5}	θ	1.21785
	σ^2	202345.83807	σ^2	69876.84537				

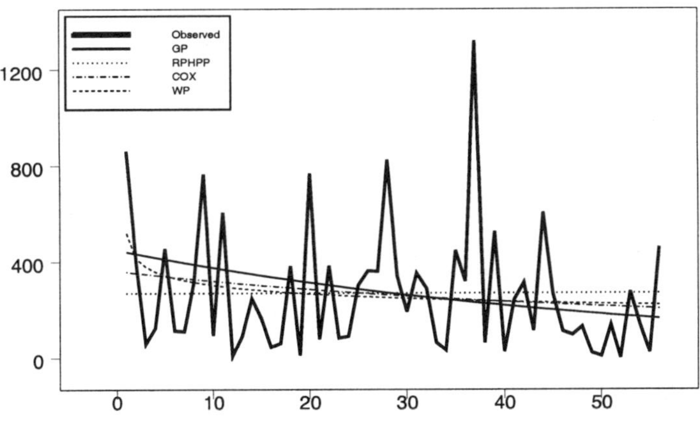

Figure 5.4.15. Observed and fitted X_i in Example 5.4.8.

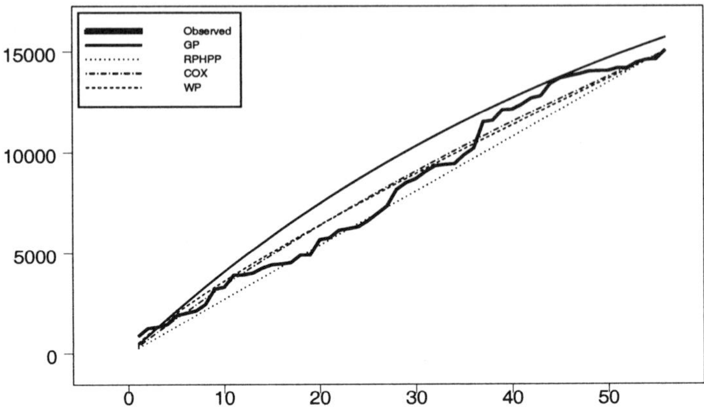

Figure 5.4.16. Observed and fitted T_i in Example 5.4.8.

Example 5.4.9. The car data.

This data set was collected and studied by Lewis (1986). The original data are the times that 41 successive vehicles travelling northwards along the $M1$ motorway in England passed a fixed point near Junction 13 in Bedfordshire on Saturday, 23rd March 1985. The adjusted data set contains 40 successive interarrival times of vehicles (also see Hand et al. (1994) for reference).

(1) Testing if the data agree with a GP.

H_0: it is a GP				H_0: it is a RP (or HPP)	
P_T^U	P_D^U	P_T^V	P_D^V	P_R	P_L
0.57812	0.25684	1.00000	0.70546	0.25581	0.23638

As in previous examples, the result show that a GP model is appropriate, a RP (or HPP) model is also acceptable.

(2) Estimation and comparison.

		GP		RP (HPP)		Cox-Lewis		WP
MSE		*58.51012		60.41000		59.11375		59.94275
MPE		0.85048		*0.41818		1.02818		1.03741
Est.	a	1.01594	a	1	α_0	-2.47994	α	3.90616×10^{-2}
	λ	10.42037	λ	7.80000	α_1	2.56010×10^{-3}	θ	1.20695
	σ^2	102.45368	σ^2	61.95897				

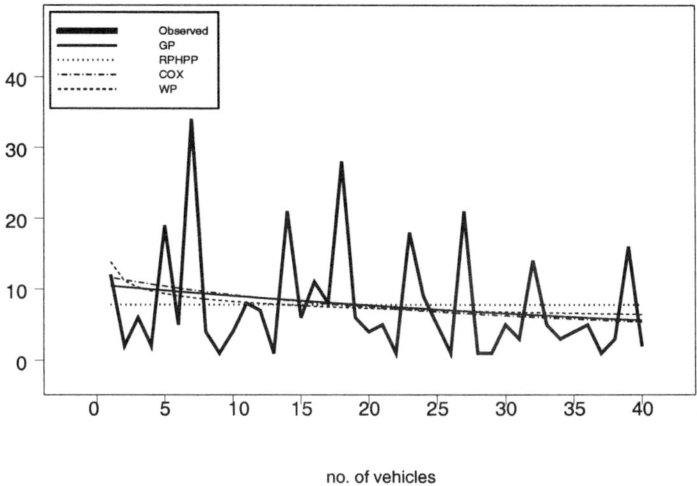

Figure 5.4.17. Observed and fitted X_i in Example 5.4.9.

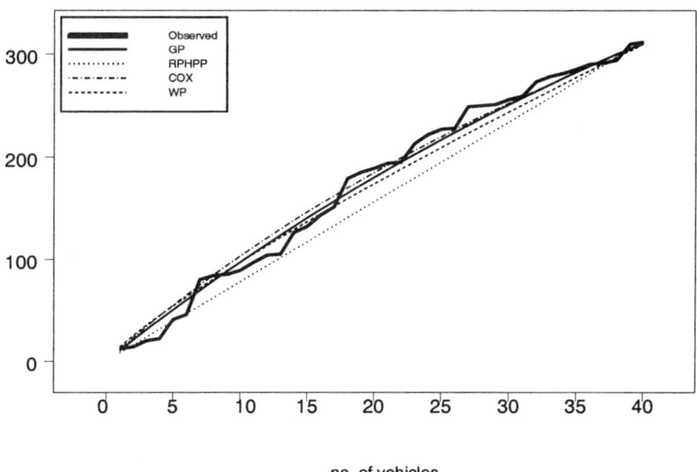

Figure 5.4.18. Observed and fitted T_i in Example 5.4.9.

Example 5.4.10. The software system failures data.

This data set originated from Musa (1979) (see also Musa et al. (1987) and Hand et al. (1994) for its source). It contains 136 failure times (in

CPU seconds, measured in terms of execution time) of a real-time command and control software system. Musa et al. (1987) suggested fitting a nonhomogeneous Poisson process model to this data set. There are three cases that the consecutive failure times are identical. As in Example 5.4.1, their interarrival times were adjusted from 0 to 0.5. The adjusted data set consists of 135 interarrival times of system failures.

(1) Testing if the data agree with a GP.

H_0: it is a GP				H_0: it is a RP (or HPP)	
P_T^U	P_D^U	P_T^V	P_D^V	P_R	P_L
0.37449	0.93216	0.07570	0.67038	*8.75589×10^{-7}	*0.00000

From the results, it is clear that the data could be modeled by a GP with $a \neq 1$.

(2) Estimation and comparison.

		GP		RP (HPP)		Cox-Lewis		WP
MSE		*7.50100×10^5		1.06327×10^6		8.18677×10^5		8.14735×10^5
MPE		*29.84097		216.36152		67.69000		1.32667
Est.	a	0.97687	a	1	α_0	-5.32116	α	0.56839
	λ	92.52291	λ	652.08456	α_1	-3.42038×10^{-5}	θ	0.48079
	σ^2	10190.64326	σ^2	1071149.05020				

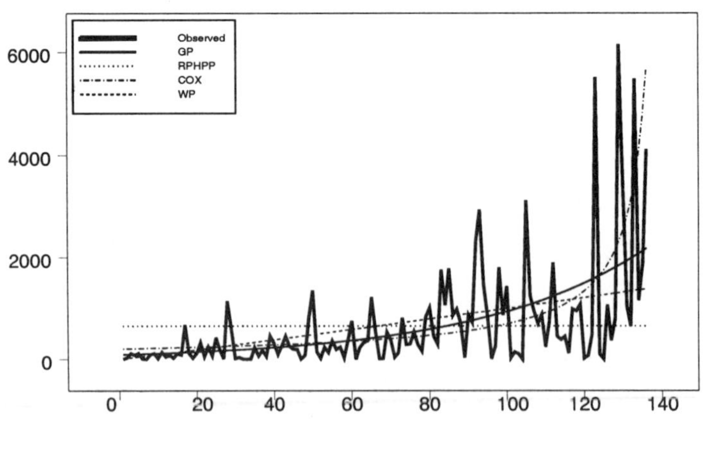

Figure 5.4.19. Observed and fitted X_i in Example 5.4.10.

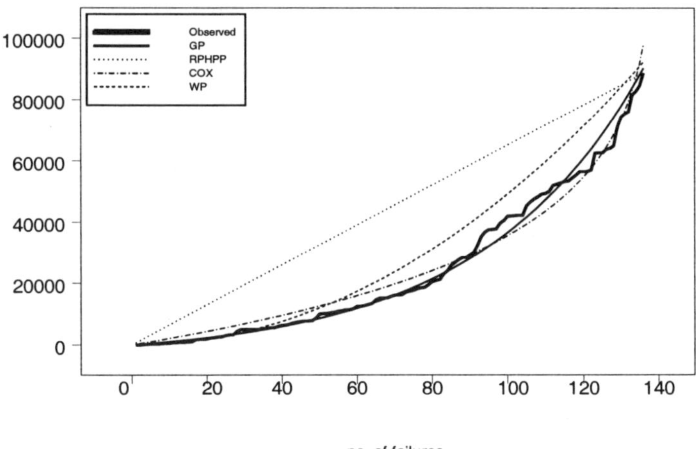

Figure 5.4.20. Observed and fitted T_i in Example 5.4.10.

From the analysis of 10 real data sets, the average values of MSE and MPE (AMSE and AMPE) for four models are summarized below. For further comparison, the values of AMSE and AMPE for the modified GP model (MGP) are also presented.

	GP	RP (HPP)	Cox-Lewis	WP	MGP
AMSE	137570	184110	146280	162320	137450
AMPE	4.4502	23.0150	8.5597	2.6598	4.3999

On the other hand, we may compare the average ranks of MSE and MPE (ARMSE and ARMPE) for the 10 real data sets.

	GP	RP (HPP)	Cox-Lewis	WP	MGP
ARMSE	1.7	3.6	1.9	2.8	1.6
ARMPE	1.9	2.7	2.5	2.7	1.7

Therefore, by comparison, the GP model has the smallest average MSE and the second smallest average MPE among these four models. The performance of GP model is even better when we compare the rank of these four models, the GP model has the smallest average ranks of MSE and MPE. The visual impression shown by the figures agrees with the results from these two summary tables.

Furthermore, from above two tables, the modified GP model is clearly better than the GP model. In fact, the average MSE and MPE for the modified GP model are respectively 1.3745×10^5 and 4.3999, and the average ranks of MSE and MPE are respectively 1.6 and 1.7. Therefore, it is suggested to fit a modified GP model to a data set. Note that under different criteria, the modified GP model may be different. For instance, in Example 5.4.5, the RP (or HPP) model is not rejected. Then according to MSE, the GP model is taken as the modified GP model, however, according to MPE, the RP model will be taken as the modified GP model.

Moreover, the range of $\hat{a}$ in the 10 data sets is from 0.96528 to 1.05009. Recall that by Theorem 4.4.2 the asymptotic variance of $\hat{a}$ is of order $O(n^{-3})$, hence $\hat{a}$ is a very accurate estimate of a, the error is of order $O_p(n^{-3/2})$. As a result, the true ratio a in these 10 data sets should be close to 1. In general, this conclusion retains the true, since in most practical situations, the trend is usually small. Thus, in the study of a GP model, one can focus his attention on a GP with ratio a close to 1, from 0.95 to 1.05 for example.

We have already seen that on average the GP model is the best model among four models. Moreover, the GP model has some other advantages. From the analysis of 10 real data sets, all of them can be fitted by a GP. The reason is probably due to the fact that the GP model is essentially a nonparametric model. Therefore, the GP model can be widely applied to analysis of data from a series of events, with trend or without trend. On the other hand, the estimation of the parameters a, λ and σ^2 for a fitted GP model is simple. Furthermore, based on Theorems 4.4.1-4.4.4, we can easily study the statistical inference problem for the GP model. In conclusion, the GP model is a simple and good model for analysis of data from a series of events with trend.

5.5 Analysis of Data by a Threshold Geometric Process Model

In analysis of data from a sequence of events, most data are not stationary but involve trend(s). In many cases, there exists just a single monotone trend, then a GP model may be applied. The ratio a of the GP measures the direction and strength of such a trend. This is what we have done in Sections 5.2 and 5.4.

In practice, many real data exhibit multiple trends. For example, in reliability engineering, many systems demonstrate that their failure rate has the shape of a bathtub curve. At the early stage of a system, as the failures of the system are due to quality-related defects, the failure rate is decreasing. During the middle stage of the system, the failure rate may be approximately constant because the failures are caused by external shocks that occur at random. In the late stage of the system, the failures are due to wearing and the failure rate is increasing. Consequently, these systems should be improving in the early stage, then become steady in the middle stage, and will be deteriorating in the late stage. In epidemiology study, the number of daily-infected cases during the outbreak of an epidemic disease often experiences a growing stage, followed by a stabilized stage and then a declining stage. The economic development of a country or a region often shows a periodic cycle, so that the gross domestic product (GDP) is increasing in the early stage of a cycle, then the GDP will be stabilized during the middle stage of the cycle, and it will be decreasing in the late stage of the cycle. All of these data exhibit multiple trends. As a GP model is appropriate for the data with a single trend, a threshold GP model should be applicable to data with multiple trends. Thus, a threshold GP model with ratio $a_1 < 1$ at the early stage, $a_2 = 1$ at the middle stage and $a_3 > 1$ at the late stage is a reasonable model for the successive operating times of a system with a bathtub shape failure rate, the number of daily-infected cases of an epidemic disease, and the GDP of a country or a region in a cycle, etc.

To demonstrate the threshold GP model, we apply the model to the study of the Severe Acute Respiratory Syndrome (SARS). The outbreak of SARS was in 2003. Many extensive researches have been done on this topic, many models were suggested for modelling the SARS data. Most of them are deterministic, for reference see Tuen et al. (2003), Hsieh et al. (2003) and Choi and Pak (2004). Recently, Chan et al. (2006) suggested a threshold GP model to analyse the SARS data. The SARS data include the daily infected cases, the daily death cases, the daily recover cases and the daily cases in treatment for four regions, namely Hong Kong, Singapore, Ontario and Taiwan. Here we focus only on modelling the number of daily infected cases in the four regions during their outbreak periods. These data sets contain some negative number of daily infected cases, possibly due to the adjustments made when some originally confirmed cases were later tested to be non-SARS cases. Therefore, these data sets are amended accordingly.

The complete data sets for all the four regions used in our analysis are attached in Appendix for reference.

A summary of the information is given in Table 5.5.1.

Table 5.5.1. Basic information of SARS data for the four regions

Regions	Hong Kong	Singapore	Ontario	Taiwan
Start date	12/3/03	13/3/03	18/3/03	1/3/03
End date	11/6/03	19/5/03	17/7/03	15/6/03
n	92	68	122	107
S_n	1,755	206	247	674
S_n/n	19.08	3.03	2.02	6.30

where start date is the reported start date, n is the number of data, S_n is the total number of cases, and S_n/n is the average number of cases.

The key problem in using a threshold GP model to fit a data set $\{X_j, j = 1, \ldots, n\}$ is to detect its thresholds or the turning points, $\{M_i, i = 1, \ldots, k\}$, each is the time for the change of the direction or strength of trends. Whenever the thresholds or the turning points are detected, for the ith piece of the threshold GP, we can estimate its ratio a_i, λ_i and σ_i^2 using the method for a GP developed in Chapter 4.

There are many methodologies for locating the turning point(s), $\{M_i, i = 1, \ldots, k\}$, a more convenient method is to plot the data set $\{X_j, j = 1, \ldots, n\}$ first. Then we can easily find some possible sets of turning points visually. For each possible set of turning points, $\{M_i, i = 1, \ldots, k\}$, we can fit a threshold GP, and evaluate the fitted values

$$\hat{X}_j = \frac{\hat{\lambda}_i}{\hat{a}_i^{j-M_i}}, \quad M_i \leq j < M_{i+1}, \ i = 1, \ldots, k, \tag{5.5.1}$$

where $M_1 = 1$ and $M_{k+1} = n + 1$.

Then define the adjusted mean squared error (ADMSE) as

$$ADMSE = \frac{1}{n} \sum_{i=1}^{k} \sum_{j=M_i}^{M_{i+1}-1} (X_j - \hat{X}_j)^2 + c(2k), \tag{5.5.2}$$

which is the sum of the mean squared error and a penalty term that is proportional to the number of thresholds k in the threshold GP model. The best threshold GP model will be chosen for minimizing the ADMSE. In practice, an appropriate value of c should be determined in advance. To do this, we may take

$$c = \frac{1}{2}\ln\bar{X}, \tag{5.5.3}$$

where $S_n = \sum_{j=1}^{n} X_j$ and $\bar{X} = S_n/n$ is the average number of infection cases per day. Then (5.5.3) means that the higher the value of $\bar{X}$ is, the heavier the penalty will be. Thus, large value c will give high penalty resulting in less number of thresholds and parameters in the model. This is an empirical choice. Intuitively, as $X_i, i = 1, \ldots, n$, are nonnegative, then large value of $\bar{X}$ may contain more fluctuations or involve more trends, in which many trends, especially small trends, are due to noise. To reduce the 'noisy' trends, a mild penalty proportional to $\ln \bar{X}$ seems plausible. We can explain this point by an example, suppose a threshold GP model is fitted to the SARS data in Singapore, and c is taken to be 0, $\frac{1}{2}\ln \bar{X}$ and 1 respectively, the number of pieces k will be 3, 2 and 1 accordingly. Visually, from Figure 5.5.3, it seems that $k = 2$ is more appropriate. In other words, $c = \frac{1}{2}\ln \bar{X}$ seems a reasonable choice.

Then, a threshold GP model is fitted to each SARS data set of the four regions using the nonparametric method developed in Chapter 4. The results are summarized in Table 5.5.2 below.

Table 5.5.2. Fitting a threshold GP models to SARS data

Region	k	M_i	ADMSE	a_i	λ_i
Hong Kong	2	1	118.98	0.9540	14.8544
		34		1.0718	41.2221
Singapore	2	1	6.338	0.9970	4.5012
		33		1.1306	6.4291
Ontario	4	1	9.110	0.9765	2.5011
		36		1.0513	0.6565
		73		1.1663	14.2766
		98		0.9999	0.0870
Taiwan	3	1	21..816	0.9681	0.3412
		46		0.9670	7.8138
		83		1.1197	8.5228

For each region, the observed and fitted values X_i and S_i are plotted against i, so that we can see the trends, locate the turning points visually, then determine the growing, stabilizing and declining stages roughly. Note that the differences between observed and fitted S_i do not necessarily reveal goodness of fit of the models as measured by ADMSE. It seems that the differences between observed and fitted X_i are quite marked at the early stage whereas they are less marked at the late stage.

(1) Hong Kong data set

Hong Kong was seriously attacked by SARS in terms of both the number of deaths and economic losses. Among the four infected regions, Hong Kong data set shows a clear growing and declining stages and has the highest number of total infections. Infection controls were enforced on March 29. The turning point T lies on April 13 of 2003.

(2) Singapore data set

Singapore data set has the lowest number of infections among the four affected regions. Like Hong Kong, Singapore data set also shows a clear growing and declining stages. Singapore invoked the infection controls on March 24. The turning point lies on April 14 of 2003.

(3) Ontario data set

Ontario data set shows clearly two phases of outbreak. Since the numbers of daily infection were mostly zero in between the two phases, the ratio between these two phases is set to 1. Three turning points on April 22, May 29 and June 23 of 2003 were identified. They marked clearly the rises and falls in each phase. On March 26, Canada declared a public health emergency and implemented infection measures.

(4) Taiwan data set

The trends for the number of daily infection in Taiwan are similar to those in Hong Kong and Singapore but with a delay in the outbreak on April 20. Two turning points, namely April 15 and May 22, were detected which marked a low-rate, a growing and then a declining stages. On March 20 of 2003, Taiwan started to implement infection controls, such as the surveillance and home quarantine system.

The daily infection data for the four regions all show trends of increasing, stabilizing and then decreasing with different region-specific patterns. Different regions have different risks of transmission: among households, hospital care workers, inpatients and community, etc. Different regions also have different checks including the precautionary measures: contact tracing, quarantine, thermal screen to outgoing and incoming passengers, stopping of hospital visitations and closing of school, etc., and the public health measures: wearing masks, frequent hand washing, avoidance of crowded places and hospitals, prompt reporting of symptoms and disinfection of living quarters, etc. On April 2, World Health Organization

(WHO) recommended that persons traveling to Hong Kong and Guangdong Province, China, would consider postponing all but essential travel. On April 23, WHO extended its SARS-related travel advice to Toronto and further to Taiwan on May 8, apart from Beijing, Shanxi Province, Guangdong Province and Hong Kong. The travel advices and precaution measures initiated by WHO helped to contain global transmissions. In the above figures, we can see the trends of outbreak in each region and evaluate it in terms of the implementation of infection controls, allowing for latency and incubation periods.

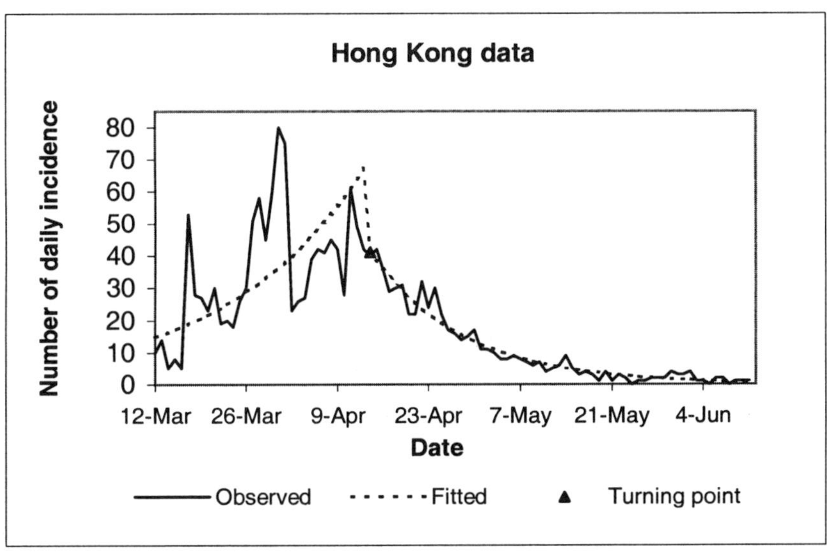

Figure 5.5.1. Hong Kong daily SARS data

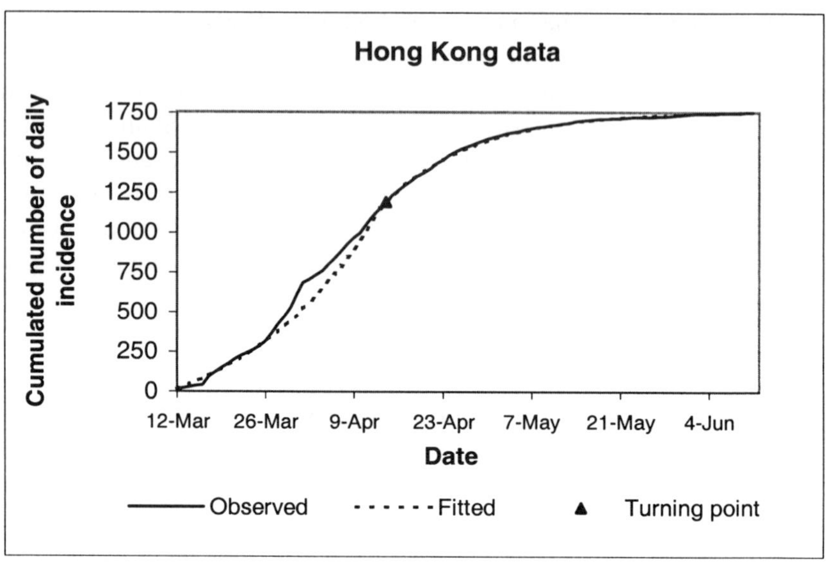

Figure 5.5.2. Hong Kong cumulative SARS data

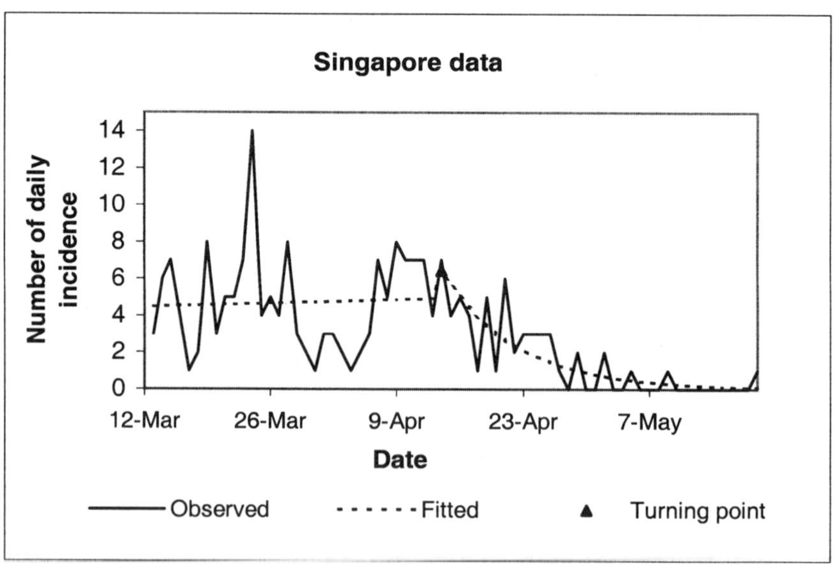

Figure 5.5.3. Singapore daily SARS data

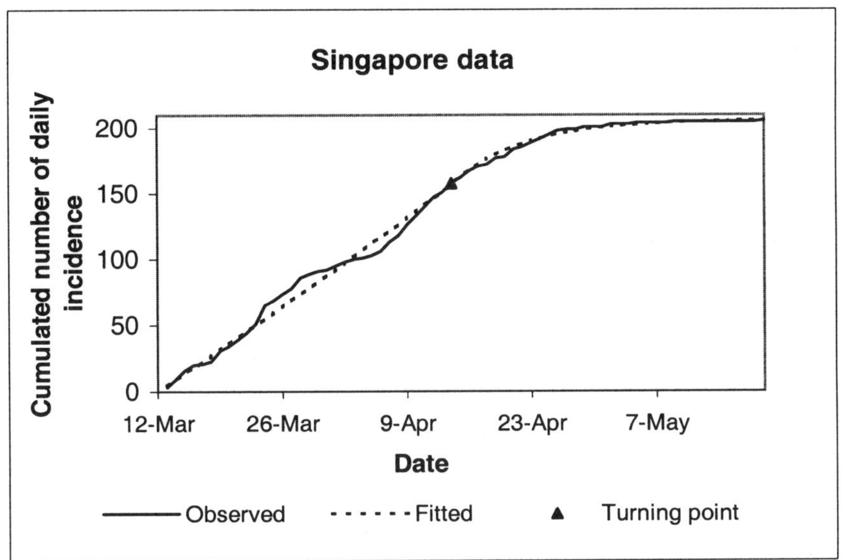

Figure 5.5.4. Singapore cumulative SARS data

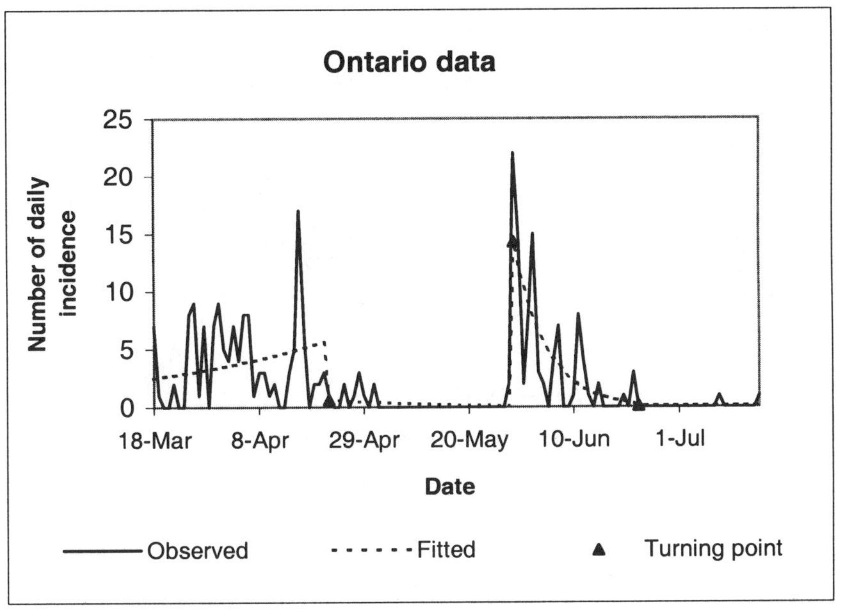

Figure 5.5.5. Ontario daily SARS data

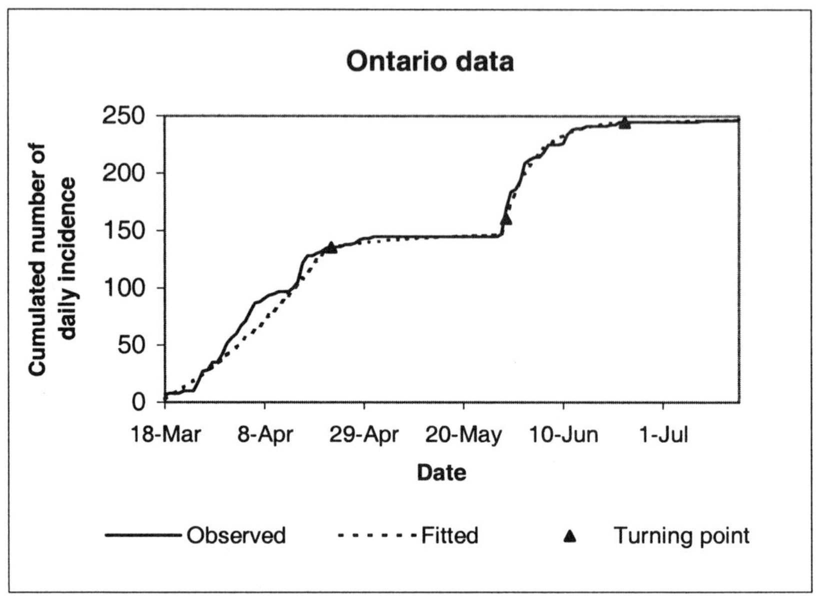

Figure 5.5.6. Ontario cumulative SARS data

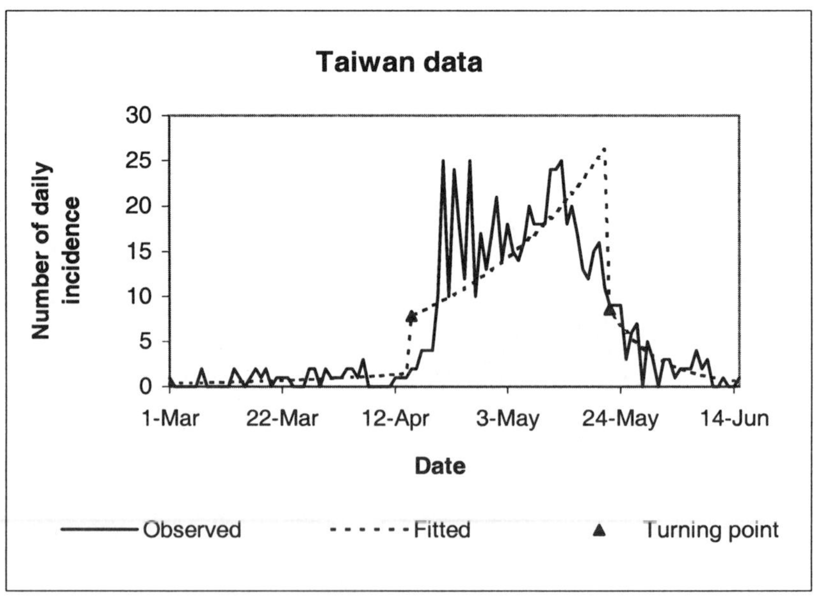

Figure 5.5.7. Taiwan daily SARS data

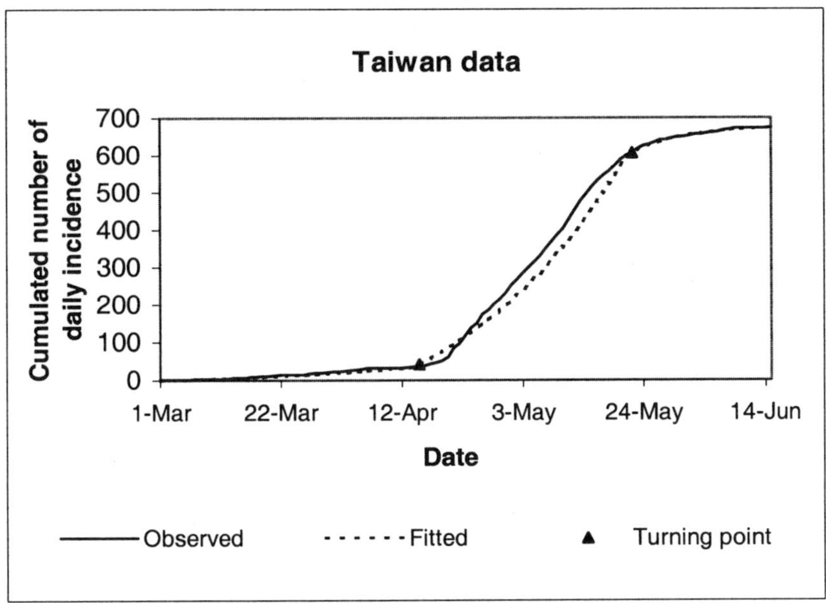

Figure 5.5.8. Taiwan cumulative SARS data

One of important objective of fitting a model is to predict the future number of infected cases. Figure 5.5.9 shows the practical significance of the GP model in prediction of the Hong Kong daily SARS data. By using the data from March 12 to April 30, a threshold GP model is refitted to the Hong Kong SARS data. Note that the turning point M_2 remains unchanged. Then the predicted number of daily infected cases after April 30 could be evaluated from (5.5.1). The results are presented in Fig. 5.5.9. One of the important factor is to predict the release date of infection controls, especially the release date of travel advice. On May 20, the Department of health predicted that the travel advice would be removed before late June. One of the requirement was the 3-day average of infected cases to be less than 5. From the predicted model, the 3-day average predicted infected cases will cut line $X = 5$ on May 19, that is very close to the observed date of May 17. By considering other criteria, such as less than 60 hospitalized cases and zero exported, the travel advice in Hong Kong was removed on May 23, 2003.

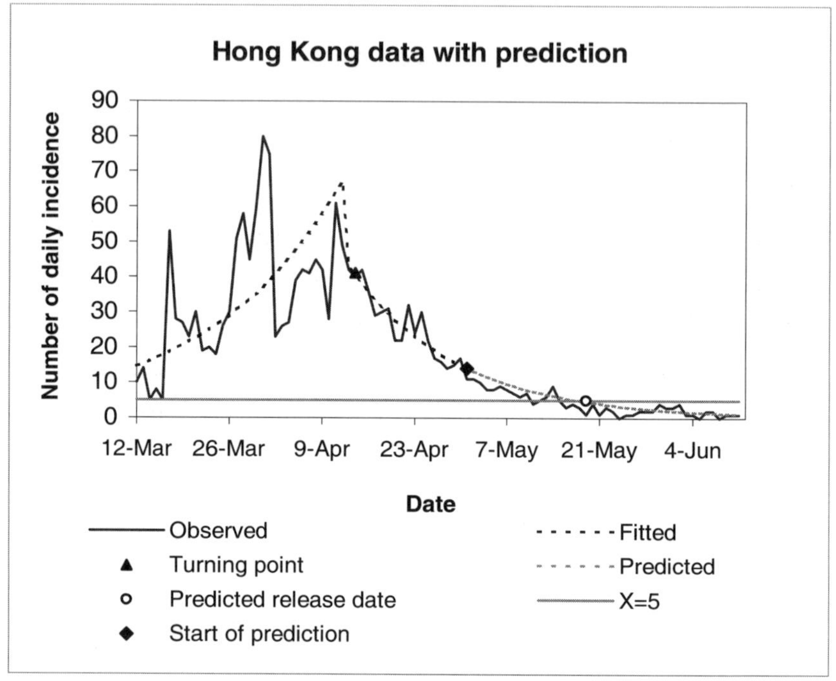

Figure 5.5.9. Hong Kong SARS data with prediction

5.6   Notes and References

In this chapter, we study the application of GP to analysis of data with trend. In Section 5.2, the methodology of using a GP model to analyze data with a single trend is considered. In Section 5.3, Poisson process models including the Cox-Lewis model and the Weibull process model, are briefly introduced. In Section 5.4, ten real data sets are fitted by four different models, including the Cox-Lewis model, the Weibull process model, RP (HPP) model and the GP model. Then the numerical results are compared by using two criteria, MSE and MPE. We can see that on average, the GP model is the best model among these four models. Besides, the GP model is simple that is an additional advantage. Sections 5.2-5.4 are due to Lam et al. (2004). The tables and figures in Section 5.4 except the last table, especially Figures 5.4.1-5.4.20, are reproduced from Lam et al. (2004) published by Springer-verlag. We greatly appreciate for the kind permission of Springer Science and Business Media for reusing the material in Lam et al. (2004) in this book.

Moreover, in epidemiology study, the daily infected cases often exhibit multiple trends. Therefore, a threshold GP model should be applicable. Section 5.5 is based on Chan et al. (2006) that applied a threshold GP model to analysis of SARS data. We are grateful to John Wiley & Sons Limited for permission to reproducing the tables and figures from Chan et al. (2006) for Section 5.5 of this book. Note that the figures in Section 5.5 are different from that in Chan et al. (2006), since another nonparametric estimates $\hat{a}_D$ and $\hat{\lambda}_D$ studied in Section 4.3 are also considered there. On the other hand, a threshold GP model is also useful in reliability study or lifetime data analysis as many systems have a bathtub shape failure rate, see Section 6.8 for a threshold GP maintenance model.

Chapter 6

Geometric Process Maintenance Model

6.1 Introduction

Lam (1988a, b) first introduced a GP model for the maintenance problem of a deteriorating system. Since then, a lot of research work on different GP maintenance models has been done. In this chapter, we shall study some GP maintenance models for a one-component system. In Section 5.2, a GP model is introduced, it is the basic model for the maintenance problem of a one-component two-state system which is either deteriorating or improving. In Section 5.3, an optimal maintenance policy is determined analytically for a deteriorating system as well as an improving system. In Section 5.4, the monotonicity of the optimal maintenance policy in a parameter for a deteriorating system is discussed. Then, a monotone process model for a one-component multistate system is introduced in Section 5.5, we show that the monotone process model is equivalent to a GP model for a one-component two-state system.

However, the above GP maintenance models just pay attention on a system that fails due to its internal cause, such as the ageing effect and accumulated wearing, but not on a system that fails due to an external cause, e.g. some shocks from the environment. In Section 6.6, a shock GP maintenance model for a system is introduced. Then in Section 6.7, a δ-shock GP maintenance model is studied.

In practice, many systems exhibit a bathtub shape failure rate. As a result, the successive operating times will have multiple trends such that they are increasing during the early stage, and stationary in the middle stage, then decreasing in the late stage. To study the maintenance problem of such a system, a threshold GP maintenance model for a system with multiple trends is considered in Section 6.8. Furthermore, it is a general knowledge that a preventive repair will usually improve the reliability or availability of a system. Based on this understanding, Section 6.9 studies a GP maintenance model with preventive repair.

6.2 A Geometric Process Maintenance Model

Now, we shall introduce a GP maintenance model that is based on Lam (2003). It is a general model, as it is not only for a deteriorating system but also for an improving system. The model is defined by making the following assumptions.

Assumption 1. At the beginning, a new system is installed. Whenever the system fails, it will be repaired. A replacement policy N is applied by which the system is replaced by a new and identical one at the time following the Nth failure.

Assumption 2. Let X_1 be the system operating time after the installation or a replacement. In general, for $n > 1$, let X_n be the system operating time after the $(n-1)$th repair, then $\{X_n, n = 1, 2, \ldots\}$ form a GP with $E(X_1) = \lambda > 0$ and ratio a . Moreover, let Y_n be the system repair time after the nth failure, then $\{Y_n, n = 1, 2, \ldots\}$ constitute a GP with $E(Y_1) = \mu \geq 0$ and ratio b. Let the replacement time be Z with $E(Z) = \tau$.

Assumption 3. The operating reward rate is r, the repair cost rate is c. The replacement cost comprises two parts: one part is the basic replacement cost R, and the other part is proportional to the replacement time Z at rate c_p.

Besides, an additional assumption is made from one of the following two assumptions.

Assumption 4. $a \geq 1$ and $0 < b \leq 1$.

Assumption 4'. $0 < a \leq 1$ and $b \geq 1$ except the case $a = b = 1$.

Then under Assumptions 1-4, the GP maintenance model is a model for a deteriorating system. However, under Assumptions 1-3 and 4', the GP maintenance model is a model for an improving system.

Remarks

At first, we shall explain the motivation of introducing the GP model. In practice, many systems are deteriorating because of the ageing effect and accumulated wearing, so that the successive operating times after repair are decreasing, while the consecutive repair times after failure are increasing.

This is not only based on our general knowledge but also on the results in real data analysis in Chapter 5. Therefore, for a deteriorating system, the successive operating times can be modelled by a decreasing GP while the consecutive repair times can be modelled by an increasing GP approximately. In this case, we should make Assumptions 1-4, that is a GP model for a deteriorating system.

On the other hand, in real life, there are some improving systems. For example, the successive operating times of a system after repair might be prolonged since the system operator can accumulate the working experience or some failed parts of the system are replaced by more advanced parts during a repair, while the consecutive repair times after failure might be shorten because the repair facility becomes more and more familiar with the system and in many cases, a repair just becomes a replacement of some parts. For an improving system, the successive operating times of the system can be modelled by an increasing GP, while the consecutive repair times of the system can be modelled by a decreasing GP approximately. In this case, we should make Assumptions 1-3 and 4', that is a GP model for an improving system.

Secondly, we shall expound the reasons of using replacement policy N. In fact, use of policy N has a long history, see Morimura (1970) and Park (1979) for reference. Nevertheless, in the literature of the maintenance problem, there are two kinds of replacement policy. One is policy T by which the system is replaced at a stopping time T, the other one is policy N by which the system is replaced at the time following the Nth failure. It is interesting to compare these two policies. For the long-run average cost per unit time case, Stadje and Zuckerman (1990) and Lam (1991b) showed that under some mild conditions, the optimal replacement policy N^* is at least as good as the optimal replacement policy T^*. Furthermore, Lam (1991c, 1992b) proved that for the expected total discounted cost case, the same conclusion holds. In addition, the implementation of a policy N is more convenient than that of a policy T, this is an extra merit of using policy N. Therefore, policy N is applied in our model.

As one part of the replacement cost, the basic replacement cost R includes the production cost of a new and identical system, the administrative cost and the transportation cost. They are clearly independent of the replacement time Z. Besides cost R, the other part of replacement cost such as the labour wages and power expenses will be proportional to the replacement time Z.

Now, we say that a cycle is completed if a replacement is completed. Thus a cycle is actually a time interval between the installation of a system and the first replacement or a time interval between two consecutive replacements. Then, the successive cycles together with the costs incurred in each cycle will constitute a renewal reward process. By applying Theorem 1.3.15, the long-run average cost per unit time (or simply the average cost) is given by

$$\frac{\text{Expected cost incurred in a cycle}}{\text{Expected length of a cycle}}. \tag{6.2.1}$$

Consequently, under Assumptions 1-3 and using policy N, the average cost is given by

$$
\begin{aligned}
C(N) &= \frac{E(c\sum_{k=1}^{N-1} Y_k - r\sum_{k=1}^{N} X_k + R + c_p Z)}{E(\sum_{k=1}^{N} X_k + \sum_{k=1}^{N-1} Y_k + Z)} \\[2mm]
&= \frac{c\mu \sum_{k=1}^{N-1}\frac{1}{b^{k-1}} - r\lambda \sum_{k=1}^{N}\frac{1}{a^{k-1}} + R + c_p\tau}{\lambda \sum_{k=1}^{N}\frac{1}{a^{k-1}} + \mu \sum_{k=1}^{N-1}\frac{1}{b^{k-1}} + \tau} \tag{6.2.2} \\[2mm]
&= A(N) - r,
\end{aligned}
$$

where

$$
A(N) = \frac{(c+r)\mu \sum_{k=1}^{N-1}\frac{1}{b^{k-1}} + R + c_p\tau + r\tau}{\lambda \sum_{k=1}^{N}\frac{1}{a^{k-1}} + \mu \sum_{k=1}^{N-1}\frac{1}{b^{k-1}} + \tau}. \tag{6.2.3}
$$

Now, our objective is to determine an optimal replacement policy N^* for minimizing $C(N)$ or $A(N)$. To do this, we may evaluate the difference of $A(N+1)$ and $A(N)$ first.

$$
\begin{aligned}
&A(N+1) - A(N) \\[2mm]
&= \frac{(c+r)\mu\{\lambda(\sum_{k=1}^{N} a^k - \sum_{k=1}^{N-1} b^k) + \tau a^N\} - (R + c_p\tau + r\tau)(\lambda b^{N-1} + \mu a^N)}{a^N b^{N-1}[\lambda \sum_{k=1}^{N}\frac{1}{a^{k-1}} + \mu \sum_{k=1}^{N-1}\frac{1}{b^{k-1}} + \tau][\lambda \sum_{k=1}^{N+1}\frac{1}{a^{k-1}} + \mu \sum_{k=1}^{N}\frac{1}{b^{k-1}} + \tau]}.
\end{aligned}
$$

Because the denominator of $A(N+1) - A(N)$ is always positive, it is clear that the sign of $A(N+1) - A(N)$ is the same as the sign of its numerator.

Then, we can define an auxiliary function

$$g(N) = g(N, a, b, \lambda, \mu, \tau, r, c, R, c_p)$$

$$= \frac{(c+r)\mu\{\lambda(\sum_{k=1}^{N} a^k - \sum_{k=1}^{N-1} b^k) + \tau a^N\}}{(R + c_p\tau + r\tau)(\lambda b^{N-1} + \mu a^N)}. \qquad (6.2.4)$$

As a result, we have the following lemma.

Lemma 6.2.1.

$$A(N+1) \overset{\geq}{\underset{<}{=}} A(N) \iff g(N) \overset{\geq}{\underset{<}{=}} 1.$$

Lemma 6.2.1 shows that the monotonicity of $A(N)$ can be determined by the value of $g(N)$.

Note that all the results above are developed under Assumptions 1-3 only. Therefore, all results including Lemma 6.2.1 hold for the models of a deteriorating system and of an improving system.

Zhang (1994) considered a GP maintenance model in which a bivariate policy (T, N) is applied. A bivariate policy (T, N) is a replacement policy such that the system will be replaced at working age T or following the time of the Nth failure, whichever occurs first.

Now, by using policy (T, N), under Assumptions 1-3 and assume that $\{X_n, n = 1, 2, \ldots\}$ and $\{Y_n, n = 1, 2, \ldots\}$ are two independent stochastic processes, we can evaluate the average cost in a similar way. To do this, let

$$U_n = \sum_{i=1}^{n} X_i, \quad V_n = \sum_{i=1}^{n} Y_i,$$

with $U_0 = 0, V_0 = 0$. Again, we say that a cycle is completed if a replacement is completed. Let W be the length of a cycle, then

$$W = \sum_{i=0}^{N-1} (T + V_i)I_{\{U_i < T \leq U_{i+1}\}} + (U_N + V_{N-1})I_{\{U_N < T\}} + Z, \qquad (6.2.5)$$

where I_A is the indicator function such that

$$I_A = \begin{cases} 1 & \text{if event } A \text{ occurs,} \\ 0 & \text{if event } A \text{ does not occur.} \end{cases} \qquad (6.2.6)$$

Assume that the operating time and repair time are continuous. Then we have

$$E(W) = E\{\sum_{i=0}^{N-1}(T+V_i)I_{\{U_i<T\le U_{i+1}\}} + (U_N + V_{N-1})I_{\{U_N<T\}} + Z\}$$

$$= E\{TI_{\{T\le U_N\}}\} + \sum_{i=1}^{N-1} E(V_i)E(I_{\{U_i<T\le U_{i+1}\}})$$

$$+E\{(U_N + V_{N-1})I_{\{U_N<T\}}\} + \tau$$

$$= E\{TI_{\{T\le U_N\}}\} + \sum_{i=1}^{N-1}(\sum_{j=1}^{i}\frac{\mu}{b^{j-1}})\{F_i(T) - F_{i+1}(T)\}$$

$$+E\{E(U_N + V_{N-1})I_{\{U_N<T\}}|U_N)\} + \tau$$

$$= T\bar{F}_N(T) + \sum_{j=1}^{N-1}(\sum_{i=j}^{N-1}\frac{\mu}{b^{j-1}})\{F_i(T) - F_{i+1}(T)\}$$

$$+E\{E[(U_N + V_{N-1})I_{\{U_N<T\}}|U_N]\} + \tau$$

$$= T\bar{F}_N(T) + \sum_{j=1}^{N-1}\frac{\mu}{b^{j-1}}\{F_j(T) - F_N(T)\}$$

$$+ \int_0^T E(U_N + V_{N-1}|U_N = u)dF_N(u) + \tau$$

$$= T\bar{F}_N(T) + \sum_{j=1}^{N-1}\frac{\mu}{b^{j-1}}F_j(T) - \sum_{j=1}^{N-1}\frac{\mu}{b^{j-1}}F_N(T)$$

$$+ \int_0^T udF_N(u) + E(V_{N-1})\int_0^T dF_N(t) + \tau$$

$$= \int_0^T \bar{F}_N(u)du + \sum_{j=1}^{N-1}\frac{\mu}{b^{j-1}}F_j(T) + \tau, \tag{6.2.7}$$

since $E(V_i) = \sum_{j=1}^{i-1}\frac{\mu}{b^{j-1}}$ and $\{X_n, n = 1, 2, \ldots\}$ and $\{Y_n, n = 1, 2, \ldots\}$ are independent. On the other hand, the expected cost incurred in a cycle is given by

$$cE(\sum_{i=1}^{N-1} V_i I_{\{U_i < T \leq U_{i+1}\}} + V_{N-1} I_{\{U_N < T\}})$$

$$- rE(TI_{\{T \leq U_N\}} + U_N I_{\{U_N < T\}}) + R + c_p E(Z)$$

$$= c \sum_{j=1}^{N-1} \frac{\mu}{b^{j-1}} F_j(T) - r \int_0^T \bar{F}_N(u)du + R + c_p\tau. \qquad (6.2.8)$$

Thus by using (6.2.1), the average cost under policy (T, N) will be given by

$$C(T, N) = \frac{c\mu \sum_{j=1}^{N-1} \frac{1}{b^{j-1}} F_j(T) - r \int_0^T \bar{F}_N(u)du + R + c_p\tau}{\int_0^T \bar{F}_N(u)du + \mu \sum_{j=1}^{N-1} \frac{1}{b^{j-1}} F_j(T) + \tau}. \qquad (6.2.9)$$

Obviously, policy N is a bivariate policy (∞, N) and policy T is a bivariate policy (T, ∞). This implies that an optimal policy of (T, N) is at least as good as an optimal policy N^* and an optimal policy T^*.

6.3 Optimal Replacement Policy

In this section, we shall determine an optimal replacement policy N^* for minimizing $C(N)$ or $A(N)$ for a deteriorating system and an improving system respectively. For this purpose, at first let $h(N) = \lambda b^{N-1} + \mu a^N$. Then it follows from (6.2.4) that

$$g(N+1) - g(N)$$

$$= \frac{(c+r)\mu}{(R + c_p\tau + r\tau)h(N)h(N+1)}\{\lambda^2 b^{N-1}(1-b)\sum_{k=1}^{N} a^k + \lambda^2 b^{N-1}(a^{N+1} - b)$$

$$+ \lambda\mu a^N(a - b^N) + \lambda\mu a^N(a-1)\sum_{k=1}^{N-1} b^k + \lambda\tau a^N b^{N-1}(a-b)\}. \qquad (6.3.1)$$

Then, two different models are considered here.

Model 1.   The model under Assumptions 1-4

Model 1 is the model for a deteriorating system. Now, from (6.3.1) and Assumption 4 the following lemma is straightforward.

Lemma 6.3.1. Function $g(N)$ is nondecreasing in N.

Consequently, Lemma 6.2.1 together with Lemma 6.3.1 gives an analytic expression for an optimal policy N^*.

Theorem 6.3.2. An optimal replacement policy N_d^* for the deteriorating system is determined by

$$N_d^* = \min\{N \mid g(N) = g(N, a, b, \lambda, \mu, \tau, r, c, R, c_p) \geq 1\}. \qquad (6.3.2)$$

The optimal replacement policy N_d^* is unique if and only if $g(N_d^*) > 1$.

We can apply Theorem 6.3.2 to determine an optimal policy for the deteriorating system. Besides, we can also determine an optimal policy numerically.

A numerical example

Now, we study a numerical example with the following parameter values: $a = 1.05, b = 0.95, R = 3000, \lambda = 40, \mu = 15, c = 10, r = 50, c_p = 10$ and $\tau = 10$.

The numerical results are presented in Table 6.3.1 and Figure 6.3.1 respectively.

Table 6.3.1. Results obtained from formulas (6.2.2) and (6.2.4)

N	C(N)	g(N)	N	C(N)	g(N)	N	C(N)	g(N)
1	22.0000	0.2354	11	-22.6116	1.5565	21	-17.6895	4.6345
2	-6.3510	0.2710	12	-22.3166	1.8050	22	-17.0663	4.9794
3	-14.8933	0.3268	13	-21.9524	2.0716	23	-16.4336	5.3236
4	-18.7671	0.4035	14	-21.5328	2.3545	24	-15.7935	5.6655
5	-20.8101	0.5020	15	-21.0682	2.6520	25	-15.1481	6.0037
6	-21.9437	0.6227	16	-20.5666	2.9622	26	-14.4991	6.3368
7	-22.5571	0.7657	17	-20.0344	3.2832	27	-13.8480	6.6638
8	-22.8432	0.9311	18	-19.4766	3.6129	28	-13.1966	6.9838
9	**-22.9089**	**1.1185**	19	-18.8976	3.9494	29	-12.5460	7.2958
10	-22.8181	1.3272	20	-18.3008	4.2905	30	-11.8978	7.5994

Model 2. The model under Assumptions 1-3 and 4'

Model 2 is the model for an improving system. Now, because of Assumption 4', instead of Lemma 6.3.1, we have the following lemma.

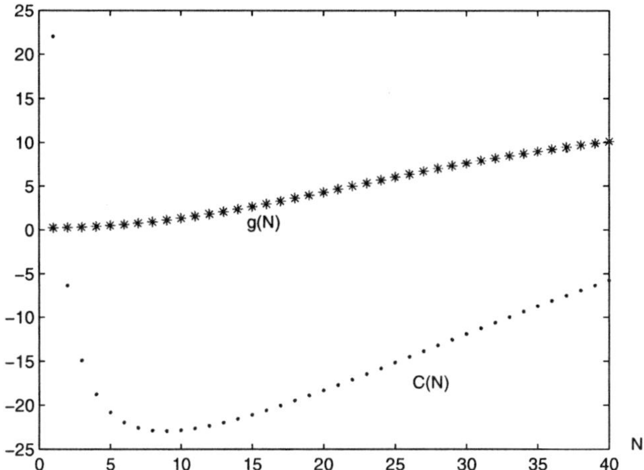

Figure 6.3.1. The plots of g(N) and C(N) against N

Lemma 6.3.3.   Function $g(N)$ is decreasing in N.

Consequently, based on Lemmas 6.2.1 and 6.3.3, we have the following result.

Theorem 6.3.4. Under Assumptions 1-3 and 4', policy $N_i^* = \infty$ is the unique optimal policy for the improving system.
Proof.

Because $g(N)$ is decreasing in N, there exists an integer N_i such that

$$N_i = \min\{N \mid g(N) \leq 1\}. \tag{6.3.3}$$

In other words, we have

$$g(N) > 1 \Longleftrightarrow N < N_i,$$

and

$$g(N) \leq 1 \Longleftrightarrow N \geq N_i.$$

Therefore, Lemma 6.2.1 implies that $C(N)$ and $A(N)$ are both unimodal of N and take their maxima at N_i. Because of (6.2.2) and Assumption 4', it is easy to check that the minimum of $C(N)$ will be given by

$$\min C(N) = \min\{C(1), C(\infty)\}$$
$$= \min\{\frac{R + c_p\tau - r\lambda}{\lambda + \tau}, -r\} = -r. \tag{6.3.4}$$

Thus, $N_i^* = \infty$ is the unique optimal replacement policy for the improving system. This completes the proof of Theorem 6.3.4.

Intuitively, it is a general knowledge that the older the improving system is, the better the system is. This means that we shall repair the system when it fails without replacement. Theorem 6.3.4 agrees with this general knowledge.

6.4 Monotonicity of the Optimal Policy for a Deteriorating System

For an improving system, Theorem 6.3.4 shows that policy $N_i^* = \infty$ is always the optimal policy. Therefore, we need only to study the monotonicity of an optimal policy N_d^* for a deteriorating system. Thus, in this section, for a deteriorating system, the optimal policy N_d^* will be denote by N^* for simplicity. Because the auxiliary function g is a function of $N, a, b, \lambda, \mu, \tau, r, c, R$ and c_p, then the optimal policy N^* will be a function of $a, b, \lambda, \mu, \tau, r, c, R$ and c_p and can be denoted by

$$N^* = N^*(a, b, \lambda, \mu, \tau, r, c, R, c_p).$$

To study the monotonicity of the optimal policy N^* in one parameter, while the others keep unchanged, we need the following lemma which can be derived from Theorem 6.3.2 directly.

Lemma 6.4.1. Auxiliary function g and optimal policy N^* possess an opposite monotonicity property in each parameter of $a, b, \lambda, \mu, \tau, r, c, R$ and c_p.

From (6.2.4), it is clear that function g is decreasing in b, R and c_p but increasing in μ and c. Therefore, from Lemma 6.4.1 the following theorem is straightforward.

Theorem 6.4.2.

(1) $N^*(a, b, \lambda, \mu, \tau, r, c, R, c_p) \leq N^*(a, b', \lambda, \mu, \tau, r, c, R, c_p), \quad \forall\, b < b'.$

(2) $N^*(a, b, \lambda, \mu, \tau, r, c, R, c_p) \geq N^*(a, b, \lambda, \mu', \tau, r, c, R, c_p), \quad \forall\, \mu < \mu'.$

(3) $N^*(a, b, \lambda, \mu, \tau, r, c, R, c_p) \geq N^*(a, b, \lambda, \mu, \tau, r, c', R, c_p), \quad \forall\, c < c'.$

(4) $N^*(a, b, \lambda, \mu, \tau, r, c, R, c_p) \leq N^*(a, b, \lambda, \mu, \tau, r, c, R', c_p), \quad \forall\, R < R'.$

(5) $N^*(a, b, \lambda, \mu, \tau, r, c, R, c_p) \leq N^*(a, b, \lambda, \mu, \tau, r, c, R, c_p'), \quad \forall\, c_p < c_p'.$

We can easily see that the results are all consistent with our practical experience. For example, as the replacement cost R increases while the other parameters keep unchanged, one would delay the replacement for saving expenses. Similarly, as the repair cost rate c raises with the others fixed, one would prefer to replace the system earlier for reducing expenditure. In other words, optimal policy N^* is increasing in R but decreasing in c.

To study the monotonicity property of N^* with respect to the other parameters, we can see from (6.2.4) that

$$\frac{\partial g}{\partial r} \gtreqless 0 \iff R + (c_p - c)\tau \gtreqless 0. \tag{6.4.1}$$

Consequently, using (6.4.1) with the help of Lemma 6.4.1, we have the following theorem.

Theorem 6.4.3.
(1) If $R + (c_p - c)\tau > 0$, then

$$N^*(a, b, \lambda, \mu, \tau, r, c, R, c_p) \geq N^*(a, b, \lambda, \mu, \tau, r', c, R, c_p), \quad \forall\, r < r'.$$

(2) If $R + (c_p - c)\tau = 0$, then

$$N^*(a, b, \lambda, \mu, \tau, r, c, R, c_p) = N^*(a, b, \lambda, \mu, \tau, r', c, R, c_p), \quad \forall\, r < r'.$$

(3) If $R + (c_p - c)\tau < 0$, then

$$N^*(a, b, \lambda, \mu, \tau, r, c, R, c_p) \leq N^*(a, b, \lambda, \mu, \tau, r', c, R, c_p), \quad \forall\, r < r'.$$

Furthermore, differentiating (6.2.4) with respect to λ yields

$$\frac{\partial g}{\partial \lambda} = \frac{(c + r)\mu a^N \left\{ \mu(\sum_{k=1}^{N} a^k - \sum_{k=1}^{N-1} b^k) - \tau b^{N-1} \right\}}{(R + c_p\tau + r\tau)(\lambda b^{N-1} + \mu a^N)^2}. \tag{6.4.2}$$

Obviously, the sign of $\frac{\partial g}{\partial \lambda}$ is determined by the sign of

$$\alpha(N) = \mu(\sum_{k=1}^{N} a^k - \sum_{k=1}^{N-1} b^k) - \tau b^{N-1},$$

since the other factors in (6.4.2) are all positive. However, a direct checking shows that $\alpha(N)$ is increasing in N. As a result, there exists an integer N_λ such that

$$N_\lambda = \min\{N \mid \alpha(N) = \mu(\sum_{k=1}^{N} a^k - \sum_{k=1}^{N-1} b^k) - \tau b^{N-1} \geq 0\}. \tag{6.4.3}$$

Note that N_λ does not depend on λ. Therefore,

$$\frac{\partial g}{\partial \lambda} < 0 \iff N < N_\lambda$$

and

$$\frac{\partial g}{\partial \lambda} \geq 0 \iff N \geq N_\lambda.$$

In other words, function g is decreasing in λ when $N < N_\lambda$ and increasing otherwise. Thus, the following theorem follows from Lemma 6.4.1 directly.

Theorem 6.4.4.

(1) If $N^*(a, b, \lambda, \mu, \tau, r, c, R, c_p) < N_\lambda$, then

$$N^*(a, b, \lambda, \mu, \tau, r, c, R, c_p) \leq N^*(a, b, \lambda', \mu, \tau, r, c, R, c_p), \quad \forall \, \lambda < \lambda'.$$

(2) If $N^*(a, b, \lambda, \mu, \tau, r, c, R, c_p) \geq N_\lambda$, then

$$N^*(a, b, \lambda, \mu, \tau, r, c, R, c_p) \geq N^*(a, b, \lambda', \mu, \tau, r, c, R, c_p), \quad \forall \, \lambda < \lambda'.$$

Furthermore, by differentiating (6.2.4) with respect to τ, we have

$$\frac{\partial g}{\partial \tau} = \frac{(c + r)\mu\{Ra^N - (c_p + r)\lambda(\sum\limits_{k=1}^{N} a^k - \sum\limits_{k=1}^{N-1} b^k)\}}{[R + (c_p + r)\tau]^2(\lambda b^{N-1} + \mu a^N)}. \tag{6.4.4}$$

Similar to the case of $\frac{\partial g}{\partial \lambda}$, the sign of $\frac{\partial g}{\partial \tau}$ is determined by the sign of

$$\beta(N) = Ra^N - (c_p + r)\lambda\{\sum\limits_{k=1}^{N} a^k - \sum\limits_{k=1}^{N-1} b^k\}. \tag{6.4.5}$$

Then we have the following result.

Lemma 6.4.5.

(1) If $(c_p + r)\lambda \leq R(1 - a^{-1})$, then $\beta(N) > 0, \quad \forall \, N$.

(2) If $R(1 - a^{-1}) < (c_p + r)\lambda \leq R$, then there exists N_τ which is independent of τ, such that $\beta(N) \geq 0$, if $N < N_\tau$ and $\beta(N) < 0$, otherwise.

(3) If $(c_p + r)\lambda > R$, then $\beta(N) < 0, \quad \forall \, N$.

Proof.

(1) If $(c_p + r)\lambda \leq R(1 - a^{-1})$, then

$$\beta(N + 1) - \beta(N) = \{[R - (c_p + r)\lambda]a - R\}a^N + (c_p + r)\lambda b^N > 0. \tag{6.4.6}$$

Inequality (6.4.6) implies that $\beta(N)$ is increasing in N with

$$\beta(1) = [R - (c_p + r)\lambda]a > 0.$$

Hence $\beta(N) > 0, \quad \forall \, N$. Therefore, part (1) follows.

(2) If $R(1 - a^{-1}) < (c_p + r)\lambda \leq R$, then

$$[R - (c_p + r)\lambda]a - R < 0 \quad \text{but} \quad R - (c_p + r)\lambda \geq 0.$$

Because $a \geq 1$ and $0 < b \leq 1$, we can write $a = \alpha b$ with $\alpha \geq 1$.

Firstly, assume $\alpha > 1$, then

$$\beta(N+1) - \beta(N) = \{[(R - (c_p + r)\lambda)a - R]\alpha^N + (c_p + r)\lambda\}b^N.$$

Because factor $\{[R - (c_p + r)\lambda]a - R\}\alpha^N + (c_p + r)\lambda$ is decreasing in N, there exists an integer N_0 such that

$$\beta(N+1) - \beta(N) > 0 \iff N < N_0,$$

and

$$\beta(N+1) - \beta(N) \leq 0 \iff N \geq N_0.$$

In other words, $\beta(N)$ is increasing if $N < N_0$ and decreasing otherwise. Note that,

$$\beta(1) = [R - (c_p + r)\lambda]a \geq 0.$$

Therefore, there exists an integer $N_\tau > N_0$ such that

$$N_\tau = \min\{N \mid \beta(N) < 0\}. \tag{6.4.7}$$

Clearly N_τ does not depend on τ. Therefore, $\beta(N) \geq 0$ for $N < N_\tau$ and $\beta(N) < 0$, for $N \geq N_\tau$.

Secondly, assume that $\alpha = 1$. In this case, $a = b = 1$. Then from (6.4.5), $\beta(N) \equiv R - (c_p + r)\lambda \geq 0$ is a constant, and (6.4.7) is still true with $N_\tau = \infty$. Thus, part (2) follows.

(3) If $(c_p + r)\lambda > R$, then

$$\beta(N+1) - \beta(N) = Ra^{N+1} - Ra^N - (c_p + r)\lambda(a^{N+1} - b^N)$$
$$\leq [R - (c_p + r)\lambda]a^N(a - 1) \leq 0.$$

Therefore, $\beta(N)$ is decreasing with $\beta(1) = [R - (c_p + r)\lambda]a < 0$. Then part (3) follows.

Recalling that the monotonicity of function g in τ is determined by the sign of $\beta(N)$, Lemmas 6.4.1 and 6.4.5 together give the following result.

Theorem 6.4.6.
(1) If $(c_p + r)\lambda \leq R(1 - a^{-1})$, then

$$N^*(a, b, \lambda, \mu, \tau, r, c, R, c_p) \geq N^*(a, b, \lambda, \mu, \tau', r, c, R, c_p), \quad \forall \, \tau < \tau'.$$

(2) If $R(1 - a^{-1}) < (c_p + r)\lambda \leq R$ and $N^*(a, b, \lambda, \mu, \tau, r, c, R, c_p) < N_\tau$, then

$$N^*(a, b, \lambda, \mu, \tau, r, c, R, c_p) \geq N^*(a, b, \lambda, \mu, \tau', r, c, R, c_p), \quad \forall \, \tau < \tau';$$

if $R(1 - a^{-1}) < (c_p + r)\lambda \leq R$ and $N^*(a, b, \lambda, \mu, \tau, r, c, R, c_p) \geq N_\tau$, then

$$N^*(a, b, \lambda, \mu, \tau, r, c, R, c_p) \leq N^*(a, b, \lambda, \mu, \tau', r, c, R, c_p), \quad \forall \, \tau < \tau',$$

where N_τ is determined by (6.4.7) and is independent of τ.

(3) If $(c_p + r)\lambda > R$, then

$$N^*(a, b, \lambda, \mu, \tau, r, c, R, c_p) \leq N^*(a, b, \lambda, \mu, \tau', r, c, R, c_p), \quad \forall \, \tau < \tau'.$$

Finally, we need to study the monotonicity of optimal policy N^* in parameter a. It is easy to see that $\frac{\partial g}{\partial a}$ is negative for large value of a. Consequently, optimal policy N^* should be increasing in parameter a for large value of a. However, in general the monotonicity of N^* in parameter a is still unknown. This is an open problem leaving for further research.

6.5   A Monotone Process Model for a Multistate System

In most existing models of maintenance problems, including the GP model, we only consider a two-state system with up and down states say. However, in many practical cases, a system may have more than 2 states. For example, an electronic component such as a diode or a transistor, may fail due to a short circuit or an open circuit. In engineering, a failure of machine may be classified by its seriousness, a slight failure or a serious failure. In man-machine system, a failure may be classified by its cause, a man-made mistake or a machine trouble. In this cases, a system will have two different failure states and one working state. On the other hand, a system may also have more than one working state, for example, a car may be driving in one out of five speeds. Lam et al. (2002) considered the case that a system has several failure states but one working state. The same problem was also studied by Zhang et al. (2002). Thereafter, Lam and Tse (2003) investigated the case that a system has several working states but one failure state. They introduced a monotone process model for these multistate deteriorating systems.

Recently, Lam (2005a) introduced a more general monotone process model for a multistate one-component system which has k working states and ℓ failure states, $k + \ell$ states in total. Furthermore, the system could be either a deteriorating or an improving system. To introduce the monotone model for the multistate system, we shall first define the system state $S(t)$

at time t by

$$S(t) = \begin{cases} i & \text{if the system is in the } i\text{th working state at time } t, \\ & i = 1, \ldots, k, \\ k+i & \text{if the system is in the } i\text{th failure state at time } t, \\ & i = 1, \ldots, \ell. \end{cases} \quad (6.5.1)$$

Therefore, the state space is $\Omega = \{1, \ldots, k, k+1, \ldots, k+\ell\}$. The set of working states is $W = \{1, \ldots, k\}$, and the set of failure states is $F = \{k+1, \ldots, k+\ell\}$. Initially, assume that a new system in working state 1 is installed. Whenever the system fails, it will be repaired. Let t_n be the completion time of the nth repair, $n = 0, 1, \ldots$ with $t_0 = 0$, and let s_n be the time of the nth failure, $n = 1, 2, \ldots$. Then clearly we have

$$t_0 < s_1 < t_1 < \ldots < s_n < t_n < s_{n+1} < \ldots$$

Furthermore, assume that the transition probability from working state i for $i = 1, \ldots, k$ to failure state $k+j$ for $j = 1, \ldots, \ell$ is given by

$$P(S(s_{n+1}) = k+j \mid S(t_n) = i) = q_j$$

with $\sum_{j=1}^{\ell} q_j = 1$. Moreover, the transition probability from failure state $k+i$ for $i = 1, \ldots, \ell$ to working state j for $j = 1, \ldots, k$ is given by

$$P(S(t_n) = j \mid S(s_n) = k+i) = p_j$$

with $\sum_{j=1}^{k} p_j = 1$.

Now let X_1 be the operating time of a system after installation. In general, let $X_n, n = 2, 3, \ldots$, be the operating time of the system after the $(n-1)$th repair and let $Y_n, n = 1, 2, \ldots$, be the repair time after the nth failure. Assume that there exists a life distribution $U(t)$ and $a_i > 0$, $i = 1, \ldots, k$, such that

$$P(X_1 \leq t) = U(t),$$

and

$$P(X_2 \leq t \mid S(t_1) = i) = U(a_i t), \quad i = 1, \ldots, k.$$

In general,

$$P(X_n \leq t \mid S(t_1) = i_1, \ldots, S(t_{n-1}) = i_{n-1})$$
$$= U(a_{i_1} \ldots a_{i_{n-1}} t), \quad i_j = 1, \ldots, k, \quad j = 1, \ldots, n-1. \quad (6.5.2)$$

Similarly, assume that there exists a life distribution $V(t)$ and $b_i > 0$, $i = 1, \ldots, \ell$, such that

$$P(Y_1 \leq t \mid S(s_1) = k+i) = V(b_i t),$$

and in general,

$$P(Y_n \leq t \mid S(s_1) = k + i_1, \ldots, S(s_n) = k + i_n)$$
$$= V(b_{i_1} \ldots b_{i_n} t), \quad i_j = 1, \ldots, \ell, \quad j = 1, \ldots, n. \tag{6.5.3}$$

We now make some additional assumptions on the model.

Assumption 1. A replacement policy N is applied by which the system is replaced by a new and identical one following the Nth failure.

Assumption 2. If a system in working state i is operating, the reward rate is r_i, $i = 1, \ldots, k$. If the system in failure state $k + i$ is under repair, the repair cost rate is c_i, $i = 1, \ldots, \ell$. The replacement cost comprises two parts, one part is the basic replacement cost R, the other part is proportional to the replacement time Z at rate c_p. In other words, the replacement cost is given by $R + c_p Z$.

Assumption 3.

$$1. \quad 1 = a_1 \leq a_2 \leq \ldots \leq a_k, \tag{6.5.4}$$

$$2. \quad 1 = b_1 \geq b_2 \geq \ldots \geq b_\ell > 0. \tag{6.5.5}$$

Under additional Assumptions 1-3, we shall argue that the model is a maintenance model for a multistate deteriorating system. For an multistate improving system, Assumption 3 will be replaced by the following assumption.

Assumption 3'.

$$1. \quad 1 = a_1 \geq a_2 \geq \ldots \geq a_k > 0, \tag{6.5.6}$$

$$2. \quad 1 = b_1 \leq b_2 \leq \ldots \leq b_\ell. \tag{6.5.7}$$

Note that not all equalities in (6.5.6) and (6.5.7) hold simultaneously.

Remarks

Since the system has k different working states, the system in different working states should have different reward rates. Similarly, the system in different failure states should have different repair cost rates. The replacement cost includes the cost of system that is a constant, it also includes the cost for dismantling the used system and installation cost of a new system, they are proportional to the replacement time.

Under Assumption 3, given two working states $0 \leq i_1 < i_2 \leq k$, we have

$$(X_2 \mid S(t_1) = i_1) \geq_{s.t.} (X_2 \mid S(t_1) = i_2). \tag{6.5.8}$$

Therefore, working state i_1 is better than working state i_2 in the sense that the system in state i_1 has a stochastically larger operating time than it has in state i_2. Consequently, the k working states are arranged in a decreasing order, such that state 1 is the best working state, ..., and state k is the worst working state. On the other hand, for two failure states $k + i_1$ and $k + i_2$ such that $k + 1 \le k + i_1 < k + i_2 \le k + \ell$,

$$(Y_1 \mid S(s_1) = k + i_1) \le_{s.t.} (Y_1 \mid S(s_1) = k + i_2). \qquad (6.5.9)$$

Therefore failure state $k + i_1$ is better than failure state $k + i_2$ in the sense that the system in state $k + i_1$ has a stochastically less repair time than it has in state $k + i_2$. Thus, ℓ failure states are also arranged in a decreasing order, such that state $k + 1$ is the best failure state, ..., and state $k + \ell$ is the worst failure state.

Under Assumption 3', given two working states $0 \le i_1 < i_2 \le k$, a similar argument shows that working state i_1 is worse than working state i_2 since the system in state i_1 has a stochastically less operating time than it has in state i_2. This means that k working states are arranged in an increasing order, such that state 1 is the worst working state, ..., and state k is the best working state. On the other hand, ℓ failure states are also arranged in an increasing order, such that state $k + 1$ is the worst failure state, ..., and state $k + \ell$ is the best failure state. Note that if all equalities in (6.5.6) and (6.5.7) hold, this will be a special case of (6.5.4) and (6.5.5). In this case Assumption 3 will hold, and it should be excluded from Assumption 3'.

In particular, if $p_1 = q_1 = 0$, $a_2 = \ldots = a_k = a$ and $b_2 = \ldots = b_\ell = b$, then the system reduces to a two-state system. In fact, (6.5.2) and (6.5.3) now become

$$P(X_n \le t) = U(a^{n-1}t),$$

and

$$P(Y_n \le t) = V(b^n t).$$

Thus, $\{X_n, n = 1, 2, \ldots\}$ will form a GP with ratio a and $X_1 \sim U$, while $\{Y_n, n = 1, 2, \ldots\}$ will constitute a GP with ratio b and $Y_1 \sim G$ where $G(t) = V(bt)$. As a result, our model reduces to the GP model for a one-component two-state system considered in Section 6.2. If in addition, $a \ge 1$ and $0 < b \le 1$, our model becomes a one-component two-state deteriorating system, if $0 < a \le 1$ and $b \ge 1$, our model becomes a one-component two-state improving system.

Now, we shall determine the distributions of X_n and Y_n respectively.

$$P\left(X_2 \leq t\right) = \sum_{i=1}^{k} P\left(X_2 \leq t \mid S\left(t_1\right) = i\right) P\left(S\left(t_1\right) = i\right)$$

$$= \sum_{i=1}^{k} p_i U\left(a_i t\right).$$

In general,

$$P\left(X_n \leq t\right)$$

$$= \sum_{i_1=1}^{k} \cdots \sum_{i_{n-1}=1}^{k} P\left(X_n \leq t \mid S\left(t_1\right) = i_1, \ldots, S\left(t_{n-1}\right) = i_{n-1}\right)$$

$$\times P\left(S\left(t_1\right) = i_1, \ldots, S\left(t_{n-1}\right) = i_{n-1}\right)$$

$$= \sum_{i_1=1}^{k} \cdots \sum_{i_{n-1}=1}^{k} p_{i_1} \ldots p_{i_{n-1}} U\left(a_{i_1} \ldots a_{i_{n-1}} t\right) \tag{6.5.10}$$

$$= \sum_{\sum_{i=1}^{k} j_i = n-1} \frac{(n-1)!}{j_1! \ldots j_k!} p_1^{j_1} \ldots p_k^{j_k} U\left(a_1^{j_1} \ldots a_k^{j_k} t\right). \tag{6.5.11}$$

By a similar way, we have

$$P\left(Y_n \leq t\right)$$

$$= \sum_{i_1=1}^{\ell} \cdots \sum_{i_n=1}^{\ell} P\left(Y_n \leq t \mid S\left(s_1\right) = k + i_1, \ldots, S\left(s_n\right) = k + i_n\right)$$

$$\times P\left(S\left(s_1\right) = k + i_1, \ldots, S\left(s_n\right) = k + i_n\right)$$

$$= \sum_{i_1=1}^{\ell} \cdots \sum_{i_n=1}^{\ell} q_{i_1} \ldots q_{i_n} V\left(b_{i_1} \ldots b_{i_n} t\right) \tag{6.5.12}$$

$$= \sum_{\sum_{i=1}^{\ell} j_i = n} \frac{n!}{j_1! \ldots j_\ell!} q_1^{j_1} \ldots q_\ell^{j_\ell} V\left(b_1^{j_1} \ldots b_\ell^{j_\ell} t\right). \tag{6.5.13}$$

Now, we shall derive the long-run average cost per unit time (or simply the average cost) under Assumptions 1 and 2 only. To do this, let $\int_0^\infty t dU\left(t\right) = \lambda$ and $\int_0^\infty t dV\left(t\right) = \mu$. Thus from (6.5.11) we have

$$E[X_n] \tag{6.5.14}$$

$$= \sum_{\sum_{i=1}^{k} j_i = n-1} \frac{(n-1)!}{j_1! \cdots j_k!} p_1^{j_1} \cdots p_k^{j_k} \int_0^\infty t\, dU\left(a_1^{j_1} \cdots a_k^{j_k} t\right)$$

$$= \sum_{\sum_{i=1}^{k} j_i = n-1} \frac{(n-1)!}{j_1! \cdots j_k!} \left(\frac{p_1}{a_1}\right)^{j_1} \cdots \left(\frac{p_k}{a_k}\right)^{j_k} \lambda$$

$$= \lambda \left(\frac{p_1}{a_1} + \ldots + \frac{p_k}{a_k}\right)^{n-1} = \frac{\lambda}{a^{n-1}}, \tag{6.5.15}$$

where $a = (\sum_{i=1}^{k} \frac{p_i}{a_i})^{-1}$. Similarly, from (6.5.13), we have

$$E[Y_n] = \mu \left(\frac{q_1}{b_1} + \ldots + \frac{q_\ell}{b_\ell}\right)^n = \frac{\mu}{b^n}, \tag{6.5.16}$$

where $b = (\sum_{i=1}^{\ell} \frac{q_i}{b_i})^{-1}$.

Then, we shall calculate the expect reward earned after the $(n-1)$th repair. For this purpose, define R_n, the reward rate after the $(n-1)$th repair, as

$$R_n = r_i \quad \text{if} \quad S(t_{n-1}) = i, \quad i = 1, \ldots, k.$$

Because $S(t_0) = 1$, then $R_1 = r_1$, the expected reward after installation or a replacement is given by

$$E[R_1 X_1] = E[r_1 X_1] = r_1 \lambda. \tag{6.5.17}$$

In general, for $n \geq 2$, the expected reward after the $(n-1)$th repair is given by

$$E[R_n X_n]$$

$$= \sum_{i_1=1}^{k} \cdots \sum_{i_{n-1}=1}^{k} E(R_n X_n \mid S(t_1) = i_1, \ldots, S(t_{n-1}) = i_{n-1})$$

$$\times P(S(t_1) = i_1, \ldots, S(t_{n-1}) = i_{n-1})$$

$$= \sum_{i_1=1}^{k} \cdots \sum_{i_{n-1}=1}^{k} p_{i_1} \cdots p_{i_{n-1}} \int_0^\infty r_{i_{n-1}} t\, dU(a_{i_1} \cdots a_{i_{n-1}} t)$$

$$= \left(\sum_{i_1=1}^{k} \cdots \sum_{i_{n-2}=1}^{k} \frac{p_{i_1} \cdots p_{i_{n-2}}}{a_{i_1} \cdots a_{i_{n-2}}}\right) \left(\sum_{i_{n-1}=1}^{k} \frac{r_{i_{n-1}} p_{i_{n-1}}}{a_{i_{n-1}}}\right) \lambda$$

$$= r\lambda \left(\frac{p_1}{a_1} + \ldots + \frac{p_k}{a_k}\right)^{n-2} = \frac{r\lambda}{a^{n-2}}, \tag{6.5.18}$$

where

$$r = \sum_{i=1}^{k} \frac{r_i p_i}{a_i}. \tag{6.5.19}$$

Thereafter, we shall evaluate the expected repair cost incurred after the nth failure. To this end, define C_n, the repair cost rate after the nth failure, as

$$C_n = c_i \quad \text{if} \quad S(s_n) = k + i, \quad i = 1, \ldots, \ell.$$

By an argument similar to the calculation of $E(R_n X_n)$, the expected repair cost after the nth failure is given by

$$E[C_n Y_n]$$
$$= c\mu \left(\frac{q_1}{b_1} + \ldots + \frac{q_k}{b_k} \right)^{n-1} = \frac{c\mu}{b^{n-1}}, \tag{6.5.20}$$

where

$$c = \sum_{i=1}^{\ell} \frac{c_i q_i}{b_i}. \tag{6.5.21}$$

Now suppose that a replacement policy N is adopted. We say a cycle is completed if a replacement is completed. Thus, a cycle is actually the time interval between the installation of the system and the first replacement or two successive replacements. Then the successive cycles and the costs incurred in each cycle form a renewal reward process. Let $\tau = E[Z]$ be the expected replacement time. Then, by applying Theorem 1.3.15, the average cost is given by

$$C(N)$$
$$= \frac{E[\sum_{n=1}^{N-1} C_n Y_n - \sum_{n=1}^{N} R_n X_n + R + c_p Z]}{E[\sum_{n=1}^{N} X_n + \sum_{n=1}^{N-1} Y_n + Z]}$$
$$= \frac{c\mu \sum_{n=1}^{N-1} \frac{1}{b^{n-1}} - \left[r_1 \lambda + r\lambda \sum_{n=2}^{N} \frac{1}{a^{n-2}} \right] + R + c_p \tau}{\lambda \sum_{n=1}^{N} \frac{1}{a^{n-1}} + \mu \sum_{n=1}^{N-1} \frac{1}{b^n} + \tau}$$
$$= \frac{bc(\frac{\mu}{b}) \sum_{n=1}^{N-1} \frac{1}{b^{n-1}} - ar\lambda \sum_{n=1}^{N} \frac{1}{a^{n-1}} + R' + c_p \tau}{\lambda \sum_{n=1}^{N} \frac{1}{a^{n-1}} + (\frac{\mu}{b}) \sum_{n=1}^{N-1} \frac{1}{b^{n-1}} + \tau}, \tag{6.5.22}$$

where

$$R' = R - (r_1 - ar)\lambda. \tag{6.5.23}$$

Clearly, a and b are two important constants in (6.5.22). To explain the meaning of a and b, we introduce the harmonic mean of a random variable here.

Definition 6.5.1. Given a random variable X with $E\left[1/X\right] \neq 0$, $m_H = 1/E[1/X]$ is the harmonic mean of X.

The harmonic mean has some nice properties.

(1) If X is a discrete random variable having a uniform distribution, such that $X = x_i$ with probability $1/n, i = 1, \ldots, n$, then the harmonic mean $m_H = n/\left[\sum_{i=1}^n (1/x_i)\right]$ of X is the harmonic mean of numbers $x_1, \ldots, x_n$.
(2) If $0 < \alpha \leq X \leq \beta$, then $\alpha \leq m_H \leq \beta$.
(3) If X is nonnegative with $E[X] > 0$, then $m_H \leq E[X]$. This is because $h\left(x\right) = 1/x$ is a convex function, from which the result follows the Jensen inequality. In fact,

$$E[\frac{1}{X}] \geq \frac{1}{E[X]}.$$

Thus, in our model, from (6.5.15) and (6.5.16), a is the harmonic mean of a random variable X with $P(X = a_i) = p_i$, $i = 1, \ldots, k$, and b is the harmonic mean of a random variable Y with $P(Y = b_i) = q_i$, $i = 1, \ldots, \ell$. For this reason, we can call respectively a and b the harmonic means of $a_1, \ldots, a_k$ and $b_1, \ldots, b_\ell$.

Then, our problem is to determine an optimal replacement policy for minimizing the average cost $C(N)$. To this end, we first observe from (6.5.22) that

$$C(N) = A(N) - ar,$$

where

$$A(N) = \frac{(bc + ar)(\frac{\mu}{b})\sum_{n=1}^{N-1}\frac{1}{b^{n-1}} + R' + c_p\tau + ar\tau}{\lambda\sum_{n=1}^{N}\frac{1}{a^{n-1}} + (\frac{\mu}{b})\sum_{n=1}^{N-1}\frac{1}{b^{n-1}} + \tau}. \qquad (6.5.24)$$

Obviously, minimizing $C(N)$ is equivalent to minimizing $A(N)$. Therefore, to determine an optimal policy for the multistate system, a similar approach as applied to a two-state system in Section 6.2 could be used. To do this, consider

$$A(N + 1) - A(N)$$

$$= \frac{1}{a^N b^N B(N) B(N+1)}\{(bc+ar)\mu\{\lambda[\sum_{n=1}^{N} a^n - \sum_{n=1}^{N-1} b^n] + \tau a^N\}$$

$$-(R' + c_p\tau + ar\tau)(\lambda b^N + \mu a^N)\}, \tag{6.5.25}$$

where

$$B(N) = \lambda \sum_{n=1}^{N} \frac{1}{a^{n-1}} + \frac{\mu}{b} \sum_{n=1}^{N-1} \frac{1}{b^{n-1}} + \tau.$$

For $R' + c_p\tau + ar\tau \neq 0$, define an auxiliary function

$$g(N) = \frac{(bc+ar)\mu\{\lambda(\sum_{n=1}^{N} a^n - \sum_{n=1}^{N-1} b^n) + \tau a^N\}}{(R' + c_p\tau + ar\tau)(\lambda b^N + \mu a^N)}. \tag{6.5.26}$$

For $R' + c_p\tau + ar\tau = 0$, another auxiliary function is defined

$$g_0(N) = (bc+ar)\mu\{\lambda(\sum_{n=1}^{N} a^n - \sum_{n=1}^{N-1} b^n) + \tau a^N\}. \tag{6.5.27}$$

Consequently, we have the following lemma.

Lemma 6.5.2.

(1) If $R' + c_p\tau + ar\tau > 0$, then

$$A(N+1) \gtreqless A(N) \iff g(N) \gtreqless 1.$$

(2) If $R' + c_p\tau + ar\tau < 0$, then

$$A(N+1) \gtreqless A(N) \iff g(N) \lesseqgtr 1.$$

(3) If $R' + c_p\tau + ar\tau = 0$, then

$$A(N+1) \gtreqless A(N) \iff g_0(N) \gtreqless 0.$$

Lemma 6.5.2 shows that the monotonicity of $A(N)$ can be determined by the value of $g(N)$ or $g_0(N)$.

For $R' + c_p\tau + ar\tau \neq 0$, let $h(N) = \lambda b^N + \mu a^N$, then from (6.5.26) we have

$$g(N+1) - g(N)$$

$$= \frac{(bc+ar)\mu}{(R' + c_p\tau + ar\tau)h(N)h(N+1)}\{\lambda^2 b^N(1-b)\sum_{n=1}^{N} a^n + \lambda^2 b^N(a^{N+1} - b)$$

$$+\lambda\mu a^N(a - b^N) + \lambda\mu a^N(a-1)\sum_{n=1}^{N-1} b^n + \lambda\tau a^N b^N(a-b)\}. \tag{6.5.28}$$

For $R' + c_p\tau + ar\tau = 0$, it follows from (6.5.27) that

$$g_0(N+1) - g_0(N)$$
$$= (bc + ar)\mu\{\lambda(a^{N+1} - b^N) + \tau(a^{N+1} - a^N)\}. \qquad (6.5.29)$$

Note that the results above are all derived under Assumptions 1 and 2 only. Therefore, all the results including Lemma 6.5.2 hold without Assumption 3 or 3'.

Now, we shall show that under Assumptions 1-3, the model is a maintenance model for a multistate deteriorating system; while under Assumptions 1, 2 and 3', it is a maintenance model for a multistate improving system.

Model 1. The model under Assumptions 1-3

Theorem 6.5.3.   Under Assumption 3, for $n = 1, 2, \ldots$, we have

$$X_n \geq_{s.t.} X_{n+1},$$

$$Y_n \leq_{s.t.} Y_{n+1}.$$

Proof.

For any $t > 0$, it follows from (6.5.10) that

$$P(X_{n+1} \leq t)$$
$$= \sum_{i_1=1}^{k} \cdots \sum_{i_n=1}^{k} p_{i_1} \ldots p_{i_n} U(a_{i_1} \ldots a_{i_n} t)$$
$$= \sum_{i_1=1}^{k} \cdots \sum_{i_{n-1}=1}^{k} p_{i_1} \ldots p_{i_{n-1}} \left[\sum_{i_n=1}^{k} p_{i_n} U(a_{i_1} \ldots a_{i_{n-1}} a_{i_n} t) \right]$$
$$\geq \sum_{i_1=1}^{k} \cdots \sum_{i_{n-1}=1}^{k} p_{i_1} \ldots p_{i_{n-1}} \left[\sum_{i_n=1}^{k} p_{i_n} U(a_{i_1} \ldots a_{i_{n-1}} t) \right]$$
$$= \sum_{i_1=1}^{k} \cdots \sum_{i_{n-1}=1}^{k} p_{i_1} \ldots p_{i_{n-1}} U(a_{i_1} \ldots a_{i_{n-1}} t)$$
$$= P(X_n \leq t).$$

Thus $X_n \geq_{s.t.} X_{n+1}$. By a similar argument, from (6.5.12) we can prove that $Y_n \leq_{s.t.} Y_{n+1}$. This completes the proof of Theorem 6.5.3.

Because $\{X_n, n = 1, 2, \ldots\}$ is stochastically decreasing and $\{Y_n, n = 1, 2, \ldots\}$ is stochastically increasing, then Theorem 6.5.3 shows that Model

1 is a monotone process model for a multistate deteriorating system.

Now we shall argue that Model 1 is equivalent to a one-component two-state GP model for a deteriorating system. In fact, under Assumption 3, we have

$$1 = a_1 \leq a \leq a_k \quad \text{and} \quad b_k \leq b \leq b_1 = 1. \tag{6.5.30}$$

Then consider a GP maintenance model for a one-component two-state system, up and down states for example. Suppose the successive operating times after repair $\{X'_n, n = 1, 2, \ldots\}$ form a GP with ratio $a \geq 1$ and $E[X'_1] = \lambda$. Suppose further the consecutive repair times after failure $\{Y'_n, n = 1, 2, \ldots\}$ constitute a GP with ratio $0 < b \leq 1$ and $E[Y'_1] = \mu/b$. The replacement time is still Z with $E[Z] = \tau$. The reward rate is ar, the repair cost rate is bc. The basic replacement cost is R' while the part of replacement cost proportional to Z is still at rate c_p. Then under policy N, following the argument in Section 6.2, we can see from (6.2.2) that the average cost for the two-state system is exactly the same as that given by (6.5.22). As a result, the multistate system and the two-state system should have the same optimal policy, since they have the same average cost. In other words, our model for the multistate deteriorating system is equivalent to a GP model for the two-state deteriorating system in the sense that they will have the same average cost and the same optimal replacement policy. In conclusion, we have proved the following theorem.

Theorem 6.5.4. The monotone process model for a $k + \ell$ multistate deteriorating system is equivalent to a GP model for a two-state deteriorating system in the sense that they will have the same average cost and the same optimal replacement policy. The successive operating times in the two-state system $\{X'_n, n = 1, 2, \ldots\}$ form a GP with ratio $a \geq 1$, the harmonic mean of $a_1, \ldots, a_k$, and $E[X'_1] = \lambda$; while its consecutive repair times after failure $\{Y'_n, n = 1, 2, \ldots\}$ constitute a GP with ratio $0 < b \leq 1$, the harmonic mean of $b_1, \ldots, b_\ell$, and $E[Y'_1] = \mu/b$. The reward rate of the two-state system is ar with r given by (6.5.19) , the repair cost rate is bc with c given by (6.5.21), the basic replacement cost is R' given by (6.5.23), but the part of replacement cost proportional to Z is still at rate c_p.

Then we shall determine an optimal policy for the multistate deteriorating system. To do this, from (6.5.28) and (6.5.29) with the help of (6.5.30), the following lemma is straightforward.

Lemma 6.5.5.

(1) If $R' + c_p \tau + a r \tau > 0$, then $g(N)$ is nondecreasing in N.
(2) If $R' + c_p \tau + a r \tau < 0$, then $g(N)$ is nonincreasing in N.
(3) Function $g_0(N)$ is nondecreasing in N.

The combination of Lemmas 6.5.2 and 6.5.5 yields the following theorem.

Theorem 6.5.6.

(1) For $R' + c_p \tau + a r \tau > 0$, an optimal replacement policy N_d^* for the multistate deteriorating system is given by

$$N_d^* = \min\{N \mid g(N) \geq 1\}. \tag{6.5.31}$$

The optimal policy N_d^* is unique if and only if $g(N_d^*) > 1$.
(2) For $R' + c_p \tau + a r \tau < 0$, an optimal replacement policy N_d^* for the multistate deteriorating system is given by

$$N_d^* = \min\{N \mid g(N) \leq 1\}. \tag{6.5.32}$$

The optimal policy N_d^* is unique if and only if $g(N_d^*) < 1$.
(3) For $R' + c_p \tau + a r \tau = 0$, an optimal replacement policy N_d^* for the multistate deteriorating system is given by

$$N_d^* = \min\{N \mid g_0(N) \geq 0\}. \tag{6.5.33}$$

The optimal replacement policy N_d^* is unique if and only if $g_0(N_d^*) > 0$.

In application of Theorem 6.5.6, we should determine the value of $R' + c_p \tau + a r \tau$ first. As an example, suppose that $R' + c_p \tau + a r \tau > 0$, then we can determine N_d^* from (6.5.31). Thus

$$g(N) < 1 \Longleftrightarrow N < N_d^*,$$

and

$$g(N) \geq 1 \Longleftrightarrow N \geq N_d^*.$$

Therefore, N_d^* is the minimum integer satisfying $g(N) \geq 1$. By Lemma 6.5.2, we have

$$A(N) > A(N_d^*) \quad \text{if} \quad N < N_d^*,$$

and

$$A(N) \geq A(N_d^*) \quad \text{if} \quad N \geq N_d^*$$

In other words, policy N_d^* is indeed an optimal replacement policy. Obviously, it is unique if $g(N_d^*) > 1$.

Model 2.   The model under Assumptions 1, 2 and 3'

Now, instead of Theorem 6.5.3, we have the following theorem.

Theorem 6.5.7. Under Assumption 3', for $n = 1, 2, \ldots$, we have

$$X_n \leq_{s.t.} X_{n+1},$$

and

$$Y_n \geq_{s.t.} Y_{n+1}.$$

The proof is similar to that of Theorem 6.5.3.

Consequently, Theorem 6.5.7 demonstrates that Model 2 is a monotone process model for a multistate improving system.

Now under Assumption 3', we have

$$a_k \leq a \leq a_1 = 1 \quad \text{and} \quad 1 = b_1 \leq b \leq b_\ell. \tag{6.5.34}$$

Also, not all equalities in (6.5.34) hold simultaneously.

As in Model 1, we can consider a GP model for a one-component two-state improving system. Then, we have the following theorem which is analogous to Theorem 6.5.4.

Theorem 6.5.8. The monotone process model for a multistate improving system is equivalent to a GP model for a two-state improving system in the sense that they have the same average cost and the same optimal replacement policy. The successive operating times after repair $\{X'_n, n = 1, 2, \ldots\}$ of the two-state system will form a GP with ratio $0 < a \leq 1$, the harmonic mean of $a_1, \ldots, a_k$, and $E[X'_1] = \lambda$; while its consecutive repair times after failure $\{Y'_n, n = 1, 2, \ldots\}$ will constitute a GP with ratio $b \geq 1$, the harmonic mean of $b_1, \ldots, b_\ell$, and $E[Y'_1] = \mu/b$. The reward rate of the two-state system is ar with r given by (6.5.19), the repair cost rate is bc with c given by (6.5.21), and the basic replacement cost R' is given by (6.5.23), but the part of replacement cost proportional to Z is still at rate c_p.

Note that, for an improving system, the average cost due to adoption of a replacement policy N is also given by (6.5.22). Moreover, we can also determine an optimal replacement policy for the improving system on the basis of Lemma 6.5.2. To do so, recall that not all equalities in (6.5.34) will hold together. Then from (6.5.28), (6.5.29) and (6.5.34), $g(N)$ and $g_0(N)$ are clearly strictly monotone. By analogy with Lemma 6.5.5, we have the following lemma.

Lemma 6.5.9.

(1) If $R' + c_p\tau + ar\tau > 0$, then $g(N)$ is decreasing in N.
(2) If $R' + c_p\tau + ar\tau < 0$, then $g(N)$ is increasing in N.
(3) Function $g_0(N)$ is decreasing in N.

The following theorem gives the optimal replacement policy for the multistate improving system.

Theorem 6.5.10.

(1) For $R' + c_p\tau + ar\tau > 0$, the unique optimal replacement policy for the multistate improving system is $N_i^* = \infty$ with

$$C(N_i^*) = C(\infty) = -ar. \tag{6.5.35}$$

(2) For $R' + c_p\tau + ar\tau < 0$, the unique optimal replacement policy for the multistate improving system $N_i^* = 1$ with

$$C(N_i^*) = C(1) = \frac{R + c_p\tau - r_1\lambda}{\lambda + \tau}. \tag{6.5.36}$$

(3) For $R' + c_p\tau + ar\tau = 0$, the optimal replacement policy N_i^* for the multistate improving system is either 1 or ∞, and

$$C(N_i^*) = C(1) = C(\infty)$$
$$= \frac{R + c_p\tau - r_1\lambda}{\lambda + \tau} = -ar. \tag{6.5.37}$$

Proof.
1. For $R' + c_p\tau + ar\tau > 0$, $g(N)$ is decreasing. Then there exists an

$$N_+ = \min\{N \mid g(N) \le 1\}. \tag{6.5.38}$$

By Lemma 6.5.2, it is easy to see that $C(N)$ and $A(N)$ are both unimodal and take their maxima at N_+. This implies that the minimum of $C(N)$ must be given by $\min\{C(1), C(\infty)\}$. Because $R' + c_p\tau + ar\tau > 0$, it follows from (6.5.22) and (6.5.23) that

$$C(1) = \frac{R + c_p\tau - r_1\lambda}{\lambda + \tau} > C(\infty) = -ar.$$

Consequently,

$$\min C(N) = \min\{C(1), C(\infty)\}$$
$$= C(\infty) = -ar. \tag{6.5.39}$$

Therefore, $N_i^* = \infty$ is the optimal replacement policy. The optimal policy is also unique because from Lemma 6.5.9, $g(N)$ is strictly decreasing.

2. For $R' + c_p\tau + ar\tau < 0$, the proof is similar.

3. For $R' + c_p\tau + ar\tau = 0$, because $g_0(N)$ is decreasing, there exists an

$$N_0 = \min\{N \mid g_0(N) \le 0\}. \tag{6.5.40}$$

A similar argument shows that $C(N)$ and $A(N)$ will both attain their maxima at N_0. This implies that the minimum of $C(N)$ must be given by $\min\{C(1), C(\infty)\}$. However, because $R' + c_p\tau + ar\tau = 0$, then

$$C(1) = \frac{R + c_p\tau - r_1\lambda}{\lambda + \tau} = C(\infty) = -ar.$$

Thus, the optimal replacement policy N_i^* is either 1 or ∞.

This completes the proof.

In practice, for most improving systems,

$$R' + c_p\tau + ar\tau = R - (r_1 - ar)\lambda + c_p\tau + ar\lambda > 0, \tag{6.5.41}$$

since R is usually large in comparison with other parameters. On the other hand, as k working states of the system are arranged in an increasing order, and state 1 is the worst working state. The worse the working state is, the lower the reward rate will be. It is natural and reasonable to assume that

$$r_1 \le r_i, \quad i = 1, 2, \ldots, k, \tag{6.5.42}$$

holds. Thus

$$r = \sum_{i=1}^{k} \frac{r_i p_i}{a_i} \ge \frac{r_1}{a},$$

and (6.5.41) holds. Thus the unique optimal policy is $N_i^* = \infty$.

Consequently, for most multistate improving systems, we should repair the system when it fails but never replace it.

6.6   A Geometric Process Shock Model

So far, we have just studied the GP models for a system that is deteriorating due to an internal cause such as the ageing effect or accumulated wearing. However, an external cause such as some random shocks produced by an environment factor may be another reason for deterioration of a system. For example, a precision instrument or meter installed in a power workshop might be affected by the random shocks due to the operation of other instruments in the environment. As a result, the operating time of the instrument or meter might be shorter. Therefore, a maintenance model with random shock is an important model in reliability. In this section, we shall study a GP shock model by making the following assumptions.

Assumption 1. A new system is installed at the beginning. Whenever the system fails, it will be repaired. A replacement policy N is adopted by which the system will be replaced by a new and identical one at the time following the Nth failure.

Assumption 2. Given that there is no random shock, let X_n be the operating time of system after the $(n-1)$th repair, then $\{X_n, \; n = 1, 2, \ldots\}$ will form a GP with ratio a and $E[X_1] = \lambda \; (> 0)$. Let Y_n be the repair time of the system after the nth failure. Then, no matter whether there is a random shock or not, $\{Y_n, \; n = 1, 2, \ldots\}$ will constitute a GP with ratio b and $E[Y_1] = \mu \; (\geq 0)$. The replacement time is a random variable Z with $E(Z) = \tau$.

Assumption 3. Let the number of random shocks by time t produced by the random environment be $N(t)$. Assume that $\{N(t), t \geq 0\}$ forms a counting process having stationary and independent increment. Whenever a shock arrives, the system operating time will be reduced. Let W_n be the reduction in the system operating time following the nth random shock. Then $\{W_n, \; n = 1, 2, \ldots\}$ are i.i.d. random variables. The successive reductions in the system operating time are additive.

If a system has failed, it will be closed in the sense that any shock arriving after failure gives no effect on the failed system.

Assumption 4. The processes $\{X_n, \; n = 1, 2, \ldots\}, \{N(t), t \geq 0\}$ and $\{W_n, \; n = 1, 2, \ldots\}$ are independent.

Assumption 5. The reward rate of the system is r, the repair cost rate of the system is c. The replacement cost comprises two parts, one part is the basic replacement cost R, the other part is proportional to the replacement time Z at rate c_p, and $E(Z) = \tau$.

Assumption 6. $a \geq 1$ and $0 < b \leq 1$.

Under Assumption 6, the model is a deteriorating system. For an improving system, we shall make another assumption.

Assumption 6'. $0 < a \leq 1$ and $b \geq 1$ except the case $a = b = 1$.

Now, we shall first study the model under the Assumptions 1-5. To do so, denote the completion time of the $(n-1)$th repair by t_{n-1}. Then the number of shocks in $(t_{n-1}, t_{n-1} + t]$ produced by the environment is given by

$$N(t_{n-1}, t_{n-1} + t] = N(t_{n-1} + t) - N(t_{n-1}), \qquad (6.6.1)$$

where $N(t_{n-1})$ and $N(t_{n-1} + t)$ are respectively the number of random shocks produced in $(0, t_{n-1}]$ and $(0, t_{n-1}+t]$. Therefore, the total reduction in the operating time in $(t_{n-1}, t_{n-1} + t]$ is given by

$$\Delta X_{(t_{n-1}, t_{n-1}+t]} = \sum_{i=N(t_{n-1})+1}^{N(t_{n-1}+t)} W_i$$
$$= \sum_{i=1}^{N(t_{n-1}, t_{n-1}+t]} W_i, \qquad (6.6.2)$$

equality (6.6.2) means that the random variables in both sides have the same distribution. Consequently, under random environment, the residual time at $t_{n-1} + t$ is given by

$$S_n(t) = X_n - t - \Delta X_{(t_{n-1}, t_{n-1}+t]}, \qquad (6.6.3)$$

subject to $S_n(t) \geq 0$. Therefore, the real system operating time after the $(n-1)$th repair is given by

$$X'_n = \inf_{t \geq 0} \{t \mid S_n(t) \leq 0\}. \qquad (6.6.4)$$

The following lemma is useful for later study, the proof is trivial.

Lemma 6.6.1.

$$P(X_n - t - \Delta X_{(t_{n-1}, t_{n-1}+t]} > 0, \ \forall t \in [0, t'])$$
$$= P(X_n - t' - \Delta X_{(t_{n-1}, t_{n-1}+t']} > 0). \qquad (6.6.5)$$

Remarks

Assumption 2 just shows that the system is deteriorating so that the consecutive repair times constitute an increasing GP, and if there is no random shock, the successive operating times form a decreasing GP.

Assumption 3 means that the effect of the random environment on the system is through a sequence of random shocks which will shorten the operating time. In practice, many examples show that the effect of a random shock is an additive reduction rather than a percentage reduction in residual operating time. In other words, $\{W_n, n = 1, 2, \ldots\}$ will act in an additive way rather than in a multiplicative way. For example, it is well known that second hand smoking is very serious, as its effect is accumulated by an additive reduction in his lifetime. Similarly, a car suffered by traffic accidents will reduce its operating time, the reduction is also additive.

Equation (6.6.3) shows that whenever the total reduction $\Delta X_{(t_{n-1}, t_{n-1}+t]}$ in system operating time in $(t_{n-1}, t_{n-1} + t]$ is greater than the residual operating time $X_n - t$, the system will fail. In other words, the chance that a shock produces an immediate failure depends on the distributions of $X_n - t$ and $\Delta X_{(t_{n-1}, t_{n-1}+t]}$. To see the reasonableness of this point, consider the following examples.

In a traffic accident, as all the passengers in the bus are suffered by the same shock, so that the reductions in their lifetimes are more or less the same, but the effect on different passengers might be quite different. An older passenger is more fragile because he has a residual lifetime less than a younger passenger has. Thus the older passenger can be injured more seriously than a younger passenger. The older passenger may even die, but the younger passenger may only suffer a slight wound. This situation also happens in engineering. Suppose many machines are installed in a workshop, all of them are suffered the same shock produced by a random environment, but the effects might be different, an old machine could be destroyed whereas a new machine might be slightly damaged. This means that the effect of a random shock depends on the residual lifetime of a system, if the reduction in the lifetime is greater than the residual time, the system will fail. Therefore result (6.6.3) is realistic. These two examples also show that $\{W_n, n = 1, 2, \ldots\}$ will act additively. In fact, if $\{W_n, n = 1, 2, \ldots\}$ act in a multiplicative way, the system could not fail after suffering a random walk.

Now, we say a cycle is completed if a replacement is completed. Then the successive cycles together with the costs incurred in each cycle form a renewal reward process. Therefore, the average cost per unit time (or simply the average cost) is also given by (1.3.36).

To begin, we shall study the distribution of X_n'. For this purpose, suppose $N(t_{n-1}, t_{n-1} + t']$, the number of random shocks occurred in $(t_{n-1}, t_{n-1} + t']$, is k. Then for $t' > 0$, by using Lemma 6.6.1, we have the following result.

$$
\begin{aligned}
&P(X_n' > t' \mid N(t_{n-1}, t_{n-1} + t'] = k) \\
&= P(\inf_{t \geq 0}\{t \mid S_n(t) \leq 0\} > t' \mid N(t_{n-1}, t_{n-1} + t'] = k) \\
&= P(S_n(t) = X_n - t - \Delta X_{(t_{n-1}, t_{n-1}+t]} > 0, \ \forall t \in [0, t'] \\
&\qquad \mid N(t_{n-1}, t_{n-1} + t'] = k) \\
&= P(X_n - t' - \Delta X_{(t_{n-1}, t_{n-1}+t']} > 0 \mid N(t_{n-1}, t_{n-1} + t'] = k) \\
&= P(X_n - \Delta X_{(t_{n-1}, t_{n-1}+t']} > t' \mid N(t_{n-1}, t_{n-1} + t'] = k) \\
&= P\left(X_n - \sum_{i=1}^{k} W_i > t'\right).
\end{aligned}
$$

Thus

$$P(X_n' > t' \mid N(t_{n-1}, t_{n-1} + t'] = k)$$
$$= \int_D \int f_n(x) h_k(w) dx dw, \tag{6.6.6}$$

where

$$D = \{(x, w) \mid x > 0, w > 0, x - w > t'\},$$

f_n is the density function of X_n and h_k is the density function of $\sum_{i=1}^{k} W_i$. Let the common density of W_i be h, and the common distribution of W_i be H. Then, h_k is the k-fold convolution of h with itself. Now, (6.6.6) yields that

$$P(X_n' > t' \mid N(t_{n-1}, t_{n-1} + t'] = k)$$
$$= \int_0^\infty [\int_{t'+w}^\infty f_n(x) dx] h_k(w) dw$$
$$= \int_0^\infty [1 - F_n(t' + w)] dH_k(w)$$
$$= 1 - \int_0^\infty F_n(t' + w) dH_k(w), \tag{6.6.7}$$

where F_n is the distribution function of X_n, H_k is the distribution of $\sum_{i=1}^{k} W_i$, it is the k-fold convolution of H with itself. Then

$$P(X_n' > t')$$
$$= \sum_{k=0}^{\infty} P(X_n' > t' \mid N(t_{n-1}, t_{n-1} + t'] = k) P(N(t_{n-1}, t_{n-1} + t'] = k)$$
$$= \sum_{k=0}^{\infty} [1 - \int_0^\infty F_n(t' + w) dH_k(w)] P(N(t_{n-1}, t_{n-1} + t'] = k)$$
$$= 1 - \sum_{k=0}^{\infty} \int_0^\infty F_n(t' + w) dH_k(w) P(N(t') = k), \tag{6.6.8}$$

where (6.6.8) is due to the fact that $\{N(t), \ t \geq 0\}$ has stationary increments. Therefore, by noting that $F_n(x) = F(a^{n-1}x)$, the distribution function I_n of X_n' is given by

$$I_n(x) = P(X_n' \leq x)$$
$$= \sum_{k=0}^{\infty} \int_0^\infty F(a^{n-1}(x + w)) dH_k(w) P(N(x) = k). \tag{6.6.9}$$

Consequently, by applying the replacement policy N, it follows from (1.3.36) that the average cost is given by

$$C(N) = \frac{E[c \sum_{n=1}^{N-1} Y_n - r \sum_{n=1}^{N} X'_n + R + c_p Z]}{E[\sum_{n=1}^{N} X'_n + \sum_{n=1}^{N-1} Y_n + Z]}$$

$$= \frac{c\mu \sum_{n=1}^{N-1} \frac{1}{b^{n-1}} - r \sum_{n=1}^{N} \lambda'_n + R + c_p \tau}{\sum_{n=1}^{N} \lambda'_n + \mu \sum_{n=1}^{N-1} \frac{1}{b^{n-1}} + \tau} \qquad (6.6.10)$$

$$= A(N) - r \qquad (6.6.11)$$

where

$$\lambda'_n = E[X'_n] = \int_0^\infty x \, dI_n(x) \qquad (6.6.12)$$

is the expected real operating time after the $(n-1)$th repair, and

$$A(N) = \frac{(c+r)\mu \sum_{n=1}^{N-1} \frac{1}{b^{n-1}} + R + c_p \tau + r\tau}{\sum_{n=1}^{N} \lambda'_n + \mu \sum_{n=1}^{N-1} \frac{1}{b^{n-1}} + \tau}. \qquad (6.6.13)$$

Now let

$$h(n) = \sum_{n=1}^{N} \lambda'_n + \mu \sum_{n=1}^{N-1} \frac{1}{b^{n-1}} + \tau.$$

Then it is easy to justify that

$$A(N+1) - A(N)$$
$$= \frac{1}{b^{N-1} h(N+1) h(N)} \{ (c+r)\mu (\sum_{n=1}^{N} \lambda'_n - \lambda'_{N+1} \sum_{n=1}^{N-1} b^n + \tau)$$
$$- (R + c_p \tau + r\tau)(\lambda'_{N+1} b^{N-1} + \mu) \}. \qquad (6.6.14)$$

Then, we introduce an auxiliary function $g(N)$.

$$g(N) = \frac{(c+r)\mu (\sum_{n=1}^{N} \lambda'_n - \lambda'_{N+1} \sum_{n=1}^{N-1} b^n + \tau)}{(R + c_p \tau + r\tau)(\lambda'_{N+1} b^{N-1} + \mu)}. \qquad (6.6.15)$$

Moreover, it is straightforward that

$$g(N+1) - g(N) = \frac{(c+r)\mu(\lambda'_{N+1} - b\lambda'_{N+2})}{(R + c_p\tau + r\tau)b^N(\lambda'_{N+1} + \frac{\mu}{b^{N-1}})(\lambda'_{N+2} + \frac{\mu}{b^N})}$$

$$\times\{\sum_{n=1}^{N+1}\lambda'_n + \mu\sum_{n=1}^{N}\frac{1}{b^{n-1}} + \tau\}. \tag{6.6.16}$$

Now, our objective is to determine an optimal replacement policy N^* for minimizing $C(N)$ or $A(N)$. To start with, from (6.6.14) and (6.6.15), we have the following lemma.

Lemma 6.6.2.

$$A(N+1) \gtreqless A(N) \iff g(N) \gtreqless 1.$$

Now, we shall consider two cases respectively.

Model 1.   The model under Assumptions 1-6

This is a GP shock model for a deteriorating system. Now, as $\{X_n, n = 1, 2, \cdots\}$ form an decreasing GP, and F_n is the distribution of X_n, then from (6.6.9) for all real t', we have

$$P(X'_n > t') \geq P(X'_{n+1} > t').$$

Thus we have the following lemma.

Lemma 6.6.3. λ'_n is nonincreasing in n.

Furthermore, it follows from (6.6.16) that

Lemma 6.6.4. $g(N)$ is nondecreasing in N.

The combination of Lemmas 6.6.2-6.6.4 yields Theorem 6.6.5.

Theorem 6.6.5.   An optimal replacement policy N^*_d is determined by

$$N^*_d = \min\{N \mid g(N) \geq 1\}. \tag{6.6.17}$$

The optimal replacement policy N^* is unique, if and only if $g(N^*) > 1$.

In particular, it follows from Theorem 6.6.5 that

Corollary 6.6.6.

(1) If

$$g(1) = \frac{(c+r)\mu(\lambda'_1 + \tau)}{(R + c_p\tau + r\tau)(\lambda'_2 + \mu)} \geq 1,$$

then $N^* = 1$. The optimal replacement policy $N^* = 1$ is unique, if and only if $g(1) > 1$.

(2) If $a > 1, 0 < b < 1$ and

$$g(\infty) = \frac{(c+r)\mu(\sum\limits_{n=1}^{\infty} \lambda'_n + \tau)}{(R + c_p\tau + r\tau)\mu} \le 1,$$

then $N^* = \infty$. The optimal replacement policy $N^* = \infty$ is unique.

Proof.

The result of part (1) is trivial. To prove part (2), note that if $a > 1$, as $E[X_n] = \frac{\lambda}{a^{n-1}}$ then

$$\sum_{n=1}^{\infty} \frac{\lambda}{a^{n-1}} < \infty.$$

Hence by noting that $\lambda'_n \le \frac{\lambda}{a^{n-1}}$, we have

$$\sum_{n=1}^{\infty} \lambda'_n < \infty.$$

Thus part (2) follows.

In other words, if $g(1) \ge 1$, the optimal policy is to replace the system whenever it fails; if $g(\infty) \le 1$, the optimal policy is to repair the system when it fails but never replace it.

Model 2. The model under Assumptions 1-5 and 6'

This is a GP shock model for an improving system. Now, instead of Lemmas 6.6.3 and 6.6.4, from (6.6.9) and (6.6.16), we have

Lemma 6.6.7. λ'_n is nondecreasing in n.

Lemma 6.6.8. $g(N)$ is nonincreasing in N.

Consequently, we have

Theorem 6.6.9. Under Assumptions 1-5 and 6', an optimal policy N_i^* is determined by

$$N_i^* = \begin{cases} 1 & A(1) < A(\infty), \\ 1 \text{ or } \infty & A(1) = A(\infty), \\ \infty & A(1) > A(\infty). \end{cases} \tag{6.6.18}$$

Proof.

Because $g(N)$ is nonincreasing in N, there exists an integer N_i such that

$$N_i = \min\{N \mid g(N) \leq 1\}. \tag{6.6.19}$$

Therefore, Lemma 6.6.2 implies that $C(N)$ and $A(N)$ are both unimodal functions of N and take their maxima at N_i. Therefore, (6.6.18) follows.

Corollary 6.6.10. If either $b > 1$ or $b = 1$ and

$$R + c_p\tau > (c + r)\lambda'_1 + c\tau, \tag{6.6.20}$$

then policy $N_i^* = \infty$ is the unique optimal policy.

Proof.

If $b > 1$, then (6.6.13) yields that $A(\infty) = 0$, since $\sum_{n=1}^{\infty} \lambda'_n$ is divergent. Now (6.6.18) implies that $N_i^* = \infty$ is the unique optimal policy.

If $b = 1$, then (6.6.13) and (6.6.20) give

$$A(\infty) \leq c + r < \frac{R + c_p\tau + r\tau}{\lambda'_1 + \tau} = A(1).$$

This also implies that $N_i^* = \infty$ is the unique optimal policy.

In practice, $b > 1$ holds in most cases, if it is not the case, R might be very large in comparison with other costs so that (6.6.20) will hold. Therefore, for an improving system, $N_i^* = \infty$ will usually be the optimal policy.

Now consider a special case that $\{N(t),\ t \geq 0\}$ is a Poisson process with rate γ. Then,

$$P(N(t) = k) = \frac{(\gamma t)^k}{k!} e^{-\gamma t}, \quad k = 0, 1, \ldots \tag{6.6.21}$$

Let the successive reduction in the system operating time caused by random shocks are $W_1, W_2, \ldots, W_k, \ldots$ They are i.i.d. each having a gamma distribution $\Gamma(\alpha, \beta)$ with density function h given by

$$h(w) = \begin{cases} \frac{\beta^\alpha}{\Gamma(\alpha)} w^{\alpha-1} e^{-\beta w} & w > 0, \\ 0 & \text{elsewhere.} \end{cases} \tag{6.6.22}$$

Therefore $\sum_{i=1}^{k} W_i$ is a gamma random variable with distribution $\Gamma(k\alpha, \beta)$. Furthermore, assume that X_1 has an exponential distribution $Exp(1/\lambda)$ with density function f given by

$$f(x) = \begin{cases} \frac{1}{\lambda} \exp(-\frac{x}{\lambda}) & x > 0, \\ 0 & \text{elsewhere.} \end{cases} \tag{6.6.23}$$

Then the distribution function F_n of X_n is given by

$$F_n(x) = F(a^{n-1}x) = \begin{cases} 1 - \exp(-a^{n-1}x/\lambda) & x > 0, \\ 0 & \text{elsewhere.} \end{cases} \quad (6.6.24)$$

Because $\{N(t), \ t \geq 0\}$ is a Poisson process, and $\{W_i, i = 1, 2, \ldots\}$ are i.i.d. random variables, then

$$\Delta X_{(0, \ t]} = \sum_{i=1}^{N(t)} W_i \quad (6.6.25)$$

forms a compound Poisson process. It follows from (6.6.9) that

$$I_n(x) = P(X'_n \leq x)$$

$$= \sum_{k=0}^{\infty} \int_0^\infty [1 - \exp(-\frac{a^{n-1}(x + w)}{\lambda})] \frac{\beta^{k\alpha}}{\Gamma(k\alpha)} w^{k\alpha-1} e^{-\beta w} dw \frac{(\gamma x)^k}{k!} e^{-\gamma x}$$

$$= 1 - \sum_{k=0}^{\infty} \exp[-(\gamma + \frac{a^{n-1}}{\lambda})x] \frac{(\gamma x)^k \beta^{k\alpha}}{k! \Gamma(k\alpha)} \int_0^\infty w^{k\alpha-1} \exp[-(\beta + \frac{a^{n-1}}{\lambda})w] dw$$

$$= 1 - \sum_{k=0}^{\infty} \exp[-(\gamma + \frac{a^{n-1}}{\lambda})x] \frac{1}{k!} [\frac{\gamma x \beta^\alpha}{(\beta + a^{n-1}/\lambda)^\alpha}]^k$$

$$= 1 - \exp\{-[\gamma(1 - (\frac{\beta}{\beta + a^{n-1}/\lambda})^\alpha) + \frac{a^{n-1}}{\lambda}]x\}.$$

Thus

$$E[X'_n] = \lambda'_n = \int_0^\infty x \, dI_n(x)$$

$$= \left(\gamma[1 - (\frac{\beta}{\beta + a^{n-1}/\lambda})^\alpha] + \frac{a^{n-1}}{\lambda}\right)^{-1}. \quad (6.6.26)$$

Clearly, if $\gamma = 0$, the system suffers no random shock, and the model reduces to the GP maintenance model introduced in Section 6.2; if $\alpha = 1$, then $W_1, W_2, \ldots, W_k, \ldots$ are i.i.d random variables each having an $Exp(\beta)$ distribution. Then (6.6.26) becomes

$$E(X'_n) = \lambda'_n = \frac{\lambda}{a^{n-1}[1 + \frac{\gamma}{\beta + a^{n-1}/\lambda}]}. \quad (6.6.27)$$

Now, we can substitute formula (6.6.26) or (6.6.27) into (6.6.15) for an explicit expression of $g(N)$. Then an optimal replacement policy N^* can be determined by using Theorem 6.6.5 or 6.6.9 accordingly.

As an explanation, a numerical example is studied here to explain how to determine an optimal replacement policy N^*. Let $c = 5, r = 25, R = $

$3600, c_p = 15, \lambda = 50, \mu = 10, \tau = 40, a = 1.01, b = 0.97, \alpha = 2, \beta = 4$ and $\gamma = 5$. Using (6.6.26), we have

$$\lambda'_n = \frac{40000 + 400a^{n-1} + a^{2n-2}}{a^{n-1}(2800 + 13a^{n-1} + 0.02a^{2n-2})}$$

Then, we can substitute λ'_n and parameter values into (6.6.10) and (6.6.15) respectively for $C(N)$ and $g(N)$. The results are tabulated in Table 6.6.1 and plotted in Figure 6.6.1. Clearly, $C(34) = -3.3612$ is the minimum of the average cost. On the other hand, $g(34) = 1.0333 > 1$, and

$$34 = \min\{N \mid g(N) \geq 1\}.$$

Thus, the unique optimal policy is $N^* = 34$, we should replace the system following the time of the 34th failure.

Table 6.6.1   Results obtained from (6.6.10) and (6.6.15)

N	C(N)	g(N)	N	C(N)	g(N)	N	C(N)	g(N)	N	C(N)	g(N)
1	70.6548	0.1295	11	3.3214	0.2399	21	-2.1982	0.4997	31	-3.3233	0.8939
2	44.9898	0.1339	12	2.2911	0.2592	22	-2.4156	0.5335	32	-3.3461	0.9394
3	31.4157	0.1398	13	1.4264	0.2800	23	-2.6020	0.5687	33	-3.3585	0.9859
4	23.0475	0.1471	14	0.6948	0.3023	24	-2.7612	0.6052	**34**	**-3.3612**	**1.0333**
5	17.3942	0.1559	15	0.0720	0.3261	25	-2.8961	0.6429	35	-3.3551	1.0816
6	13.3355	0.1661	16	-0.4610	0.3514	26	-3.0095	0.6819	36	-3.3410	1.1306
7	10.2932	0.1779	17	-0.9187	0.3782	27	-3.1036	0.7221	37	-3.3195	1.1805
8	7.9386	0.1911	18	-1.3127	0.4064	28	-3.1804	0.7634	38	-3.2913	1.2310
9	6.0710	0.2058	19	-1.6525	0.4361	29	-3.2417	0.8058	39	-3.2568	1.2823
10	4.5610	0.2221	20	-1.9456	0.4672	30	-3.2888	0.8493	40	-3.2167	1.3342

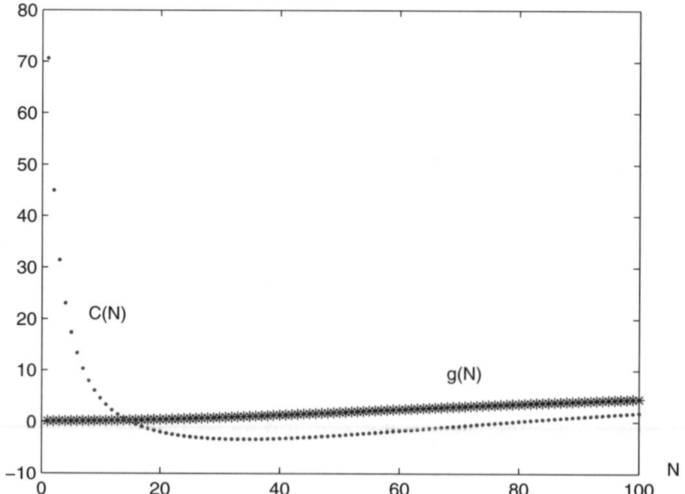

Figure 6.6.1. The plots of g(N) and C(N) against N

6.7 A Geometric Process δ-Shock Model

In Section 6.6, we study a GP shock model for a system under a random environment. Whenever a shock arrives, the system operating time will be reduced, the successive reductions in operation time are i.i.d. random variables and are additive. A shock is a deadly shock if the system will fail after suffering the shock. Then a deadly shock is a shock when it arrives the accumulated reduction in operating time will exceed the residual operating time. A δ-shock is different from the above shock. If the interarrival time of two successive shocks is smaller than a specified threshold δ, then the system will fail, and the latter shock is a deadly δ-shock. Now we shall introduce a GP δ-shock model for a repairable system by making the following assumptions.

Assumption 1. At the beginning, a new system is installed. Whenever the system fails, it will be repaired. A replacement policy N is adopted by which the system will be replaced by a new and identical one at the time following the Nth failure.

Assumption 2. The system is subject to a sequence of shocks. The shocks will arrive according to a Poisson process with rate θ. If the system has been repaired for n times, $n = 0, 1, \ldots$, the threshold of deadly shock will be $\alpha^n \delta$ where α is the rate and δ is the threshold of deadly shock for a new system. This means that whenever the time to the first shock is less than δ or the interarrival time of two successive shocks after the nth repair is less than $\alpha^n \delta$, the system will fail. During the repair time, the system is closed, so that any shock arriving when the system is under repair is ineffective. The successive repair times of the system after failures form a GP with ratio b. The mean repair time after the first failure is $\mu \geq 0$. The replacement time is a random variable Z with $E(Z) = \tau$.

Assumption 3. The repair cost rate is c, the reward rate when the system is operating is r. The replacement cost comprises two parts, one part is the basic replacement cost R, the other part is proportional to the replacement time Z at rate c_p.

Assumption 4. The Poisson process and the GP are independent.

Assumption 5. $\alpha \geq 1$ and $0 < b \leq 1$.

Then under Assumptions 1-5, the GP δ-shock model is a maintenance model for a deteriorating system. For an improving system, Assumption 5 will be replaced by Assumption 5'.

Assumption 5'. $0 < \alpha \leq 1$ and $b \geq 1$ except the case $\alpha = b = 1$.

Then under Assumptions 1-4 and 5', the GP δ-shock model is a maintenance model for an improving system.

Remarks

In this model, we assume there is one repair facility. Therefore, the repair facility will repair the system immediately when it fails until the system is recovered from failure. Therefore, the repair facility will be free if the system is operating.

In many cases, the interarrival times of shocks are i.i.d. random variables. Then a Poisson process will be an adequate approximation of the real arrival process of the shocks. On the other hand, Assumption 4 is natural, as the Poisson process is due to an external cause, the effect of random environment, while the GP is determined by the system itself.

In practice, most systems are deteriorating. For a deteriorating system, it will be more fragile and easier to break down after repair. As a result, the threshold of a deadly shock of the system will be increasing in n, the number of repairs taken. In other words, if the number of repairs n increases, the threshold of a deadly shock will increase accordingly. As an approximation, we may assume that the threshold value increases in n geometrically at rate $\alpha \geq 1$. Furthermore, for a deteriorating system, the consecutive repair times of the system will be longer and longer. In conclusion, Assumptions 2 and 5 just indicate that the system is deteriorating from different phases. Therefore, under Assumptions 1-5, the GP δ-shock model is a maintenance model for a deteriorating system.

However, in real life, there do have some improving systems. For an improving system, the older the system is, the more solid the system will be. Thus, the threshold of a deadly shock could be decreasing geometrically, while the successive repair times of the system will constitute a decreasing GP. Consequently, under Assumptions 1-4 and 5', the GP δ-shock model is a maintenance model for an improving system.

Now, we shall say that a cycle is completed if a replacement is completed. Thus a cycle is actually a time interval between the installation of a system and the first replacement or an interarrival time interval between two consecutive replacements. Then, the successive cycles together with the costs incurred in each cycle will constitute a renewal reward process. The average cost per unit time (or simply the average cost) is again given by (1.3.36). Let X_n be the operating time of a system following the $(n-1)$th repair in a cycle, and let Y_n be the repair time after the nth failure of the system in the cycle. Thus under replacement policy N, from (1.3.36) the average cost $C(N)$ is given by

$$C(N) = \frac{E(c \sum_{n=1}^{N-1} Y_n - r \sum_{n=1}^{N} X_n + R + c_p Z)}{E(\sum_{n=1}^{N} X_n + \sum_{n=1}^{N-1} Y_n + Z)}$$

$$= \frac{c\mu \sum_{n=1}^{N-1} \frac{1}{b^{n-1}} - r \sum_{n=1}^{N} \lambda_n + R + c_p \tau}{\sum_{n=1}^{N} \lambda_n + \mu \sum_{n=1}^{N-1} \frac{1}{b^{n-1}} + \tau} \tag{6.7.1}$$

$$= A(N) - r, \tag{6.7.2}$$

where $\lambda_n = E(X_n)$ is the expected operating time following the $(n-1)$th repair, and

$$A(N) = \frac{(c+r)\mu \sum_{n=1}^{N-1} \frac{1}{b^{n-1}} + R + (c_p + r)\tau}{\sum_{n=1}^{N} \lambda_n + \mu \sum_{n=1}^{N-1} \frac{1}{b^{n-1}} + \tau}. \tag{6.7.3}$$

Now, we need to evaluate the values of $\lambda_n, n = 1, 2, \ldots$ To do this, let U_{n1} be the arrival time of the first shock following the $(n-1)$th repair. In general, let U_{nk} be the interarrival time between the $(k-1)$th and kth shocks following the $(n-1)$th repair. Let $M_n, n = 1, 2, \ldots$, be the number of shocks following the $(n-1)$th repair until the first deadly shock occurred. Then,

$$M_n = \min\{m \mid U_{n1} \geq \alpha^{n-1}\delta, \ldots, U_{n\ m-1} \geq \alpha^{n-1}\delta, U_{nm} < \alpha^{n-1}\delta\}. \tag{6.7.4}$$

Let U_n be a random variable with exponential distribution $Exp(\theta)$ with mean $1/\theta$. Then, M_n will have a geometric distribution $G(p_n)$ with

$$p_n = P(U_n < \alpha^{n-1}\delta) = \int_0^{\alpha^{n-1}\delta} \theta e^{-\theta x} dx$$

$$= 1 - \exp(-\theta\alpha^{n-1}\delta) \tag{6.7.5}$$

and $q_n = 1 - p_n$. Thus

$$X_n = \sum_{i=1}^{M_n} U_{ni}. \tag{6.7.6}$$

Now, suppose that $M_n = m$, then

$$X_n = X_{nm} + U_{nm} \tag{6.7.7}$$

with

$$X_{nm} = \sum_{i=1}^{m-1} U_{ni}$$

and

$$U_{n1} \geq \alpha^{n-1}\delta, \ldots, U_{n\,m-1} \geq \alpha^{n-1}\delta, \quad \text{but} \quad U_{nm} < \alpha^{n-1}\delta. \tag{6.7.8}$$

Consequently,

$$X_{nm} = \sum_{i=1}^{m-1} (U_{ni} - \alpha^{n-1}\delta) + (m-1)\alpha^{n-1}\delta.$$

Because exponential distribution is memoryless, $U_{ni} - \alpha^{n-1}\delta$, $i = 1, 2, \ldots, m-1$, are i.i.d. random variables, each has the same exponential distribution $Exp(\theta)$ as U_n has. This implies that $X_{nm} - (m-1)\alpha^{n-1}\delta$ will have a gamma distribution $\Gamma(m-1, \theta)$. Thus, the density g_{nm} of X_{nm} is given by

$$g_{nm}(x) = \begin{cases} \frac{\theta^{m-1}}{(m-2)!}(x-K)^{m-2}e^{-\theta(x-K)} & x > K, \\ 0 & \text{elsewhere,} \end{cases} \tag{6.7.9}$$

where $K = (m-1)\alpha^{n-1}\delta$. As a result

$$E(X_{nm}) = \frac{m-1}{\theta} + (m-1)\alpha^{n-1}\delta. \tag{6.7.10}$$

On the other hand, because $U_{nm} < \alpha^{n-1}\delta$, we have

$$E(U_{nm}) = E(U_n \mid U_n < \alpha^{n-1}\delta)$$

$$= \int_0^{\alpha^{n-1}\delta} u\theta e^{-\theta u} / [1 - \exp(-\theta\alpha^{n-1}\delta)]du$$

$$= \frac{1}{\theta} - \frac{\alpha^{n-1}\delta \exp(-\theta\alpha^{n-1}\delta)}{1 - \exp(-\theta\alpha^{n-1}\delta)}. \tag{6.7.11}$$

Then (6.7.7) with the help of (6.7.10) and (6.7.11) yields

$$\lambda_n = E(X_n)$$

$$= \sum_{m=1}^{\infty} E(X_n \mid M_n = m) P(M_n = m)$$

$$= \sum_{m=1}^{\infty} E(X_{nm} + U_{nm}) q_n^{m-1} p_n$$

$$= \sum_{m=1}^{\infty} \left\{ \frac{m-1}{\theta} + (m-1)\alpha^{n-1}\delta + \frac{1}{\theta} - \frac{\alpha^{n-1}\delta \exp(-\theta\alpha^{n-1}\delta)}{1 - \exp(-\theta\alpha^{n-1}\delta)} \right\} q_n^{m-1} p_n$$

$$= \frac{1-p_n}{p_n}\left(\frac{1}{\theta} + \alpha^{n-1}\delta\right) + \frac{1}{\theta} - \frac{\alpha^{n-1}\delta \exp(-\theta\alpha^{n-1}\delta)}{1 - \exp(-\theta\alpha^{n-1}\delta)}$$

$$= \frac{1}{\theta[1 - \exp(-\theta\alpha^{n-1}\delta)]}. \tag{6.7.12}$$

Consequently, from (6.7.1), the average cost is given by

$$C(N) = \frac{c\mu \sum_{n=1}^{N-1} \frac{1}{b^{n-1}} - r \sum_{n=1}^{N} \frac{1}{\theta[1-\exp(-\theta\alpha^{n-1}\delta)]} + R + c_p\tau}{\sum_{n=1}^{N} \frac{1}{\theta[1-\exp(-\theta\alpha^{n-1}\delta)]} + \mu \sum_{n=1}^{N-1} \frac{1}{b^{n-1}} + \tau} \tag{6.7.13}$$

$$= A(N) - r, \tag{6.7.14}$$

where $A(N)$ from (6.7.3) is given by

$$A(N) = \frac{(c+r)\mu \sum_{n=1}^{N-1} \frac{1}{b^{n-1}} + R + (c_p+r)\tau}{\sum_{n=1}^{N} \frac{1}{\theta[1-\exp(-\theta\alpha^{n-1}\delta)]} + \mu \sum_{n=1}^{N-1} \frac{1}{b^{n-1}} + \tau}. \tag{6.7.15}$$

Our objective is to find an optimal policy for minimizing $C(N)$ or $A(N)$. For this purpose, first of all, we consider the difference of $A(N+1)$ and $A(N)$,

$$A(N+1) - A(N)$$

$$= \frac{(c+r)\mu\left(\sum_{n=1}^{N} \lambda_n - \lambda_{N+1} \sum_{n=1}^{N-1} b^n + \tau\right) - (R + (c_p+r)\tau)(\lambda_{N+1}b^{N-1} + \mu)}{b^{N-1}\left[\sum_{n=1}^{N} \lambda_n + \mu \sum_{n=1}^{N-1} \frac{1}{b^{n-1}} + \tau\right]\left[\sum_{n=1}^{N+1} \lambda_n + \mu \sum_{n=1}^{N} \frac{1}{b^{n-1}} + \tau\right]}.$$

Then define the following auxiliary function

$$g(N) = \frac{(c+r)\mu(\sum_{n=1}^{N}\lambda_n - \lambda_{N+1}\sum_{n=1}^{N-1}b^n + \tau)}{(R+(c_p+r)\tau)(\lambda_{N+1}b^{N-1} + \mu)} \qquad (6.7.16)$$

$$= \frac{(c+r)\mu\{\sum_{n=1}^{N}\frac{1}{\theta[1-\exp(-\theta\alpha^{n-1}\delta)]} - \frac{1}{\theta[1-\exp(-\theta\alpha^N\delta)]}\sum_{n=1}^{N-1}b^n + \tau\}}{(R+(c_p+r)\tau)\{\frac{1}{\theta[1-\exp(-\theta\alpha^N\delta)]}b^{N-1} + \mu\}}. \qquad (6.7.17)$$

As the denominator of $A(N+1) - A(N)$ is always positive, it is clear that the sign of $A(N+1) - A(N)$ is the same as the sign of its numerator. Consequently, we have the following lemma.

Lemma 6.7.1.

$$A(N+1) \mathop{\gtreqless}^{>}_{<} A(N) \iff g(N) \mathop{\gtreqless}^{>}_{<} 1.$$

Lemma 6.7.1 shows that the monotonicity of $A(N)$ can be determined by the value of $g(N)$.

Note that the results in Section 6.7 so far are developed under Assumptions 1-4 only. Therefore, all the results including Lemma 6.7.1 hold for both the deteriorating and improving systems.

To determine an optimal replacement policy analytically, we shall consider two cases.

Model 1. The model under Assumptions 1-5

This is a GP δ-shock model for a deteriorating system. Now, we have the following lemma.

Lemma 6.7.2. Under Assumption 1-5, we have
(1) λ_n is nonincreasing in n,
(2) $g(N)$ is nondecreasing in N.
Proof.

Because $\alpha \geq 1$, then from (6.7.12), λ_n is clearly nonincreasing in n. Moreover, for any integer N, from (6.7.16) and Assumption 5, we have

$$g(N+1) - g(N)$$

$$= \frac{(c+r)\mu(\lambda_{N+1} - b\lambda_{N+2})(\sum_{n=1}^{N+1}\lambda_n + \mu\sum_{n=1}^{N}\frac{1}{b^{n-1}} + \tau)}{(R+(c_p+r)\tau)b^N(\lambda_{N+1} + \frac{\mu}{b^{N-1}})(\lambda_{N+2} + \frac{\mu}{b^N})} \geq 0. \qquad (6.7.18)$$

Therefore, part (2) of Lemma 6.7.2 follows.

Then, from Lemmas 6.7.1 and 6.7.2, the following theorem gives an analytic expression of an optimal policy.

Theorem 6.7.3. Under Assumptions 1-5, an optimal replacement policy N_d^* for a deteriorating system is determined by

$$N_d^* = \min\{N \mid g(N) \geq 1\}. \tag{6.7.19}$$

Furthermore, the optimal policy is unique if and only if $g(N_d^*) > 1$.

In particular, if

$$g(1) = \frac{(c+r)\mu(\lambda_1 + \tau)}{(R + (c_p + r)\tau)(\lambda_2 + \mu)} \geq 1$$

then $N_d^* = 1$. This means that an optimal replacement policy is to replace the system immediately whenever it fails. If $g(\infty)$ exists and $g(\infty) \leq 1$, then $N_d^* = \infty$. This means that the optimal policy is to continually repair when it fails but never replace the system.

Model 2. The model under Assumptions 1-4 and 5'

This model is a GP δ-shock model for an improving system. Now, instead of Lemma 6.7.2, Assumption 5' and (6.7.18) yield the following lemma.

Lemma 6.7.4. Under Assumption 1-4 and 5', we have
(1) λ_n is nondecreasing in n,
(2) $g(N)$ is decreasing in N.

Consequently, Lemmas 6.7.1 and 6.7.4 give the following result.

Theorem 6.7.5. Under Assumptions 1-4 and 5', policy $N_i^* = \infty$ is the unique optimal replacement policy for the improving system.
Proof.
In fact, because $g(N)$ is decreasing in N, there exists an integer N_i such that

$$N_i = \min\{N \mid g(N) \leq 1\}. \tag{6.7.20}$$

Therefore, Lemma 6.7.1 yields that $A(N)$ and $C(N)$ are both unimodal functions of N and both take maximum at N_i. This implies that the minimum of $C(N)$ will be given by

$$\min C(N) = \min\{C(1), C(\infty)\}$$
$$= \min\{\frac{R + c_p\tau - r\lambda_1}{\lambda_1 + \tau}, -r\} = -r. \qquad (6.7.21)$$

Consequently, $N_i^* = \infty$ is the unique optimal replacement policy for the improving system. This completes the proof of Theorem 6.7.5.

As the optimal policy for the improving system is always $N_i^* = \infty$, we shall study a numerical example for a deteriorating system here. The parameters are: $\alpha = 1.04, b = 0.95, \theta = 0.04, \mu = 15, c = 10, r = 20, R = 5000, c_p = 8, \delta = 1$ and $\tau = 40$. The results of $C(N)$ and $g(N)$ against N are tabulated in Table 6.7.1 and plotted in Figure 6.7.1. From Table 6.7.1, it is clear that $C(17) = -8.6911$ is the unique minimum of the average cost. Consequently, policy $N_d^* = 17$ is the unique optimal replacement policy. The same conclusion can be obtained by using Theorem 6.7.3. In fact, from Table 6.7.1 we can see that $g(17) = 1.1219 > 1$ and

$$17 = \min\{N \mid g(N) \geq 1\}.$$

Therefore, $N^* = 17$ is the unique optimal replacement policy. This means that we should replace the system at the time following the 17th failure.

In conclusion, an optimal replacement policy obtained by using Theorem 6.7.3 is the same as that obtained by calculating values of $C(N)$. However, we note here that Theorem 6.7.3 is a theoretical result for determination of an optimal replacement policy for a deteriorating system. Furthermore, by using Theorem 6.7.3 we can always stop the search whenever $g(N)$ crosses over 1.

Table 6.7.1 Results obtained from (6.7.13) and (6.7.14)

N	C(N)	g(N)	N	C(N)	g(N)	N	C(N)	g(N)	N	C(N)	g(N)
1	-1.5582	0.0424	11	-8.5842	0.3971	21	-8.6541	1.9777	31	-8.3540	6.0935
2	-5.4484	0.0499	12	-8.6230	0.4815	22	-8.6361	2.2518	32	-8.2991	6.6911
3	-6.7674	0.0618	13	-8.6512	0.5789	23	-8.6150	2.5530	33	-8.2505	7.3241
4	-7.4247	0.0787	14	-8.6709	0.6903	24	-8.5909	2.8827	34	-8.1991	7.9921
5	-7.8137	0.1012	15	-8.6834	0.8170	25	-8.5639	3.2424	35	-8.1449	8.6943
6	-8.0673	0.1301	16	-8.6898	0.9604	26	-8.5342	3.6333	36	-8.0880	9.4295
7	-8.2429	0.1660	**17**	**-8.6911**	**1.1219**	27	-8.5017	4.0568	37	-8.0281	10.1963
8	-8.3693	0.2097	18	-8.6877	1.3028	28	-8.4666	4.5138	38	-7.9653	10.9929
9	-8.4625	0.2622	19	-8.6802	1.5048	29	-8.4287	5.0051	39	-7.8996	11.8170
10	-8.5321	0.3243	20	-8.6689	1.7292	30	-8.3882	5.5316	40	-7.8308	12.6661

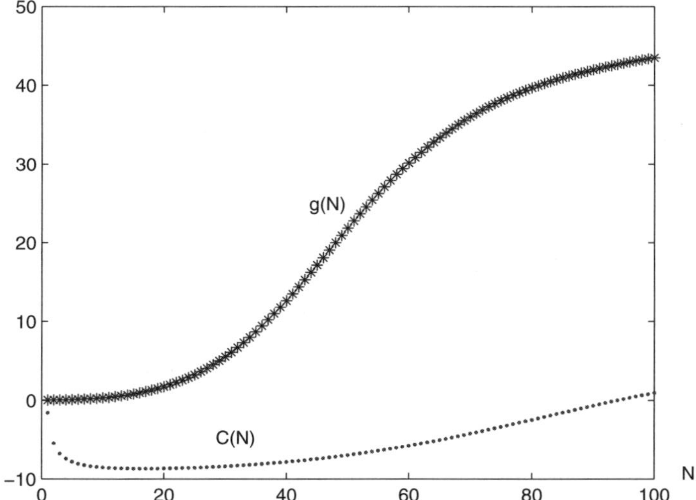

Figure 6.7.1. The plots of g(N) and C(N) against N

6.8   A Threshold Geometric Process Maintenance Model

In practice, many systems demonstrate that their failure rate has the shape
of a bathtub curve. In other words, in the early stage, the failure rate of a
system is decreasing; during the middle stage, the failure rate may be approximately a constant; and in the late stage, the failure rate is increasing.
Consequently, the system should be improving in the early stage, steady
in the middle stage, and deteriorating in the late stage. A maintenance
model for such a system is clearly important in practice. Accordingly, Lam

(2007b) introduced a threshold GP model by making the following assumptions.

Assumption 1. At the beginning, a new system is installed. Whenever the system fails, it will be repaired. A replacement policy N is applied by which the system is replaced by a new and identical one at the time following the Nth failure.

Assumption 2. Let X_1 be the operating time after the installation or a replacement. In general, for $n > 1$, let X_n be the operating time of the system after the $(n-1)$th repair. Assume that $\{X_n, n = 1, 2, \ldots\}$ form a threshold GP having ratios $\{a_i, i = 1, \ldots, k\}$ and thresholds $\{M_i, i = 1, \ldots, k\}$ with $M_1 = 1$ and $E[X_{M_i}] = \lambda_i > 0, i = 1, \ldots, k$. Moreover, let Y_n be the repair time after the nth failure, then $\{Y_n, n = 1, 2, \ldots\}$ constitute a threshold GP having ratios $\{b_i, i = 1, \ldots, k\}$ and thresholds $\{M_i, i = 1, \ldots, k\}$ with $E[Y_{M_i}] = \mu_i \geq 0, i = 1, \ldots, k$. Let the replacement time be Z with $E[Z] = \tau$.

Assumption 3. The operating reward rate after the $(n-1)$th repair is

$$r(n) = r_i \quad M_i \leq n < M_{i+1}, \quad i = 1, \ldots, k, \tag{6.8.1}$$

with $M_{k+1} = \infty$. The repair cost rate after the nth failure is

$$c(n) = c_i \quad M_i \leq n < M_{i+1}, \quad i = 1, \ldots, k. \tag{6.8.2}$$

The replacement cost comprises two parts, one is the basic replacement cost R, the other one is proportional to the replacement time Z at rate c_p.

Now, we say that a cycle is completed if a replacement is completed. Since a cycle is actually a time interval between the installation of a system and the first replacement or a time interval between two consecutive replacements of the system, the successive cycles will form a renewal process. Furthermore, the successive cycles together with the costs incurred in each cycle will constitute a renewal reward process. It follows from (1.3.36) that the long-run average cost per unit time (or simply the average cost) is given by

$$C(N) = \frac{E\{\sum_{n=1}^{N-1} c(n)Y_n - \sum_{n=1}^{N} r(n)X_n + R + c_p Z\}}{E\{\sum_{n=1}^{N} X_n + \sum_{n=1}^{N-1} Y_n + Z\}} \tag{6.8.3}$$

Then the average cost $C(N)$ is given as follows.

For $1 = M_1 \leq N < M_2$, let

$$R_1 = R + c_p \tau \quad \text{and} \quad \tau_1 = \tau.$$

Then (6.8.3) gives

$$C(N) = \frac{c_1 \mu_1 \sum\limits_{n=1}^{N-1} \frac{1}{b_1^{n-1}} - r_1 \lambda_1 \sum\limits_{n=1}^{N} \frac{1}{a_1^{n-1}} + R_1}{\lambda_1 \sum\limits_{n=1}^{N} \frac{1}{a_1^{n-1}} + \mu_1 \sum\limits_{n=1}^{N-1} \frac{1}{b_1^{n-1}} + \tau_1}. \qquad (6.8.4)$$

In general, for $M_i \leq N < M_{i+1}, i = 2, \ldots, k$, let

$$R_i = R_{i-1} + c_{i-1} \mu_{i-1} \sum_{n=1}^{M_i - M_{i-1}} \frac{1}{b_{i-1}^{n-1}} - r_{i-1} \lambda_{i-1} \sum_{n=1}^{M_i - M_{i-1}} \frac{1}{a_{i-1}^{n-1}}$$

and

$$\tau_i = \tau_{i-1} + \lambda_{i-1} \sum_{n=1}^{M_i - M_{i-1}} \frac{1}{a_{i-1}^{n-1}} + \mu_{i-1} \sum_{n=1}^{M_i - M_{i-1}} \frac{1}{b_{i-1}^{n-1}}.$$

Then by induction, (6.8.3) yields

$$C(N) = \frac{c_i \mu_i \sum\limits_{n=M_i}^{N-1} \frac{1}{b_i^{n-M_i}} - r_i \lambda_i \sum\limits_{n=M_i}^{N} \frac{1}{a_i^{n-M_i}} + R_i}{\lambda_i \sum\limits_{n=M_i}^{N} \frac{1}{a_i^{n-M_i}} + \mu_i \sum\limits_{n=M_i}^{N-1} \frac{1}{b_i^{n-M_i}} + \tau_i}$$

$$= \frac{c_i \mu_i \sum\limits_{n=1}^{N-M_i} \frac{1}{b_i^{n-1}} - r_i \lambda_i \sum\limits_{n=1}^{N-M_i+1} \frac{1}{a_i^{n-1}} + R_i}{\lambda_i \sum\limits_{n=1}^{N-M_i+1} \frac{1}{a_i^{n-1}} + \mu_i \sum\limits_{n=1}^{N-M_i} \frac{1}{b_i^{n-1}} + \tau_i}. \qquad (6.8.5)$$

Note that although R_1 is positive, R_i, $i = 2, \ldots, k$, may be negative. Furthermore, from (6.8.4) and (6.8.5), we can see that for different ranges of N, the structure of average cost $C(N)$ is the same. Now, for the ith piece of the threshold GP, assume that policy N is applied. Then the average cost is given by

$$C_i(N) = \frac{c_i \mu_i \sum\limits_{n=1}^{N-1} \frac{1}{b_i^{n-1}} - r_i \lambda_i \sum\limits_{n=1}^{N} \frac{1}{a_i^{n-1}} + R_i}{\lambda_i \sum\limits_{n=1}^{N} \frac{1}{a_i^{n-1}} + \mu_i \sum\limits_{n=1}^{N-1} \frac{1}{b_i^{n-1}} + \tau_i} \qquad (6.8.6)$$

$$= A_i(N) - r_i, \qquad (6.8.7)$$

where

$$A_i(N) = \frac{(c_i + r_i)\mu_i \sum\limits_{n=1}^{N-1} \frac{1}{b_i^{n-1}} + R_i + r_i\tau_i}{\lambda_i \sum\limits_{n=1}^{N} \frac{1}{a_i^{n-1}} + \mu_i \sum\limits_{n=1}^{N-1} \frac{1}{b_i^{n-1}} + \tau_i}. \tag{6.8.8}$$

Obviously, the average cost $C(N)$ is equal to

$$C(N) = C_i(N - M_i + 1) \quad M_i \leq N < M_{i+1}, \quad i = 1,\ldots,k. \tag{6.8.9}$$

Consequently, $C(N)$ could be minimized if we could minimize each of $C_i(N)$, $i = 1,\ldots,k$. As in each piece of the threshold GP, we study a GP maintenance model, we can apply the method developed in Section 6.2. To do so, we shall first study the difference of $A_i(N + 1)$ and $A_i(N)$.

$$A_i(N + 1) - A_i(N)$$
$$= \frac{(c_i + r_i)\mu_i\{\lambda_i(\sum\limits_{n=1}^{N} a_i^n - \sum\limits_{n=1}^{N-1} b_i^n) + \tau_i a_i^N\} - (R_i + r_i\tau_i)(\lambda_i b_i^{N-1} + \mu_i a_i^N)}{a_i^N b_i^{N-1}[\lambda_i \sum\limits_{n=1}^{N} \frac{1}{a_i^{n-1}} + \mu_i \sum\limits_{n=1}^{N-1} \frac{1}{b_i^{n-1}} + \tau_i][\lambda_i \sum\limits_{n=1}^{N+1} \frac{1}{a_i^{n-1}} + \mu_i \sum\limits_{n=1}^{N} \frac{1}{b_i^{n-1}} + \tau_i]}.$$

Then, if $R_i + r_i\tau_i \neq 0$, define an auxiliary function

$$g_i(N) = \frac{(c_i + r_i)\mu_i\{\lambda_i(\sum\limits_{n=1}^{N} a_i^n - \sum\limits_{n=1}^{N-1} b_i^n) + \tau_i a_i^N\}}{(R_i + r_i\tau_i)(\lambda_i b_i^{N-1} + \mu_i a_i^N)}. \tag{6.8.10}$$

If $R_i + r_i\tau_i = 0$, define another auxiliary function

$$g_{i0}(N) = (c_i + r_i)\mu_i\{\lambda_i(\sum\limits_{n=1}^{N} a_i^n - \sum\limits_{n=1}^{N-1} b_i^n) + \tau_i a_i^N\}. \tag{6.8.11}$$

Accordingly, we can define functions

$$g(N) = g_i(N - M_i + 1) \quad M_i \leq N < M_{i+1}, \quad i = 1,\ldots,k \tag{6.8.12}$$

and

$$g_0(N) = g_{i0}(N - M_i + 1) \quad M_i \leq N < M_{i+1}, \quad i = 1,\ldots,k$$

$$\tag{6.8.13}$$

respectively.

Since the denominator of $A_i(N + 1) - A_i(N)$ is always positive, the sign of $A_i(N + 1) - A_i(N)$ is determined by the sign of its numerator. Consequently, we have the following lemma.

Lemma 6.8.1.

(1) If $R_i + r_i\tau_i > 0$, then
$$A_i(N+1) \gtreqless A_i(N) \iff g_i(N) \gtreqless 1.$$

(2) If $R_i + r_i\tau_i < 0$, then
$$A_i(N+1) \gtreqless A_i(N) \iff g_i(N) \lesseqgtr 1.$$

(3) If $R_i + r_i\tau_i = 0$, then
$$A_i(N+1) \gtreqless A_i(N) \iff g_{i0}(N) \gtreqless 0.$$

Thus, the monotonicity of $A_i(N)$ can be determined by the value of $g_i(N)$ or $g_{i0}(N)$. Therefore, from (6.8.7), the following result is straightforward.

Lemma 6.8.2.

(1) If $R_i + r_i\tau_i > 0$, then
$$C_i(N+1) \gtreqless C_i(N) \iff g_i(N) \gtreqless 1.$$

(2) If $R_i + r_i\tau_i < 0$, then
$$C_i(N+1) \gtreqless C_i(N) \iff g_i(N) \lesseqgtr 1.$$

(3) If $R_i + r_i\tau_i = 0$, then
$$C_i(N+1) \gtreqless C_i(N) \iff g_{i0}(N) \gtreqless 0.$$

Now, let $h_i(N) = \lambda_i b_i^{N-1} + \mu_i a_i^N$. Then from (6.8.10), we have

$$g_i(N+1) - g_i(N)$$
$$= \frac{(c_i + r_i)\mu_i}{(R_i + r_i\tau_i)h_i(N)h_i(N+1)}\{\lambda_i^2 b_i^{N-1}(1 - b_i)\sum_{n=1}^{N} a_i^n + \lambda_i^2 b_i^{N-1}(a_i^{N+1} - b_i)$$
$$+ \lambda_i\mu_i a_i^N(a_i - b_i^N) + \lambda_i\mu_i a_i^N(a_i - 1)\sum_{n=1}^{N-1} b_i^n + \lambda_i\tau_i a_i^N b_i^{N-1}(a_i - b_i)\}.$$
$$(6.8.14)$$

On the other hand, (6.8.11) yields

$$g_{i0}(N+1) - g_{i0}(N)$$
$$= (c_i + r_i)\mu_i\{\lambda_i(a_i^{N+1} - b_i^N) + \tau_i(a_i^{N+1} - a_i^N)\}. \qquad (6.8.15)$$

Now we can determine an optimal policy N_i^* for the ith piece of the threshold GP. If $a_i \geq 1$ and $0 < b_i \leq 1$, then in the ith piece of the threshold GP, the successive operating times after repair will be stochastically decreasing, while the consecutive repair times after failure will be stochastically increasing. Thus, the system is in a deteriorating stage. However, if $0 < a_i \leq 1$ and $b_i \geq 1$, then in the ith piece of the threshold GP, the successive operating times after repair will be stochastically increasing, while their consecutive repair times after failure will be stochastically decreasing. Therefore, the system is in an improving stage. In both cases, by using Lemma 6.8.1 with the help of (6.8.10) or (6.8.11), an optimal policy N_i^* can also be determined. Finally, by comparison of average costs $C_i(N_i^*)$ for $i = 1, \ldots, k$, an overall optimal policy N^* for the system could be determined.

As a particular and important example, we shall study the optimal replacement policy for a system with a bathtub shape intensity function. To do this, an additional assumption is made here.

Assumption 4. Assume that $k = 3$ and the ratios satisfy the following conditions:

$$1.\ 0 < a_1 \leq 1,\ b_1 \geq 1, \tag{6.8.16}$$

$$2.\ a_2 = 1,\ b_2 = 1, \tag{6.8.17}$$

$$3.\ a_3 \geq 1,\ 0 < b_3 \leq 1, \tag{6.8.18}$$

not all equalities in (6.8.16) or (6.8.18) hold together.

According to Assumption 4, the threshold GP has three thresholds, and the system has three stages or phases. The early stage or the infant failure phase of the system is from the beginning to the completion of the $(M_2 - 1)$th repair. Since $0 < a_1 \leq 1$ and $b_1 \geq 1$, the system is in an improving stage. The middle stage or the useful life phase of the system is from the completion of the $(M_2 - 1)$th repair to the completion of the $(M_3 - 1)$th repair. Because $a_2 = b_2 = 1$, the system is in a stationary stage so that the successive operating times are identically distributed, so are the consecutive repair times. The late stage or the wear out phase of the system starts from the completion of the $(M_3 - 1)$th repair. As $a_3 \geq 1$ and $0 < b_3 \leq 1$, the system is in a deteriorating stage. Therefore, Assumption 4 means that the system has a bathtub shape intensity function.

Now we shall discuss three different stages respectively.

Stage 1. For $1 \leq N < M_2$, then $0 < a_1 \leq 1$ and $b_1 \geq 1$. The system is in the improving stage.

Obviously, $R_1 + r_1\tau_1 > 0$, it follows from (6.8.14) that function $g_1(N)$ is nonincreasing in N. Therefore, there exists an integer N_i such that

$$N_i = \min\{N \mid g_1(N) \leq 1\}. \tag{6.8.19}$$

Then, Lemma 6.8.2 yields that $C_1(N)$ is a unimodal function of N and takes its maximum at N_i. Although N_i may be greater than $M_2 - 1$, the minimum of $C_1(N)$ is always given by

$$\min_{1 \leq N < M_2} C(N) = \min_{1 \leq N < M_2} C_1(N)$$
$$= \min\{C_1(1), C_1(M_2 - 1)\} = \min\{C(1), C(M_2 - 1)\}. \tag{6.8.20}$$

Stage 2. For $M_2 \leq N < M_3$, then $a_2 = 1$ and $b_2 = 1$. The system is in the stationary stage.

Now, from (6.8.10) and (6.8.11), $g_2(N)$ and $g_{20}(N)$ are both constants. Therefore, Lemma 6.8.2 implies that $C_2(N)$ will be either monotone or a constant. As a result, the minimum of $C_2(N)$ will be taken at one of two end points. Then, we have

$$\min_{M_2 \leq N < M_3} C(N) = \min_{1 \leq N < M_3 - M_2 + 1} C_2(N)$$
$$= \min\{C_2(1), C_2(M_3 - M_2)\} = \min\{C(M_2), C(M_3 - 1)\}. \tag{6.8.21}$$

Stage 3. For $N \geq M_3$, then $a_3 \geq 1$ and $0 < b_3 \leq 1$. The system is in the deterioraing stage.

If $R_3 + r_3\tau_3 > 0$, then (6.8.14) shows that $g_3(N)$ is nondecreasing in N. Therefore, there exists

$$N_3^* = \min\{N \mid g_3(N) \geq 1\}. \tag{6.8.22}$$

Consequently, Lemma 6.8.2 shows that $C_3(N)$ is minimized at N_3^*.

If $R_3 + r_3\tau_3 < 0$, a similar argument shows that there exists

$$N_3^* = \min\{N \mid g_3(N) \leq 1\}, \tag{6.8.23}$$

N_3^* is the optimal policy.

Now, assume that $R_3 + r_3\tau_3 = 0$. Because

$$g_{30}(1) = \mu_3 a_3(c_3 + r_3)(\lambda_3 + \tau_3) > 0,$$

and g_{30} is nondecreasing in N, then $g_{30}(N) > 0$ for all integer $N \geq 1$. Lemma 6.6.2 yields that $C_3(N)$ is minimized at $N_3^* = 1$.

Consequently, we have

$$\min_{N \geq M_3} C(N) = \min_{N \geq 1} C_3(N)$$
$$= C_3(N_3^*) = C(M_3 + N_3^* - 1). \tag{6.8.24}$$

In conclusion, it follows from (6.8.20), (6.8.21) and (6.8.24) that an optimal replacement policy N^* for a system with a bathtub shape intensity function is given by

$$C(N^*) = \min_{N \geq 1} C(N)$$
$$= \min\{C(1), C(M_2 - 1), C(M_2), C(M_3 - 1), C(M_3 + N_3^* - 1)\},$$
$$\tag{6.8.25}$$

where N_3^* is determined by one of (6.8.22) and (6.8.23) if $R_3 + r_3\tau_3 \neq 0$, and $N_3^* = 1$ otherwise.

Thereafter, we can state the conditions for the uniqueness of the optimal policy N^*. The proof is straightforward.

1. For $N^* = 1$ and $g_1(1) > 1$, then the optimal policy $N^* = 1$ is unique.

2. For $N^* = M_2 - 1$ and $g_1(M_2 - 2) < 1$, then the optimal policy $N^* = M_2 - 1$ is unique.

3. For $N^* = M_2$, then

(1) If $R_2 + r_2\tau_2 > 0$ and $g_2(1) > 1$, the optimal policy $N^* = M_2$ is unique.

(2) If $R_2 + r_2\tau_2 < 0$, the optimal policy $N^* = M_2$ is always unique, since $g_2(1) < 0$ by (6.8.10).

(3) If $R_2 + r_2\tau_2 = 0$, the optimal policy $N^* = M_2$ is always unique, since $g_{20}(1) > 0$ by (6.8.11).

4. For $N^* = M_3 - 1$, if $g_2(1) < 1$, the optimal policy $N^* = M_3 - 1$ is unique, since $R_2 + r_2\tau_2 > 0$. Otherwise, if $R_2 + r_2\tau_2 \leq 0$ then Lemma 6.8.2 with the help of (6.8.10) or (6.8.11) implies that $C_2(M_3 - 1) > C_2(M_2)$, then $N^* \neq M_3 - 1$, this is impossible.

5. For $N^* = M_3 + N_3^* - 1$, then

(1) If $R_3 + r_3\tau_3 > 0$ and $g_3(N_3^*) > 1$, the optimal policy $N^* = M_3 + N_3^* - 1$ is unique.

(2) If $R_3 + r_3\tau_3 < 0$ and $g_3(N_3^*) < 1$, the optimal policy $N^* = M_3 + N_3^* - 1$ is unique.

(3) If $R_3 + r_3\tau_3 = 0$, the optimal policy $N^* = M_2 + N_3^* - 1 = M_3$ is always unique.

Three particular cases are then discussed below.

1. If $M_2 = \infty$, then the threshold GP model becomes a GP model for an improving system. It follows from (6.8.20) that

$$C(N^*) = \min_{N \geq 1} C(N) = \min\{C(1), C(\infty)\}$$

$$= \{\frac{R_1 - r_1 \lambda_1}{\lambda_1 + \tau_1}, -r_1\} = -r_1. \tag{6.8.26}$$

Therefore, $N^* = \infty$ is the unique optimal policy for the improving system. This particular model has been studied in Section 6.2.

2. If $M_2 = 1$ but $M_3 = \infty$, then the threshold GP model reduced to a RP model for a stationary system. Then from (6.8.21), the optimal policy N^* is determined by

$$N^* = \begin{cases} 1 & \text{if } C(1) \leq C(\infty), \\ \infty & \text{if } C(1) \geq C(\infty). \end{cases} \tag{6.8.27}$$

This is a perfect repair model with i.i.d. operating times and repair times.

3. If $M_3 = 1$, then the threshold GP model becomes a GP model for a deteriorating system. From (6.8.22), the optimal policy $N^* = N_3^*$. This particular model has also been considered in Section 6.2.

To demonstrate the threshold GP model, consider an example of a 3-piece threshold GP model for a system with a bathtub shape intensity function. The parameter values are: $a_1 = 0.95, b_1 = 1.05, a_2 = 1, b_2 = 1, a_3 = 1.03, b_3 = 0.97$, and $M_2 = 6, M_3 = 26$. Therefore, the system is in the infant failure phase for $1 \leq N < 6$, it is in the useful life phase for $6 \leq N < 26$, and is in the wear out phase for $N \geq 26$. The cost parameters are: $c_1 = 50, r_1 = 20, c_2 = 10, r_2 = 30, c_3 = 20, r_3 = 40, R = 600, c_p = 10$. The other parameter values are: $\lambda_1 = 40, \mu_1 = 15, \lambda_2 = 50, \mu_2 = 10, \lambda_3 = 48, \mu_3 = 12$ and $\tau = 10$.

Then, we can evaluate the values of $C(N)$ according to (6.8.5) and $g(N)$ from (6.8.12). The results are tabulated in Table 6.8.1.

According to (6.8.5), for $1 \leq N < M_2 = 6$, we can see from Table 6.8.1 and (6.8.19) that $C_1(N)$ is maximized at $N_i = 2$ and is minimized at $M_2 - 1 = 5$. For $6 = M_2 \leq N < M_3 = 26$, as $R_2 + r_2 \tau_2 = 8677.0419 > 0$ and $g_2(N) = 0.3235 < 1$ is a constant, $C_2(N)$ is decreasing and is minimized at $M_3 - M_2 = 20$, and $C(N)$ is minimized at $M_3 - 1 = 25$. Finally, for $N \geq M_3 = 26$, since $R_3 + r_3 \tau_3 = 33843.2742 > 0$, then from Table 6.8.1 and (6.8.22), $C_3(N)$ is minimized at $N_3^* = 11$. Consequently for $N \geq M_3 = 26$, $C(N)$ is minimized at $M_3 + N_3^* - 1 = 36$.

Table 6.8.1. Results obtained from (6.8.5) and (6.8.12)

N	C(N)	g(N)	N	C(N)	g(N)	N	C(N)	g(N)
1	**-2.0000**	**1.0215**	16	-15.8782	0.3235	31	-19.6556	0.7739
2	-1.7936	0.8639	17	-16.2431	0.3235	32	-19.7728	0.8147
3	-2.1981	0.6323	18	-16.5667	0.3235	33	-19.8624	0.8578
4	-2.7307	0.3332	19	-16.8557	0.3235	34	-19.9267	0.9032
5	**-3.3029**	**-0.0264**	20	-17.1153	0.3235	35	-19.9678	0.9508
6	**-5.1722**	**0.3235**	21	-17.3499	0.3235	**36**	**-19.9875**	**1.0007**
7	-7.6836	0.3235	22	-17.5628	0.3235	37	-19.9872	1.0528
8	-9.5417	0.3235	23	-17.7569	0.3235	38	-19.9684	1.1071
9	-10.9720	0.3235	24	-17.9346	0.3235	39	-19.9323	1.1637
10	-12.1070	0.3235	**25**	**-18.0980**	**0.3235**	40	-19.8800	1.2223
11	-13.0296	0.3235	26	-18.5403	0.6028	41	-19.8123	1.2830
12	-13.7943	0.3235	27	-18.8477	0.6327	42	-19.7301	1.3457
13	-14.4385	0.3235	28	-19.1081	0.6647	43	-19.6432	1.4102
14	-14.9885	0.3235	29	-19.3267	0.6989	44	-19.5253	1.4767
15	-15.4636	0.3235	30	-19.5079	0.7353	45	-19.4040	1.5448

Finally, it follows from (6.8.25) that the minimum average cost $C(N^*)$ is given by

$$C(N^*) = \min\{C(1), C(M_2 - 1), C(M_2), C(M_3 - 1), C(M_3 + N_3^* - 1)\}$$
$$= \min\{C(1), C(5), C(6), C(25), C(36)\}$$
$$= \min\{-2.0000, -3.3029, -5.1722, -18.0980, -19.9875\}$$
$$= -19.9875. \tag{6.8.28}$$

Therefore, an optimal replacement policy is $N^* = 36$. The system should be replaced by a new and identical one following the 36th failure. The optimal policy is unique because

$$g(36) = g_3(36 - 26 + 1) = g_3(11) = g_3(N_3^*) = 1.0007 > 1.$$

6.9 A Geometric Process Preventive Maintenance Model

So far, we have just studied a GP model in which a system will be repaired after failure. In many cases, such as in a hospital or a steel manufacturing complex, a cut off in electricity supply may cause a serious catastrophe. To maintain their electric generator with a high reliability is important for production efficiency, economic profits and even life safety. A preventive repair is a very powerful measure, since a preventive repair will extend system lifetime and raise the reliability at a lower cost rate. Therefore, we shall introduce a GP preventive maintenance model.

Assume that a new system is installed at the beginning. A maintenance policy (T, N) is applied, by which whenever the system fails or its operating time reaches T whichever occurs first, the system will be repaired. A repair when a system fails is a failure repair, while a repair when the system operating time reaches T is a preventive repair. The system will be replaced by a new and identical one at the time following the Nth failure.

The time interval between the installation and the first replacement or time interval between two successive replacements is called a cycle. For $n = 1, 2, \ldots, N-1$, the time interval between the completion of the $(n-1)$th failure repair and the nth failure repair in a cycle is called the nth period in the cycle. However, the Nth period is the time interval between the completion of the $(N-1)$th failure repair in the cycle and the replacement. Let $X_n^{(i)}, n = 1, 2, \ldots, i = 1, 2, \ldots$ be the operating time of the system after the ith preventive repair in the nth period. Let $Y_n^{(i)}, n = 1, 2, \ldots, i = 1, 2, \ldots$ be the ith preventive repair time of the system in the nth period. Denote the failure repair time of the system in the nth period by Z_n. Given that the number of preventive repairs in the nth period is $M_n, n = 1, 2, \ldots, N$, a diagram of one possible realization of the system process in a cycle may be given in Figure 6.9.1.

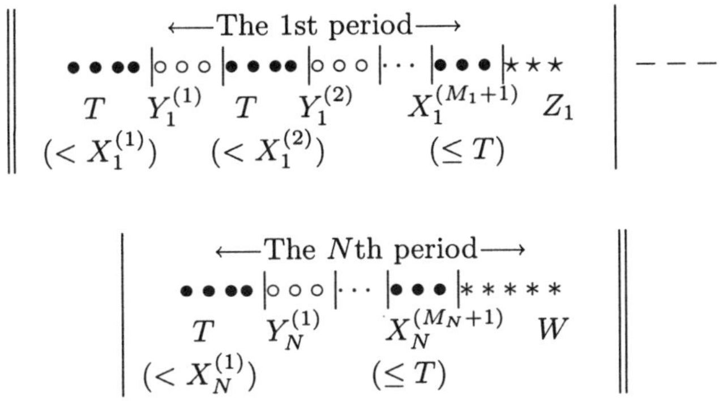

$$\longleftarrow\text{The 1st period}\longrightarrow$$

$$\bullet\bullet\bullet\bullet \mid \circ\circ\circ \mid \bullet\bullet\bullet\bullet \mid \circ\circ\circ \mid \cdots \mid \bullet\bullet\bullet \mid \star\star\star \qquad ---$$

$$T \quad Y_1^{(1)} \quad T \quad Y_1^{(2)} \qquad X_1^{(M_1+1)} \quad Z_1$$

$$(< X_1^{(1)}) \qquad (< X_1^{(2)}) \qquad (\leq T)$$

$$\longleftarrow\text{The }N\text{th period}\longrightarrow$$

$$\bullet\bullet\bullet\bullet \mid \circ\circ\circ \mid \cdots \mid \bullet\bullet\bullet \mid \ast\ast\ast\ast\ast$$

$$T \quad Y_N^{(1)} \qquad X_N^{(M_N+1)} \quad W$$

$$(< X_N^{(1)}) \qquad (\leq T)$$

$\bullet\bullet$: working state, $\circ\circ$: preventive repair state, $\star\star$: failure repair state, $\ast\ast$: replacement state

Figure 6.9.1. The diagram of a realization of the system process in a cycle

Lam (2007a) introduced a GP preventive maintenance model by making the following assumptions.

Assumption 1. The system after preventive repair is as good as new so that $\{X_n^{(i)}, i = 1, 2, \ldots\}$ are i.i.d. random variables, while $\{Y_n^{(i)}, i = 1, 2, \ldots\}$ are also i.i.d. random variables. However, the system after failure repair is not as good as new so that the successive operating times $\{X_n^{(1)}, n = 1, 2, \ldots\}$ after failure repair form a GP with ratio a and $E[X_1^{(1)}] = \lambda$, while the consecutive failure repair times $\{Z_n, n = 1, 2, \ldots\}$ constitute a GP with ratio b and $E[Z_1] = \mu$. On the other hand, the preventive repair times $\{Y_n^{(1)}, n = 1, 2, \ldots\}$ in successive periods form a GP with ratio b_p and $E[Y_1^{(1)}] = \nu$. The replacement time W is a random variable with $E[W] = \tau$.

Assumption 2. The processes $\{X_n^{(i)}, n = 1, 2, \ldots, i = 1, 2, \ldots\}$, $\{Y_n^{(i)}, n = 1, 2, \ldots, i = 1, 2, \ldots\}$ and $\{Z_n, n = 1, 2, \ldots\}$ are independent and are all independent of W.

Assumption 3. The failure repair cost rate is c, the preventive repair rate is c_p, the reward rate when the system is operating is r. The replacement cost comprises two parts, one is the basic replacement cost R, and the other one is proportional to the replacement time W at rate c_r.

Assumption 4. Assume that T has a lower bound $T_0 > 0$, i.e. $T \geq T_0$.

Remarks

Under Assumptions 1-4, the model is a GP preventive maintenance model. As a preventive repair is adopted when the system is operating, it is reasonable to assume that the system after preventive repair is as good as new. Therefore, in Assumption 1, we assume that the successive operating times after preventive repair are i.i.d. random variables. On the other hand, we still assume that the system after failure repair is not as good as new such that the successive operating times after failure repair will form a GP and the consecutive failure repair times will constitute a GP. This is the difference between two kinds of repair. It is also the advantage of using the preventive repair. Furthermore, the GP preventive maintenance model is a model for a deteriorating system, if $a \geq 1$ and $0 < b, b_p \leq 1$, it is a model for an improving system if $0 < a \leq 1$ and $b, b_p \geq 1$. Therefore, the GP model is a general model that can be flexibly applied to a deteriorating system as well as an improving system.

Assumption 4 means that time T has a lower bound. In practice, there should have a break interval T_0 between two consecutive repairs for recovering the repair facility and preparing next repair. Then, a repair can only start after break interval T_0 that can be taken as a lower bound of T. On the other hand, we may choose a lower bound T_0 such that

$$P(X_1^{(1)} \leq T_0) \leq \theta_0$$

with a small θ_0 equal to 0.05 for example. Note that if $T_0 = \infty$, our model will reduce to the GP maintenance model without preventive repair studied in Section 6.2.

Our problem is to determine an optimal maintenance policy (T^*, N^*) for minimizing the long-run average cost per unit time (or simply the average cost) $C(T, N)$. Once again, as the successive cycles form a renewal process, the successive cycles together with the costs incurred in each cycle will constitute a renewal reward process.

Now, suppose a maintenance policy (T, N) is adopted. Let X_n be the total operating time of the system in the nth period of a cycle, and M_n be the number of preventive repairs taken in the period. Then M_n is the number of operating times longer than T in the nth period. Denote the distribution function and density function of $X_n^{(i)}$ by $F_n(x) = F(a^{n-1}x)$ and $f_n(x) = a^{n-1}f(a^{n-1}x)$ respectively. By using (1.3.36), the average cost $C(T, N)$ is given by

$$C(T, N) = \frac{E[c_p \sum\limits_{n=1}^{N} \sum\limits_{i=1}^{M_n} Y_n^{(i)} + c \sum\limits_{n=1}^{N-1} Z_n - r \sum\limits_{n=1}^{N} X_n + R + c_r W]}{E[\sum\limits_{n=1}^{N} X_n + \sum\limits_{n=1}^{N} \sum\limits_{i=1}^{M_n} Y_n^{(i)} + \sum\limits_{n=1}^{N-1} Z_n + W]}$$

$$= \frac{c_p \sum\limits_{n=1}^{N} \frac{\nu}{b_p^{n-1}} E[M_n] + c \sum\limits_{n=1}^{N-1} \frac{\mu}{b^{n-1}} - r \sum\limits_{n=1}^{N} E[X_n] + R + c_r \tau}{\sum\limits_{n=1}^{N} E[X_n] + \sum\limits_{n=1}^{N} \frac{\nu}{b_p^{n-1}} E[M_n] + \sum\limits_{n=1}^{N-1} \frac{\mu}{b^{n-1}} + \tau}. \qquad (6.9.1)$$

To derive an explicit expression for $C(T, N)$, we shall first study the distribution of M_n. Recall that M_n is the number of preventive repairs taken in the nth period. Thus, $Y_n = M_n + 1$ will have a geometric distribution $G(p_n)$. Then (1.2.10) yields that

$$P(M_n = i) = P(Y_n = i + 1) = p_n q_n^i, \quad i = 0, 1, \ldots$$

with

$$p_n = P(X_n^{(1)} \le T) = F(a^{n-1}T) \tag{6.9.2}$$

and $q_n = P(X_n^{(1)} > T) = 1 - p_n$. Consequently,

$$E[M_n] = \frac{q_n}{p_n}. \tag{6.9.3}$$

Furthermore, from (1.4.21), the expected total operating time in the nth period is given by

$$E[X_n] = \lambda(T, n) = \frac{1}{F(a^{n-1}T)} \int_0^T \{1 - F(a^{n-1}t)\} dt. \tag{6.9.4}$$

Now, (6.9.1) becomes

$$
C(T, N)
$$
$$
= \frac{c_p \sum_{n=1}^{N} \frac{\nu}{b_p^{n-1}} \frac{q_n}{p_n} + c \sum_{n=1}^{N-1} \frac{\mu}{b^{n-1}} - r \sum_{n=1}^{N} \lambda(T, n) + R + c_r \tau}{\sum_{n=1}^{N} \lambda(T, n) + \sum_{n=1}^{N} \frac{\nu}{b_p^{n-1}} \frac{q_n}{p_n} + \sum_{n=1}^{N-1} \frac{\mu}{b^{n-1}} + \tau}. \tag{6.9.5}
$$

Then

$$C(T, N) = A(T, N) + c_p, \tag{6.9.6}$$

where

$$A(T, N) = \frac{I(T, N)}{J(T, N)} \tag{6.9.7}$$

with

$$
I(T, N)
$$
$$
= (c - c_p) \sum_{n=1}^{N-1} \frac{\mu}{b^{n-1}} - (r + c_p) \sum_{n=1}^{N} \lambda(T, n) + R + (c_r - c_p)\tau, \tag{6.9.8}
$$

and

$$J(T, N) = \sum_{n=1}^{N} \lambda(T, n) + \sum_{n=1}^{N} \frac{\nu}{b_p^{n-1}} \frac{q_n}{p_n} + \sum_{n=1}^{N-1} \frac{\mu}{b^{n-1}} + \tau. \tag{6.9.9}$$

Therefore, to minimize $C(T, N)$ is equivalent to minimize $A(T, N)$. In general, an optimal maintenance policy (T^*, N^*) could be obtained numerically by minimizing $C(T, N)$ or $A(T, N)$. However, under some additional conditions, it is easy to determine the optimal maintenance policy (T^*, N^*).

Lemma 6.9.1.

(1) $\frac{q_n}{p_n}$ is nonincreasing in T.

(2) $\frac{q_n}{p_n}$ is nonincreasing in n if $a \geq 1$ and nondecreasing in n if $a < 1$.

(3) $\lambda(T, n)$ is nonincreasing in n if $a \geq 1$ and increasing in n if $a < 1$.

Proof.

Parts (1) and (2) follow from (6.9.2) directly. To prove part (3), by considering n as a continuous variable, then (6.9.4) gives

$$\frac{\partial \lambda(T, n)}{\partial n} = -\frac{a^{n-1}\ell na}{F(a^{n-1}T)^2}$$

$$\times \{F(a^{n-1}T)\int_0^T tf(a^{n-1}t)dt + f(a^{n-1}T)T\int_0^T (1 - F(a^{n-1}t))dt\}.$$

$$(6.9.10)$$

Obviously, it is nonpositive if $a \geq 1$, and positive if $a < 1$. This completes the proof of part (3).

Lemma 6.9.2. If life distribution $F(x)$ is ERBLE, then life distribution $F(a^{n-1}x)$ is also ERBLE and $\lambda(T, n)$ is nonincreasing in T.

Proof.

If life distribution $F(x)$ is ERBLE, then by Definition 1.4.10,

$$\lambda(T, 1) = \frac{1}{F(T)} \int_0^T \{1 - F(x)\}dx \qquad (6.9.11)$$

is nonincreasing in T. Therefore from (6.9.4), the expected total operating time in the nth period is given by

$$\lambda(T, n) = \frac{1}{F(a^{n-1}T)} \int_0^T \{1 - F(a^{n-1}t)\}dt$$

$$= \frac{1}{a^{n-1}F(a^{n-1}T)} \int_0^{a^{n-1}T} \{1 - F(u)\}du$$

$$= \lambda(a^{n-1}T, 1)/a^{n-1}. \qquad (6.9.12)$$

It is clearly nonincreasing in T. Thus life distribution $F(a^{n-1}t)$ is also ERBLE. This completes the proof of Lemma 6.9.2.

Now, we can prove the following theorem.

Theorem 6.9.3. Assume that the life distribution $F \in$ ERBLE, and

$$I(T_0, N) \geq 0 \qquad \text{for all} \quad N \geq 1, \qquad (6.9.13)$$

then

$$\min_{T,N} A(T, N) = \min_N A(T_0, N). \qquad (6.9.14)$$

Proof.

Because the distribution F is ERBLE, it follows from Lemma 6.9.2 that $\lambda(T, n)$ is nonincreasing in T. On the other hand, by Lemma 6.9.1, $\frac{q_n}{p_n}$ is nonincreasing in T. Consequently, $I(T, N)$ is nondecreasing in T, while $J(T, N)$ is nonincreasing in T and positive. Then Assumption 4 and (6.9.13) yield that

$$\min_{T,N} A(T, N) = \min_{N} \min_{T} \frac{I(T, N)}{J(T, N)}$$

$$= \min_{N} \frac{I(T_0, N)}{J(T_0, N)} = \min_{N} A(T_0, N).$$

The following corollary is a special case of Theorem 6.9.3.

Corollary 6.9.4. Assume that life distribution $F \in$ ERBLE, $a \geq 1, 0 < b \leq 1$ and $c \geq c_p$. In addition, there exists an integer k such that

$$I(T_0, n) \geq 0 \quad \text{for} \quad n = 1, 2, \ldots, k, \tag{6.9.15}$$

and

$$I(T_0, k + 1) - I(T_0, k)$$
$$= (c - c_p)\frac{\mu}{b^{k-1}} - (r + c_p)\lambda(T_0, k + 1) \geq 0. \tag{6.9.16}$$

Then

$$\min_{T,N} A(T, N) = \min_{N} A(T_0, N).$$

Proof.

Actually, we need only check condition (6.9.13). Because of (6.9.15), (6.9.13) holds for $N \leq k$. For $N > k$, by noting that $a \geq 1, 0 < b \leq 1$ and $c \geq c_p$, it follows from (6.9.15), (6.9.16) and Lemma 6.9.1 that

$$I(T_0, N)$$

$$= I(T_0, k) + \sum_{i=k}^{N-1} \{I(T_0, i + 1) - I(T_0, i)\}$$

$$= I(T_0, k) + \sum_{i=k}^{N-1} \{(c - c_p)\frac{\mu}{b^{i-1}} - (r + c_p)\lambda(T_0, i + 1)\}$$

$$\geq I(T_0, k) + (N - k)\{(c - c_p)\frac{\mu}{b^{k-1}} - (r + c_p)\lambda(T_0, k + 1)\}$$

$$\geq 0. \tag{6.9.17}$$

Thus, from Theorem 6.9.3, Corollary 6.9.4 holds.

As a result, if a system is deteriorating with an ERBLE distribution, Corollary 6.9.4 may be applicable. In using Corollary 6.9.4, essentially we need to check conditions (6.9.15) and (6.9.16) only, since in practice, the failure repair cost rate should be higher than the preventive repair cost rate, and $c \geq c_p$ should be true.

Now, we consider a special case by assuming that the operating time of a new system is an random variable with exponential distribution $Exp(1/\lambda)$ having density

$$f(t) = \begin{cases} \frac{1}{\lambda}e^{-t/\lambda} & t > 0, \\ 0 & \text{elsewhere.} \end{cases} \tag{6.9.18}$$

Then, it is straightforward that

$$p_n = P(X_n^{(1)} \leq T) = \int_0^T \frac{a^{n-1}}{\lambda} \exp\{-a^{n-1}t/\lambda\}dt$$
$$= 1 - \exp\{-a^{n-1}T/\lambda\}. \tag{6.9.19}$$

Consequently

$$q_n = P(X_n^{(1)} > T) = \exp\{-a^{n-1}T/\lambda\}. \tag{6.9.20}$$

Furthermore, from (6.9.4) we have

$$\lambda(T,n) = \frac{\lambda}{a^{n-1}}. \tag{6.9.21}$$

Then, (6.9.8) and (6.9.9) become

$$I(T,N) = (c - c_p)\sum_{n=1}^{N-1} \frac{\mu}{b^{n-1}} - (r + c_p)\sum_{n=1}^{N} \frac{\lambda}{a^{n-1}} + R + (c_r - c_p)\tau, \tag{6.9.22}$$

and

$$J(T,N) = \sum_{n=1}^{N} \frac{\lambda}{a^{n-1}} + \sum_{n=1}^{N} \frac{\nu}{b_p^{n-1}}\frac{q_n}{p_n} + \sum_{n=1}^{N-1} \frac{\mu}{b^{n-1}} + \tau. \tag{6.9.23}$$

Hence we have the following theorem.

Theorem 6.9.5. Assume that the operating time of a new system has an exponential distribution. Then an optimal maintenance policy (T^*, N^*) is determined by

$$A(T^*, N^*) = \min\{\min_N A(T_0, N), \min_N A(\infty, N)\}. \tag{6.9.24}$$

Proof.

For any policy (T, N), (6.9.21) shows that $I(T, N)$ does not depend on T, then

$$I(T, N) = I(T_0, N) = I(\infty, N).$$

Furthermore, $J(T, N) > 0$ and is nonincreasing in T by Lemma 6.9.1. Then if $I(T, N) \geq 0$, we have

$$
\begin{aligned}
A(T, N) &= \frac{I(T, N)}{J(T, N)} = \frac{I(T_0, N)}{J(T, N)} \\
&\geq \frac{I(T_0, N)}{J(T_0, N)} = A(T_0, N) \\
&\geq \min_N A(T_0, N).
\end{aligned}
\tag{6.9.25}
$$

On the other hand, if $I(T, N) < 0$, then

$$
\begin{aligned}
A(T, N) &= \frac{I(T, N)}{J(T, N)} = \frac{I(\infty, N)}{J(T, N)} \\
&\geq \frac{I(\infty, N)}{J(\infty, N)} = A(\infty, N) \\
&\geq \min_N A(\infty, N).
\end{aligned}
\tag{6.9.26}
$$

This completes the proof of Theorem 6.9.5.

In general, a numerical method should be applied for finding an optimal policy. To start with, an upper bound T_1 of T is chosen in the following way.

$$P(X_1^{(1)} \geq T_1) \leq \theta_1$$

with small $\theta_1 = 0.05$ for example. Then a grid method could be adopted as follows:

Step 1: Divide interval (T_0, T_1) by points $T_0 = t_0 < t_1 < \cdots < t_k = T_1$ with large k so that the maximum length of subintervals $(t_i, t_{i+1}), i = 0, \ldots, k-1$, is small enough.

Step 2: For each $i = 0, 1, \ldots, k$, evaluate the average cost $A(t_i, n)$ and choose

$$A(t_i, n_i) = \min_n \{A(t_i, n)\}.$$

Step 3: Determine

$$A(t^*, n^*) = \min_{0 \leq i \leq k} \{A(t_i, n_i)\}.$$

Then the policy (t^*, n^*) is an approximate optimal maintenance policy for the system.

Now, for explaining the GP model and the methodology developed in this section, we consider three numerical examples.

Example 6.9.6.   Assume that the operating time of a new system has a Weibull distribution $W(\alpha, \beta)$ with parameters $\alpha = 2.2, \beta = 1$. The other parameters are $a = 1.02, b = 0.96, b_p = 0.98, \mu = 0.8, \nu = 0.2, \tau = 1.0, r = 40, c = 30, c_p = 5, R = 500, c_r = 25$.

As the life distribution F is $W(2.2, 1)$, Corollary 1.4.12 implies that $F \in \mathrm{ERBLE}$. Moreover, $a = 1.02, b = 0.96, b_p = 0.98$, the system is a deteriorating system. However, Theorem 6.9.3 and Corollary 6.9.4 are not applicable, since

$$I(0.1, 1) = -194.0462 < 0.$$

Therefore, we shall apply the grid method for finding an approximate optimal policy. As the value of T_0 is not specified in this example, we may determine a lower bound T_0 and an upper bound T_1 first by a probability consideration. To do so, we note that

$$P(X_1^{(1)} \geq T) = \exp\{-\beta T^\alpha\}.$$

Then

$$P(0.1 \leq X_1^{(1)} \leq 2.2) = \exp(-0.1^{2.2}) - \exp(-2^{2.2}) = 0.9836.$$

Thus, it is reasonable to take $T_0 = 0.1$ and $T_1 = 2.0$. Then divide interval $(T_0, T_1) = (0.1, 2.0)$ into 20 subintervals with equal length by $t_i = 0.1(i + 1), i = 0, \ldots, 19$. Afterward, we can find

$$A(t_i, n_i) = \min_n A(t_i, n) \tag{6.9.27}$$

by comparison of values $A(t_i, n)$. Then an approximate optimal policy (t^*, n^*) could be determined by

$$A(t^*, n^*) = \min_{0 \leq i \leq k} A(t_i, n_i). \tag{6.9.28}$$

The numerical results are given in Table 6.9.1.

From the first six columns of Table 6.9.1, we can see that minimum value of $A(t_i, n_i)$ is -14.6362, thus an approximate optimal policy is $(t^*, n^*) = (0.2, 13)$.

Table 6.9.1. The values of $A(t_i, n_i)$ and n_i

t_i	$A(t_i, n_i)$	n_i	t_i	$A(t_i, n_i)$	n_i	t_i	$A(t_i, n_i)$	n_i	t_i	$A(t_i, n_i)$	n_i
0.1	-12.1717	9	1.1	7.6696	22	0.11	-12.7113	10	0.21	-14.6124	13
0.2	**-14.6362**	**13**	1.2	8.5506	23	0.12	-13.1669	10	0.22	-14.5528	13
0.3	-13.1251	14	1.3	9.1675	23	0.13	-13.5531	11	0.23	-14.4601	13
0.4	-9.9919	15	1.4	9.5922	24	0.14	-13.8709	11	0.24	-14.3371	13
0.5	-6.4019	16	1.5	9.8773	24	0.15	-14.1237	11	0.25	-14.1881	14
0.6	-2.9301	17	1.6	10.0723	25	0.16	-14.3255	12	0.26	-14.0179	14
0.7	0.1545	18	1.7	10.1907	25	0.17	-14.4728	12	0.27	-13.8246	14
0.8	2.7480	19	1.8	10.2662	25	0.18	-14.5697	12	0.28	-13.6101	14
0.9	4.8368	20	1.9	10.3130	25	0.19	-14.6209	13	0.29	-13.3764	14
1.0	6.4565	21	2.0	10.3413	25	**0.20**	**-14.6362**	**13**	0.30	-13.1251	14

Thus an optimal policy should lay in interval $(0.1, 0.3)$. To obtain a more precise approximate optimal policy, we can divide $(0.1, 0.3)$ further, 20 subintervals with equal length 0.01 for example, and repeat the above procedure. The numerical results are given in the last six columns of Table 6.9.1. Afterward, we can find minimum value $A(t_i, n_i) = A(0.20, 13) = -14.6362$. Therefore a more precise approximate optimal policy is $(t^*, n^*) = (0.20, 13)$. Clearly, we can partition interval $(0.19, 0.21)$ further for obtaining a better approximate optimal solution.

Example 6.9.7.   Assume that the operating time of a system has a Weibull distribution $W(\alpha, \beta)$ with density function

$$f(x) = \begin{cases} \alpha\beta x^{\alpha-1}\exp(-\beta x^\alpha) & t > 0, \\ 0 & \text{elsewhere,} \end{cases} \tag{6.9.29}$$

and $\alpha = 2, \beta = 1$. The other parameters are $a = 1.05, b = 0.95, b_p = 0.97, \mu = 0.3, \nu = 0.2, \tau = 0.4, r = 30, c = 25, c_p = 10, R = 2000, c_r = 20, T_0 = 0.5$.

Because the lifetime distribution F has a Weibull distribution $W(\alpha, \beta)$ with $\alpha = 2 > 1$, Corollary 1.4.12 implies that $F \in$ ERBLE. Moreover, since $a = 1.05 > 1, b = 0.95, b_p = 0.97$, the system is deteriorating. As $c = 25 > c_p = 10$, to apply Corollary 6.9.4, we need check conditions (6.9.15) and (6.9.16). To do this, the values of $I(T_0, N)$ and $A(T_0, N)$ are tabulated in Table 6.9.2. Moreover, the values of $A(T_0, N)$ are plotted in Figure 6.9.2.

Table 6.9.2. The values of $I(T_0, N)$ and $A(T_0, N)$

N	$I(T_0, N)$	$A(T_0, N)$	N	$I(T_0, N)$	$A(T_0, N)$	N	$I(T_0, N)$	$A(T_0, N)$
1	1920.5854	602.1537	11	1456.8149	59.2853	21	1332.5433	35.0326
2	1849.1004	306.2005	12	1433.4250	54.9077	**22**	**1330.9632**	33.8745
3	1784.5895	205.4929	13	1412.9526	51.2392	23	1330.9714	32.8196
4	1726.4372	154.8939	14	1395.1807	48.1250	24	1332.5145	31.8534
5	1674.0869	124.5468	15	1379.9152	45.4517	25	1335.5503	30.9641
6	1627.0352	104.3807	16	1366.9832	43.1341	30	1372.3450	27.3720
7	1584.8278	90.0496	17	1356.2310	41.1070	50	1942.0809	19.8209
8	1547.0543	79.3748	18	1347.5230	39.3195	100	14546.4130	15.4142
9	1513.3444	71.1276	19	1340.7395	37.7316	200	231804.902	15.0025
10	1483.3648	64.5880	20	1335.7763	36.3113	∞	∞	**15.0000**

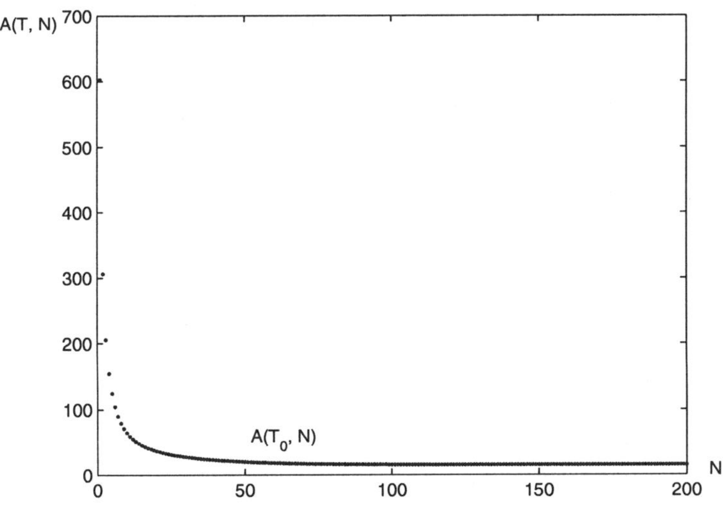

Figure 6.9.2. Plots of $A(T_0, N)$ against N.

It is easy to see from Table 6.9.2, conditions (6.9.15) and (6.9.16) hold for $k = 22$. Consequently, by Corollary 6.9.4, an optimal maintenance policy (T^*, N^*) is determined by

$$A(T^*, N^*) = \min_N A(T_0, N) = A(0.5, \infty). \tag{6.9.30}$$

 Geometric Process and Its Applications

Note that series $\sum_{n=1}^{N} \lambda(T, n)$ and $\sum_{n=1}^{N} \frac{\nu}{b_p^{n-1}} \frac{q_n}{p_n}$ are both convergent, but series $\sum_{n=1}^{N-1} \frac{\mu}{b^{n-1}}$ is divergent. Therefore, (6.9.7) yields that

$$A(0.5, \infty) = \lim_{N \to \infty} A(0.5, N) = c - c_p = 15. \qquad (6.9.31)$$

Therefore, the optimal maintenance policy is $(T^*, N^*) = (0.5, \infty)$, i.e. a preventive repair is taken when the system age reaches 0.5 and a failure repair is taken when the system fails whichever occurs first without replacement. Correspondingly, the minimum average cost is

$$C(T^*, N^*) = C(0.5, \infty) = A(0.5, \infty) + c_p = 25.$$

Example 6.9.8. Assume that the operating time of a system has an exponential distribution $Exp(1/\lambda)$ with parameter $\lambda = 120$. The other parameters are $a = 1.01, b = 0.95, b_p = 0.98, \mu = 25, \nu = 10, \tau = 20, r = 40, c = 30, c_p = 10, R = 8000, c_r = 20, T_0 = 20$. As the operating time distribution is exponential, Theorem 6.9.5 is applicable. The optimal maintenance policy (T^*, N^*) is determined by

$$A(T^*, N^*) = \min\{\min_N A(T_0, N), \min_N A(\infty, N)\}. \qquad (6.9.32)$$

The values of $A(T_0, N)$ and $A(\infty, N)$ are tabulated in Table 6.9.3 and plotted together in Figure 6.9.3.

From Table 6.9.3, it is easy to see that

$$A(T^*, N^*) = \min\{\min_N A(T_0, N), \min_N A(\infty, N)\}$$
$$= \min\{A(20, 11), A(\infty, 12)\} = A(\infty, 12) = -30.2739. \qquad (6.9.33)$$

Table 6.9.3. The values of $A(T_0, N)$ and $A(\infty, N)$

N	$A(T_0, N)$	$A(\infty, N)$	N	$A(T_0, N)$	$A(\infty, N)$	N	$A(T_0, N)$	$A(\infty, N)$
1	11.2740	15.7143	**11**	**-21.6341**	-30.2598	21	-20.1125	-28.2280
2	-8.2121	-11.4181	**12**	-21.6302	**-30.2739**	22	-19.8475	-27.8538
3	-14.4533	-20.0954	13	-21.5764	-30.2164	23	-19.5698	-27.4593
4	-17.4271	-24.2433	14	-21.4823	-30.1004	24	-19.2802	-27.0457
5	-19.0988	-26.5885	15	-21.3548	-29.9356	25	-18.9794	-26.6143
6	-20.1192	-28.0318	16	-21.1993	-29.7292	30	-17.3267	-24.2180
7	-20.7662	-28.9572	17	-21.0196	-29.4866	40	-13.4163	-18.4840
8	-21.1783	-29.5560	18	-20.8189	-29.2121	50	-8.9056	-11.9621
9	-21.4325	-29.9343	19	-20.5996	-28.9092	70	0.7990	0.9945
10	-21.5747	-30.1555	20	-20.3636	-28.5804	100	12.3635	13.6820

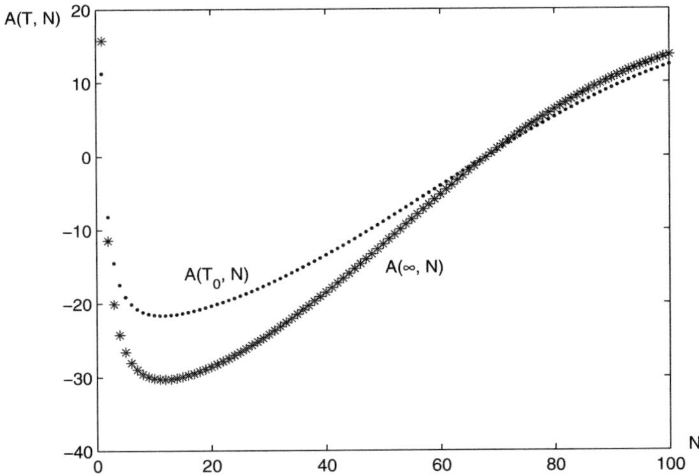

Figure 6.9.3. Plots of $A(T_0, N)$ and $A(\infty, N)$ against N

Therefore, the optimal maintenance policy is $(T^*, N^*) = (\infty, 12)$. Correspondingly, the minimum average cost is

$$C(T^*, N^*) = C(\infty, 12) = A(\infty, 12) + c_p = -20.2739.$$

6.10 Notes and References

In this chapter, we study the GP maintenance models. Assume that the successive operating times of a system after repair form a GP with ratio a, while the consecutive repair times of the system after failure constitute a GP with ratio b. By giving different values of a and b, the GP models can be applied to a deteriorating system as well as an improving system. In Section 6.2, the GP model for a two-state system using policy N is formulated. It was first time introduced by Lam (1988a, b) and then generalized by Lam (2003). In Section 6.3, an optimal replacement policy is determined analytically for a deteriorating system, while for an improving system, we show that policy $N^* = \infty$ is always the optimal policy. Then, the monotonicity properties of the optimal policy for a deteriorating system are considered in Section 6.4. The materials in Sections 6.2-6.4 are due to Lam (2003). On the other hand, Section 6.2 also studies a GP maintenance model in which a bivariate policy (T, N) is applied. This model was considered by Zhang (1994). The formulas presented here are based on Leung (2001) and Zhang

(2007). It is easy to see that the results of Leung (2001) and Zhang (1994) are equivalent.

In Section 6.5, a monotone process model for a multistate system is introduced, it is shown that the model is equivalent to a GP model for a two-state system. Furthermore, an optimal policy for the multistate system is also determined analytically. Section 6.5 is on the basis of Lam (2005a), it was first published in Journal of Applied Probability by The Applied Probability Trust. However, some special cases were considered earlier by Lam et al. (2002), Zhang et al. (2002) and Lam and Tse (2003). On the other hand, Lam (1991a) considered a monotone process maintenance model for a two-state system.

In Sections 6.6, a GP shock model for a system under a random environment is introduced. It is originally due to Lam and Zhang (2003). However, the GP shock model for an improving system and hence Theorem 6.6.9 are new. This model is actually a GP cumulative damage model. A cumulative damage model was initially considered by Barlow and Proschan (1975). Later on, Shanthikumar and Sumita (1983, 1984) studied a more general shock model under the assumption that the magnitude of the nth shock and the time interval between the $(n-1)$th shock and the nth shock are correlated.

A δ-shock model was studied by Li (1984) and Li et al. (1999), they derived the life distribution of a system without repair and the economic design of the components in the system. In Section 6.7, a GP δ-shock model for a repairable system is considered in which the interarrival times of successive shocks are i.i.d. with exponential distribution. This is a generalization of Li et al. (1999) by considering a repairable system that is either deteriorating or improving. Section 6.7 is on the basis of Lam and Zhang (2004). Afterward, Tang and Lam (2006a, b) studied more general δ-shock GP models in which the interarrival times of successive shocks are also i.i.d. but with gamma, Weibull and lognormal distributions respectively, also see Liang et al. (2006) for further reference.

Then, a threshold GP maintenance model is studied in Section 6.8. that is based on Lam (2007b). On the other hand, Zhang (2002) considered a GP model with preventive repair and using a replacement policy N. Thereafter, Wang and Zhang (2006) studied a GP model with periodic preventive repair. However, in Section 6.9, we study a more general GP model with preventive repair by using a maintenance policy (T, N), it is due to Lam's paper (2007a).

Finally, we note here all the numerical examples in Chapter 6 are new.

For more reference about GP maintenance model or related work, see Rangon and Esther Grace (1989), Leung and Lee (1998), Sheu (1999), Péréz-Ocon and Torres-Castro (2002), Wang and Zhang (2005), Zhang et al. (2001, 2002), Zhang et al. (2007), among others.

Chapter 7

Application to Analysis of System Reliability

7.1 Introduction

In this chapter, we shall study a GP model for a two-component system with one repairman. The main concern is to analyze the system reliability. In Sections 7.2-7.4, we shall consider respectively the series, parallel and cold standby systems. By introducing some supplementary variables, a set of differential equations for the probability distribution of the system state are derived. The equations can be solved analytically or numerically. Then some important reliability indices of the system such as the availability, the rate of occurrence of failure, the reliability and the mean time to the first failure are determined. In Section 7.5, as a particular case of a two-component system, we shall study the maintenance problem for a cold standby system.

7.2 Reliability Analysis for a Series System

Now we consider a GP model for a two-component series system with one repairman. For convenience, we shall use 'service' to represent 'either repair or replacement' here. Then we make the following assumptions.

Assumption 1.   At the beginning, a new two-component series system is installed. The system is up if and only if two components are both operating. Whenever a component fails, it will be repaired and the system is down. Assume that two components shut off each other, i.e. the behaviour of one component does not affect the other one.

Assumption 2.   For $i = 1, 2$, let X_{i1} be the operating time of com-

227

ponent i after the installation or a replacement. In general, for $n > 1$, let X_{in} be the operating time of component i after its $(n-1)$th repair, then $\{X_{in}, n = 1, 2, \ldots\}$ form a GP with ratio $a_i \geq 1$, and X_{i1} has an exponential distribution $Exp(\lambda_i)$ with distribution function

$$P(X_{i1} \leq t) = 1 - e^{-\lambda_i t}, \quad t \geq 0,$$

and 0 otherwise. Moreover, let Y_{in} be the repair time of the ith component after its nth failure. Then $\{Y_{in}, n = 1, 2, \ldots\}$ constitute a GP with ratio $0 < b_i \leq 1$, and Y_{i1} has an exponential distribution $Exp(\mu_i)$ with distribution function

$$P(Y_{i1} \leq t) = 1 - e^{-\mu_i t}, \quad t \geq 0,$$

and 0 otherwise.

Assumption 3. For $i = 1, 2$, assume that component i will be replaced by a new and identical one following the time of its N_ith failure. Moreover, for $i = 1, 2$, let Z_i be the replacement time of component i. Assume that Z_i has an exponential distribution $Exp(\tau_i)$ with distribution function

$$P(Z_i \leq t) = 1 - e^{-\tau_i t}, \quad t \geq 0,$$

and 0 otherwise.

Assumption 4. Assume further $\{X_{in}, n = 1, 2, \ldots\}$, $\{Y_{in}, n = 1, 2, \ldots\}$ and Z_i are all independent.

Under Assumption 2, for $i = 1, 2$, the successive operating times $\{X_{in}, n = 1, 2, \ldots\}$ form a decreasing GP, while the consecutive repair times $\{Y_{in}, n = 1, 2, \ldots\}$ constitute an increasing GP. Therefore our model is a GP model for a deteriorating two-component series system.

Now, the system state at time t can be defined by

$$I(t) = \begin{cases} 0 & \text{if both components are operating at time } t, \\ 1 & \text{if component 1 is under service at time } t, \\ 2 & \text{if component 2 is under service at time } t. \end{cases}$$

Obviously, $\{I(t), t \geq 0\}$ is a stochastic process with state space $S = \{0, 1, 2\}$, working state set $W = \{0\}$ and failure state set $F = \{1, 2\}$. Thus, the system is up at time t if and only if $I(t) = 0$. Clearly, $\{I(t), t \geq 0\}$ is not a Markov process. However, a Markov process may be constructed by applying the method of supplementary variable. To do this, two supplementary variables $I_1(t)$ and $I_2(t)$ are introduced, where $I_i(t), i = 1, 2$, is the number of failures of component i by time t since the installation or the last replacement. Then, the system state at time t is $(I(t), I_1(t), I_2(t))$, and process $\{(I(t), I_1(t), I_2(t)), t \geq 0\}$ will form a Markov process. Note that the system state

$(I(t), I_1(t), I_2(t)) = (0, j, k), \quad j = 0, \ldots, N_1 - 1, \ k = 0, \ldots, N_2 - 1,$

means that at time t both components are operating, hence the system is up, but component 1 has failed for j times while component 2 has failed for k times. On the other hand, the system state

$(I(t), I_1(t), I_2(t)) = (1, j, k), \quad j = 0, \ldots, N_1 - 1, \ k = 0, \ldots, N_2 - 1,$

means that component 1 has failed for j times and is under repair and component 2 has failed for k times, hence the system is down. Moreover, the system state

$(I(t), I_1(t), I_2(t)) = (1, N_1, k), \quad k = 0, \ldots, N_2 - 1,$

means that component 1 has failed for N_1 times and is being replaced and component 2 has failed for k times, hence the system is down. A similar explanation can be applied to system state

$(I(t), I_1(t), I_2(t)) = (2, j, k), j = 0, \ldots, N_1 - 1, \ k = 0, \ldots, N_2.$

Now, the probability distribution of the system state at time t is given by

$$p_{ijk}(t) = P\{(I(t), I_1(t), I_2(t)) = (i, j, k) \mid (I(0), I_1(0), I_2(0)) = (0, 0, 0)\},$$
$$i = 0, \ j = 0, \ldots, N_1 - 1; \ k = 0, \ldots, N_2 - 1;$$
$$i = 1, \ j = 1, \ldots, N_1; \ k = 0, \ldots, N_2 - 1;$$
$$i = 2, \ j = 0, \ldots, N_1 - 1; \ k = 1, \ldots, N_2.$$

Then, we can derive the following differential equations.

$$(\frac{d}{dt} + a_1^j \lambda_1 + a_2^k \lambda_2) p_{0jk}(t) = b_1^{j-1} \mu_1 p_{1jk}(t) + b_2^{k-1} \mu_2 p_{2jk}(t),$$
$$j = 1, \ldots, N_1 - 1, k = 1, \ldots, N_2 - 1; \quad (7.2.1)$$

$$(\frac{d}{dt} + \lambda_1 + a_2^k \lambda_2) p_{00k}(t) = b_2^{k-1} \mu_2 p_{20k}(t) + \tau_1 p_{1N_1 k}(t),$$
$$k = 1, \ldots, N_2 - 1; \quad (7.2.2)$$

$$(\frac{d}{dt} + a_1^j \lambda_1 + \lambda_2) p_{0j0}(t) = b_1^{j-1} \mu_1 p_{1j0}(t) + \tau_2 p_{2jN_2}(t),$$
$$j = 1, \ldots, N_1 - 1; \quad (7.2.3)$$

$$(\frac{d}{dt} + \lambda_1 + \lambda_2) p_{000}(t) = \tau_1 p_{1N_10}(t) + \tau_2 p_{20N_2}(t); \quad (7.2.4)$$

$$(\frac{d}{dt} + b_1^{j-1} \mu_1) p_{1jk}(t) = a_1^{j-1} \lambda_1 p_{0j-1k}(t),$$
$$j = 1, \ldots, N_1 - 1, k = 0, \ldots, N_2 - 1; \quad (7.2.5)$$

$$(\frac{d}{dt} + \tau_1)p_{1N_1k}(t) = a_1^{N_1-1}\lambda_1 p_{0N_1-1k}(t),$$
$$k = 0,\ldots,N_2-1; \tag{7.2.6}$$

$$(\frac{d}{dt} + b_2^{k-1}\mu_2)p_{2jk}(t) = a_2^{k-1}\lambda_2 p_{0jk-1}(t),$$
$$j = 0,\ldots,N_1-1, k = 1,\ldots,N_2-1; \tag{7.2.7}$$

$$(\frac{d}{dt} + \tau_2)p_{2jN_2}(t) = a_2^{N_2-1}\lambda_2 p_{0jN_2-1}(t),$$
$$j = 0,\ldots,N_1-1. \tag{7.2.8}$$

Equations (7.2.1)-(7.2.8) are actually the Kolmogorov forward equations. They can be derived by applying a classical probability analysis. As an example, we shall derive equation (7.2.1) here. For $j = 1,\ldots,N_1-1, k = 1,\ldots,N_2-1$, we have

$$p_{0jk}(t+\Delta t)$$

$$= P(\text{no component fails in } (t,t+\Delta t] \mid (I(t),I_1(t),I_2(t)) = (0,j,k))p_{0jk}(t)$$

$$+ P(\text{component 1 completes repair in } (t,t+\Delta t]$$

$$\mid (I(t),I_1(t),I_2(t)) = (1,j,k))p_{1jk}(t)$$

$$+ P(\text{component 2 completes repair in } (t,t+\Delta t]$$

$$\mid (I(t),I_1(t),I_2(t)) = (2,j,k))p_{2jk}(t) + o(\Delta t)$$

$$= (1 - a_1^j\lambda_1\Delta t)(1 - a_2^k\lambda_2\Delta t)p_{0jk}(t) + b_1^{j-1}\mu_1 p_{1jk}(t)\Delta t$$

$$+ b_2^{k-1}\mu_2 p_{2jk}(t)\Delta t + o(\Delta t).$$

Then

$$\frac{p_{0jk}(t+\Delta t) - p_{0jk}(t)}{\Delta t}$$

$$= -(a_1^j\lambda_1 + a_2^k\lambda_2)p_{0jk}(t) + b_1^{j-1}\mu_1 p_{1jk}(t) + b_2^{k-1}\mu_2 p_{2jk}(t) + o(1).$$

Therefore, (7.2.1) follows by letting $\Delta t \to 0$. The other equations can be derived in a similar way. The initial condition is

$$p_{ijk}(0) = \begin{cases} 1 & (i,j,k) = (0,0,0), \\ 0 & \text{elsewhere.} \end{cases} \tag{7.2.9}$$

In general, the Kolmogorov forward equations (7.2.1)-(7.2.8) with initial condition (7.2.9) may be solved by using the Laplace transform. After obtaining the probability distribution $p_{ijk}(t)$ by inversion, we can determine the reliability indices straightforwardly.

In fact, the availability of the system at time t is determined by

$$A(t) = P(\text{the system is up at time } t) = P(I(t) = 0)$$

$$= \sum_{j=0}^{N_1-1} \sum_{k=0}^{N_2-1} p_{0jk}(t). \tag{7.2.10}$$

If

$$\lim_{t\to\infty} p_{0jk}(t) = A_{0jk},$$

then the equilibrium availability is given by

$$A = \lim_{t\to\infty} A(t) = \sum_{j=0}^{N_1-1}\sum_{k=0}^{N_2-1} A_{0jk}. \qquad (7.2.11)$$

Let

$$p_{ijk}^{*}(s) = \int_{0}^{\infty} e^{-st} p_{ijk}(t)\,dt$$

be the Laplace transform of $p_{ijk}(t)$. Then, an alternative way to determine the equilibrium availability is to apply the Tauberian theorem

$$A = \lim_{t\to\infty} A(t) = \lim_{s\to0} sA^{*}(s)$$

$$= \sum_{j=0}^{N_1-1}\sum_{k=0}^{N_2-1} \lim_{s\to0} sp_{0jk}^{*}(s). \qquad (7.2.12)$$

The rate of occurrence of failures (ROCOF) is another important reliability index. Let the ROCOF at time t be $m_f(t)$, then Theorem 1.6.5 gives

$$m_f(t) = \sum_{j=0}^{N_1-1}\sum_{k=0}^{N_2-1} p_{0jk}(t)(a_1^j\lambda_1 + a_2^k\lambda_2). \qquad (7.2.13)$$

Moreover,

$$m_f = \lim_{t\to\infty} m_f(t) = \sum_{j=0}^{N_1-1}\sum_{k=0}^{N_2-1} A_{0jk}(a_1^j\lambda_1 + a_2^k\lambda_2). \qquad (7.2.14)$$

Consequently, $M_f(t)$, the expected number of failures by time t, will have an asymptotic line with slope m_f.

The reliability of the system at time t is given by

$$R(t) = P(\text{the system operating time} \geq t)$$

$$= P(\min(X_{11}, X_{21}) > t) = \exp\{-(\lambda_1 + \lambda_2)t\}. \qquad (7.2.15)$$

Then the mean time to the first failure (MTTFF) is determined by

$$MTTFF = \int_{0}^{\infty} R(t)\,dt$$

$$= \int_{0}^{\infty} \exp\{-(\lambda_1 + \lambda_2)t\}\,dt = \frac{1}{\lambda_1 + \lambda_2}. \qquad (7.2.16)$$

In particular, consider the case that $N_1 = N_2 = 1$. This is a series system with two different components in which a failed component will be replaced by a new and identical one without repair. Because $\{I(t), t \geq 0\}$ form a Markov process, we need not introduce the supplementary variables. Therefore, we can define the probability distribution as

$$p_i(t) = P(I(t) = i), \quad i = 0, 1, 2.$$

Now instead of equations (7.2.1)-(7.2.8), we have

$$(\frac{d}{dt} + \lambda_1 + \lambda_2)p_0(t) = \tau_1 p_1(t) + \tau_2 p_2(t), \tag{7.2.17}$$

$$(\frac{d}{dt} + \tau_1)p_1(t) = \lambda_1 p_0(t), \tag{7.2.18}$$

$$(\frac{d}{dt} + \tau_2)p_2(t) = \lambda_2 p_0(t). \tag{7.2.19}$$

The initial condition (7.2.9) becomes

$$p_i(0) = \begin{cases} 1 & i = 0, \\ 0 & i = 1, 2. \end{cases} \tag{7.2.20}$$

Then for $i = 0, 1, 2$, let $p_i^*(s)$ be the Laplace transform of $p_i(t)$ defined by

$$p_i^*(s) = \int_0^\infty e^{-st} p_i(t)dt.$$

By taking the Laplace transform on both sides of (7.2.17)-(7.2.19) with the help of (7.2.20), it follows that

$$(s + \lambda_1 + \lambda_2)p_0^*(s) - 1 = 1 + \tau_1 p_1^*(s) + \tau_2 p_2^*(s),$$
$$(s + \tau_1)p_1^*(s) = \lambda_1 p_0^*(s),$$
$$(s + \tau_2)p_2^*(s) = \lambda_2 p_0^*(s).$$

Therefore

$$\begin{aligned}
p_0^*(s) &= \frac{(s + \tau_1)(s + \tau_2)}{s[s^2 + (\lambda_1 + \lambda_2 + \tau_1 + \tau_2)s + (\lambda_1\tau_2 + \lambda_2\tau_1 + \tau_1\tau_2)]} \\
&= \frac{(s + \tau_1)(s + \tau_2)}{s(s + \alpha)(s + \beta)},
\end{aligned} \tag{7.2.21}$$

where $-\alpha$ and $-\beta$ are two roots of the quadratic equation

$$s^2 + (\lambda_1 + \lambda_2 + \tau_1 + \tau_2)s + (\lambda_1\tau_2 + \lambda_2\tau_1 + \tau_1\tau_2) = 0. \tag{7.2.22}$$

The discriminant of (7.2.22) is

$$\begin{aligned}
\Delta &= (\lambda_1 + \lambda_2 + \tau_1 + \tau_2)^2 - 4(\lambda_1\tau_2 + \lambda_2\tau_1 + \tau_1\tau_2) \\
&= (\lambda_1 - \lambda_2 + \tau_1 - \tau_2)^2 + 4\lambda_1\lambda_2 > 0.
\end{aligned}$$

Therefore, both α and β are positive numbers.

Consequently, we have

$$p_1^*(s) = \frac{\lambda_1(s + \tau_2)}{s(s + \alpha)(s + \beta)} \qquad (7.2.23)$$

and

$$p_2^*(s) = \frac{\lambda_2(s + \tau_1)}{s(s + \alpha)(s + \beta)}. \qquad (7.2.24)$$

It follows by partial fraction

$$p_i^*(s) = \frac{A_i}{s} + \frac{B_i}{s + \alpha} + \frac{C_i}{s + \beta}, \quad i = 0, 1, 2. \qquad (7.2.25)$$

Now

$$A_0 = \frac{\tau_1\tau_2}{\alpha\beta} = \frac{\tau_1\tau_2}{\lambda_1\tau_2 + \lambda_2\tau_1 + \tau_1\tau_2},$$

$$B_0 = \frac{(\tau_1 - \alpha)(\tau_2 - \alpha)}{\alpha(\alpha - \beta)},$$

$$C_0 = \frac{(\tau_1 - \beta)(\tau_2 - \beta)}{\beta(\beta - \alpha)},$$

and for $i = 1, 2$,

$$A_i = \frac{\lambda_i\tau_{3-i}}{\alpha\beta} = \frac{\lambda_i\tau_{3-i}}{\lambda_1\tau_2 + \lambda_2\tau_1 + \tau_1\tau_2},$$

$$B_i = \frac{\lambda_i(\tau_{3-i} - \alpha)}{\alpha(\alpha - \beta)},$$

$$C_i = \frac{\lambda_i(\tau_{3-i} - \beta)}{\beta(\beta - \alpha)}.$$

By inversion, (7.2.25) yields that

$$p_i(t) = A_i + B_i e^{-\alpha t} + C_i e^{-\beta t}, \quad i = 0, 1, 2. \qquad (7.2.26)$$

Thus, the availability at time t is

$$A(t) = p_0(t) = A_0 + B_0 e^{-\alpha t} + C_0 e^{-\beta t}. \qquad (7.2.27)$$

Because $\alpha > 0$ and $\beta > 0$, the equilibrium availability exists and is given by

$$\lim_{t \to \infty} A(t) = A_0.$$

Alternatively, it follows from (7.2.12) that

$$\lim_{t \to \infty} A(t) = \lim_{s \to 0} s p_0^*(s) = A_0.$$

On the other hand, from (7.2.13), the ROCOF at time t is given by

$$m_f(t) = p_0(t)(\lambda_1 + \lambda_2)$$
$$= (\lambda_1 + \lambda_2)(A_0 + B_0 e^{-\alpha t} + C_0 e^{-\beta t}).$$

Furthermore

$$M_f(t) = (\lambda_1 + \lambda_2)\{A_0 t + \frac{B_0}{\alpha}(1 - e^{-\alpha t}) + \frac{C_0}{\beta}(1 - e^{-\beta t})\}.$$

The equilibrium ROCOF is

$$m_f = \lim_{t \to \infty} m_f(t) = A_0(\lambda_1 + \lambda_2).$$

The asymptotic line of $M_f(t)$ is

$$y = (\lambda_1 + \lambda_2)(A_0 t + \frac{B_0}{\alpha} + \frac{C_0}{\beta}).$$

Furthermore, the reliability and MTTFF are given by (7.2.15) and (7.2.16) respectively.

7.3   Reliability Analysis for a Parallel System

Now we consider a GP model for a two-component parallel system with one repairman under the following assumptions. As in Section 7.2, we shall use 'service' to represent 'either repair or replacement' here. Now, we make the following assumptions.

Assumption 1. At the beginning, a new two-component parallel system is installed such that the system is up if and only if at least one component is operating. A failed component will be repaired by the repairman. Then component i, $i = 1, 2$, will be replaced by a new and identical one following the time of its N_ith failure. If one component fails but the other component is being service, it must wait for service and the system is down.

Assumption 2. For $i = 1, 2$, let X_{i1} be the operating time of component i after installation or a replacement. In general, for $n > 1$, let X_{in} be the operating time of component i after its $(n - 1)$th repair, then $\{X_{in}, n = 1, 2, \ldots\}$ form a GP with ratio $a_i \geq 1$, and X_{i1} has an exponential distribution $Exp(\lambda_i)$ with distribution function

$$P(X_{i1} \leq t) = 1 - e^{-\lambda_i t}, \quad t \geq 0,$$

and 0 otherwise. Moreover, let Y_{in} be the repair time of component i after its nth failure. Then $\{Y_{in}, n = 1, 2, \ldots\}$ constitute a GP with ratio $0 <$

$b_i \leq 1$, and Y_{i1} has an exponential distribution $Exp(\mu_i)$ with distribution function

$$P(Y_{i1} \leq t) = 1 - e^{-\mu_i t}, \quad t \geq 0,$$

and 0 otherwise.

Assumption 3. For $i = 1, 2$, let Z_i be the replacement time of component i. Assume that Z_i has an exponential distribution $Exp(\tau_i)$ with distribution function

$$P(Z_i \leq t) = 1 - e^{-\tau_i t}, \quad t \geq 0,$$

and 0 otherwise.

Assumption 4. Assume further $\{X_{in}, n = 1, 2, \ldots\}$, $\{Y_{in}, n = 1, 2, \ldots\}$ and Z_i are all independent.

Under Assumption 2, for $i = 1, 2$, the successive operating times $\{X_{in}, n = 1, 2, \ldots\}$ form a decreasing GP, while the consecutive repair times $\{Y_{in}, n = 1, 2, \ldots\}$ constitute an increasing GP. Therefore our model is a GP model for a deteriorating two-component parallel system.

Then the system state at time t can be defined by

$$I(t) = \begin{cases} 0 & \text{if both components are operating at time } t, \\ 1 & \text{if component 1 is under service and component 2} \\ & \text{is operating at time } t, \\ 2 & \text{if component 2 is under service and component 1} \\ & \text{is operating at time } t, \\ 3 & \text{if component 1 is under service and component 2} \\ & \text{is waiting for service at time } t, \\ 4 & \text{if component 2 is under service and component 1} \\ & \text{is waiting for service at time } t. \end{cases}$$

Now, the working state set is $W = \{0, 1, 2\}$, the failure state set is $F = \{3, 4\}$, and the state space is $S = W \cup F$. Once again, $\{I(t), t \geq 0\}$ is not a Markov process. However, we can also introduce two supplementary variables $I_1(t)$ and $I_2(t)$, where $I_i(t)$, $i = 1, 2$, is the number of failures of component i by time t since installation or the last replacement. Then, the system state at time t will be $(I(t), I_1(t), I_2(t))$, and process $\{(I(t), I_1(t), I_2(t)), t \geq 0\}$ will become a Markov process. Note that the system state
$$(I(t), I_1(t), I_2(t)) = (0, j, k), \quad j = 0, \ldots, N_1 - 1, \ k = 0, \ldots, N_2 - 1,$$
means that at time t both component are operating, hence the system is up, but component 1 has failed for j times while component 2 has failed for

k times. We can also explain the meaning of other state $(I(t), I_1(t), I_2(t))$ accordingly. Now, the probability distribution of the system state at time t is given by

$$p_{ijk}(t) = P\{(I(t), I_1(t), I_2(t)) = (i, j, k) \mid (I(0), I_1(0), I_2(0)) = (0, 0, 0)\},$$

$$i = 0, \ j = 0, \ldots, N_1 - 1; \ k = 0, \ldots, N_2 - 1;$$
$$i = 1, \ j = 1, \ldots, N_1; \ k = 0, \ldots, N_2 - 1;$$
$$i = 2, \ j = 0, \ldots, N_1 - 1; \ k = 1, \ldots, N_2;$$
$$i = 3, \ j = 1, \ldots, N_1; \ k = 1, \ldots, N_2;$$
$$i = 4, \ j = 1, \ldots, N_1; \ k = 1, \ldots, N_2.$$

Consequently, by using classical probability analysis, we have the following Kolmogorov forward equations.

$$(\frac{d}{dt} + a_1^j \lambda_1 + a_2^k \lambda_2)p_{0jk}(t) = b_1^{j-1}\mu_1 p_{1jk}(t) + b_2^{k-1}\mu_2 p_{2jk}(t),$$
$$j = 1, \ldots, N_1 - 1, k = 1, \ldots, N_2 - 1; \qquad (7.3.1)$$

$$(\frac{d}{dt} + \lambda_1 + a_2^k \lambda_2)p_{00k}(t) = b_2^{k-1}\mu_2 p_{20k}(t) + \tau_1 p_{1N_1 k}(t),$$
$$k = 1, \ldots, N_2 - 1; \qquad (7.3.2)$$

$$(\frac{d}{dt} + a_1^j \lambda_1 + \lambda_2)p_{0j0}(t) = b_1^{j-1}\mu_1 p_{1j0}(t) + \tau_2 p_{2jN_2}(t),$$
$$j = 1, \ldots, N_1 - 1; \qquad (7.3.3)$$

$$(\frac{d}{dt} + \lambda_1 + \lambda_2)p_{000}(t) = \tau_1 p_{1N_1 0}(t) + \tau_2 p_{20N_2}(t); \qquad (7.3.4)$$

$$(\frac{d}{dt} + a_2^k \lambda_2 + b_1^{j-1}\mu_1)p_{1jk}(t) = a_1^{j-1}\lambda_1 p_{0j-1k}(t) + b_2^{k-1}\mu_2 p_{4jk}(t),$$
$$j = 1, \ldots, N_1 - 1, k = 1, \ldots, N_2 - 1; \qquad (7.3.5)$$

$$(\frac{d}{dt} + \lambda_2 + b_1^{j-1}\mu_1)p_{1j0}(t) = a_1^{j-1}\lambda_1 p_{0j-10}(t) + \tau_2 p_{4jN_2}(t),$$
$$j = 1, \ldots, N_1 - 1; \qquad (7.3.6)$$

$$(\frac{d}{dt} + a_2^k \lambda_2 + \tau_1)p_{1N_1 k}(t) = a_1^{N_1-1}\lambda_1 p_{0N_1-1k}(t) + b_2^{k-1}\mu_2 p_{4N_1 k}(t),$$
$$k = 1, \ldots, N_2 - 1; \qquad (7.3.7)$$

$$(\frac{d}{dt} + \lambda_2 + \tau_1)p_{1N_1 0}(t) = a_1^{N_1-1}\lambda_1 p_{0N_1-10}(t) + \tau_2 p_{4N_1 N_2}(t); \quad (7.3.8)$$

$$(\frac{d}{dt} + a_1^j \lambda_1 + b_2^{k-1}\mu_2)p_{2jk}(t) = a_2^{k-1}\lambda_2 p_{0jk-1}(t) + b_1^{j-1}\mu_1 p_{3jk}(t),$$
$$j = 1, \ldots, N_1 - 1, k = 1, \ldots, N_2 - 1; \qquad (7.3.9)$$

$$(\frac{d}{dt} + \lambda_1 + b_2^{k-1}\mu_2)p_{20k}(t) = a_2^{k-1}\lambda_2 p_{00k-1}(t) + \tau_1 p_{3N_1k}(t),$$
$$k = 1,\ldots,N_2 - 1; \tag{7.3.10}$$

$$(\frac{d}{dt} + a_1^j\lambda_1 + \tau_2)p_{2jN_2}(t) = a_2^{N_2-1}\lambda_2 p_{0jN_2-1}(t) + b_1^{j-1}\mu_1 p_{3jN_2}(t),$$
$$j = 1,\ldots,N_1 - 1; \tag{7.3.11}$$

$$(\frac{d}{dt} + \lambda_1 + \tau_2)p_{20N_2}(t) = a_2^{N_2-1}\lambda_2 p_{00N_2-1}(t) + \tau_1 p_{3N_1N_2}(t); \tag{7.3.12}$$

$$(\frac{d}{dt} + b_1^{j-1}\mu_1)p_{3jk}(t) = a_2^{k-1}\lambda_2 p_{1jk-1}(t),$$
$$j = 1,\ldots,N_1 - 1, k = 1,\ldots,N_2; \tag{7.3.13}$$

$$(\frac{d}{dt} + \tau_1)p_{3N_1k}(t) = a_2^{k-1}\lambda_2 p_{1N_1k-1}(t),$$
$$k = 1,\ldots,N_2; \tag{7.3.14}$$

$$(\frac{d}{dt} + b_2^{k-1}\mu_2)p_{4jk}(t) = a_1^{j-1}\lambda_1 p_{2j-1k}(t),$$
$$j = 1,\ldots,N_1, k = 1,\ldots,N_2 - 1; \tag{7.3.15}$$

$$(\frac{d}{dt} + \tau_2)p_{4jN_2}(t) = a_1^{j-1}\lambda_1 p_{2j-1N_2}(t),$$
$$j = 1,\ldots,N_1. \tag{7.3.16}$$

The initial condition is

$$p_{ijk}(0) = \begin{cases} 1 & (i,j,k) = (0,0,0), \\ 0 & \text{elsewhere.} \end{cases} \tag{7.3.17}$$

Although we may solve equations (7.3.1)-(7.3.16) with initial condition (7.3.17) by using the Laplace transform, it might be better to apply a numerical method. For reference see Lam (1995) in which a numerical example is studied by using the Runge-Kutta method. As for the series case, after obtaining the probability distribution $p_{ijk}(t)$, we can determine the reliability indices of the system.

The availability of the system at time t is given by

$$A(t) = P(\text{the system is up at time } t)$$
$$= \sum_{j=0}^{N_1-1}\sum_{k=0}^{N_2-1} p_{0jk}(t) + \sum_{j=1}^{N_1}\sum_{k=0}^{N_2-1} p_{1jk}(t) + \sum_{j=0}^{N_1-1}\sum_{k=1}^{N_2} p_{2jk}(t). \tag{7.3.18}$$

Furthermore, let the ROCOF at time t be $m_f(t)$, then Theorem 1.6.5 gives

$$m_f(t) = \sum_{j=1}^{N_1}\sum_{k=0}^{N_2-1} p_{1jk}(t)a_2^k\lambda_2 + \sum_{j=0}^{N_1-1}\sum_{k=1}^{N_2} p_{2jk}(t)a_1^j\lambda_1. \tag{7.3.19}$$

If there exists a limiting distribution, let

$$\lim_{t\to\infty} p_{ijk}(t) = A_{ijk}.$$

Then the equilibrium availability is given by

$$A = \lim_{t\to\infty} A(t)$$

$$= \sum_{j=0}^{N_1-1}\sum_{k=0}^{N_2-1} A_{0jk} + \sum_{j=1}^{N_1}\sum_{k=0}^{N_2-1} A_{1jk} + \sum_{j=0}^{N_1-1}\sum_{k=1}^{N_2} A_{2jk}. \qquad (7.3.20)$$

The equilibrium ROCOF is given by

$$m_f = \lim_{t\to\infty} m_f(t) = \sum_{j=1}^{N_1}\sum_{k=0}^{N_2-1} A_{1jk} a_2^k \lambda_2 + \sum_{j=0}^{N_1-1}\sum_{k=1}^{N_2} A_{2jk} a_1^j \lambda_1. \qquad (7.3.21)$$

Consequently, as in the series case, $M_f(t)$ will have an asymptotic line with slope m_f.

To study the reliability of the system, we can treat the failure states 3 and 4 as absorbing states, and denote the system state at time t by $\tilde{I}(t)$. Then after introducing the same supplementary variables $I_1(t)$ and $I_2(t)$, process $\{(\tilde{I}(t), I_1(t), I_2(t)), t \geq 0\}$ will form a Markov process with the probability mass function

$$q_{ijk}(t) = P\{(\tilde{I}(t), I_1(t), I_2(t)) = (i,j,k) \mid (\tilde{I}(0), I_1(0), I_2(0)) = (0,0,0)\},$$

$$i = 0, \quad j = 0,\ldots,N_1 - 1; \quad k = 0,\ldots,N_2 - 1;$$

$$i = 1, \quad j = 1,\ldots,N_1; \quad k = 0,\ldots,N_2 - 1;$$

$$i = 2, \quad j = 0,\ldots,N_1 - 1; \quad k = 1,\ldots,N_2.$$

The Kolmogorov forward equations now are as follows.

$$(\frac{d}{dt} + a_1^j\lambda_1 + a_2^k\lambda_2)q_{0jk}(t) = b_1^{j-1}\mu_1 q_{1jk}(t) + b_2^{k-1}\mu_2 q_{2jk}(t),$$

$$j = 1,\ldots,N_1 - 1, k = 1,\ldots,N_2 - 1; \qquad (7.3.22)$$

$$(\frac{d}{dt} + \lambda_1 + a_2^k\lambda_2)q_{00k}(t) = b_2^{k-1}\mu_2 q_{20k}(t) + \tau_1 q_{1N_1 k}(t),$$

$$k = 1,\ldots,N_2 - 1; \qquad (7.3.23)$$

$$(\frac{d}{dt} + a_1^j\lambda_1 + \lambda_2)q_{0j0}(t) = b_1^{j-1}\mu_1 q_{1j0}(t) + \tau_2 q_{2jN_2}(t),$$

$$j = 1,\ldots,N_1 - 1; \qquad (7.3.24)$$

$$(\frac{d}{dt} + \lambda_1 + \lambda_2)q_{000}(t) = \tau_1 q_{1N_10}(t) + \tau_2 q_{20N_2}(t); \qquad (7.3.25)$$

$$(\frac{d}{dt} + a_2^k\lambda_2 + b_1^{j-1}\mu_1)q_{1jk}(t) = a_1^{j-1}\lambda_1 q_{0j-1k}(t),$$

$$j = 1,\ldots,N_1 - 1, k = 0,\ldots,N_2 - 1; \qquad (7.3.26)$$

$$(\frac{d}{dt} + a^k\lambda_2 + \tau_1)q_{1N_1k}(t) = a_1^{N_1-1}\lambda_1 q_{0N_1-1k}(t),$$

$$k = 0 , \ldots, N_2 - 1; \tag{7.3.27}$$

$$(\frac{d}{dt} + a_1^j\lambda_1 + b_2^{k-1}\mu_2)q_{2jk}(t) = a_2^{k-1}\lambda_2 q_{0jk-1}(t),$$

$$j = 0 , \ldots, N_1 - 1, k = 1, \ldots, N_2 - 1; \tag{7.3.28}$$

$$(\frac{d}{dt} + a_1^j\lambda_1 + \tau_2)q_{2jN_2}(t) = a_2^{N_2-1}\lambda_2 q_{0jN_2-1}(t),$$

$$j = 0 , \ldots, N_1 - 1; \tag{7.3.29}$$

The initial condition is

$$q_{ijk}(0) = \begin{cases} 1 & (i,j,k) = (0,0,0), \\ 0 & \text{elsewhere.} \end{cases} \tag{7.3.30}$$

After solving the equations, the reliability at time t of the system is now given by

$$R(t) = P(\text{the system operating time} \geq t)$$

$$= \sum_{j=0}^{N_1-1}\sum_{k=0}^{N_2-1} q_{0jk}(t) + \sum_{j=1}^{N_1}\sum_{k=0}^{N_2-1} q_{1jk}(t) + \sum_{j=0}^{N_1-1}\sum_{k=1}^{N_2} q_{2jk}(t). \tag{7.3.31}$$

Moreover, the mean time to the first failure (MTTFF) is determined by

$$MTTFF = \int_0^\infty R(t)dt = \lim_{s\to 0} R^*(s). \tag{7.3.32}$$

7.4 Reliability Analysis for a Cold Standby System

In practice, for improving the reliability or availability of a one-component system, a standby component is usually installed, such a two-component system is called a standby system. For example, in a nuclear plant, to reduce the risk of the 'scram' of a reactor in case of a coolant pipe breaking or some other failure happening, a standby diesel generator should be installed. In a hospital or a steel manufacturing complex, if the power supply suddenly suspends when required, the consequences might be catastrophic such as a patient may die in an operating room, a standby generator is then installed. Therefore, it is interesting to study the reliability of a standby system. In practice, many standby systems are cold standby systems in the sense that the standby component will neither fail nor deteriorate. A GP model for a cold standby system with one repairman is considered under the following assumptions.

Assumption 1. At the beginning, a new system of two components is installed in which component 1 is operating and component 2 is in cold standby. Whenever an operating component fails, it will be repaired by the repairman, and the standby component is operating. A failed component after repair will be operating if the other one fails, and standby otherwise. If a component fails when the other one is under repair it will wait for repair, and the system is down. Assume that the shift of switch is reliable and the change of system state is instantaneous. Replacement policy N is applied by which the system will be replaced by a new and identical one following the Nth failure of component 2.

Assumption 2. For $i = 1, 2$, let X_{i1} be the operating time of component i after the installation or a replacement. In general, for $n > 1$, let X_{in} be the operating time of component i after its $(n-1)$th repair, then $\{X_{in}, n = 1, 2, \ldots\}$ form a GP with ratio $a_i \geq 1$. Assume that X_{i1} has an exponential distribution $Exp(\lambda_i)$ with distribution function

$$P(X_{i1} \leq t) = 1 - e^{-\lambda_i t}, \quad t \geq 0,$$

and 0 otherwise. Let Y_{in} be the repair time after the nth failure. Then $\{Y_{in}, n = 1, 2, \ldots\}$ constitute a GP with ratio $0 < b_i \leq 1$. Assume that Y_{i1} has an exponential distribution $Exp(\mu_i)$ with distribution function

$$P(Y_{i1} \leq t) = 1 - e^{-\mu_i t}, \quad t \geq 0,$$

and 0 otherwise.

Assumption 3. Let Z be the replacement time of the system. Assume that Z has an exponential distribution $Exp(\tau)$ with distribution function

$$P(Z \leq t) = 1 - e^{-\tau t}, \quad t \geq 0,$$

and 0 otherwise.

Assumption 4. Assume that $\{X_{in}, n = 1, 2, \ldots\}$, $\{Y_{in}, n = 1, 2, \ldots\}$ and Z are all independent.

Now, the system state at time t can be defined by

$$
I(t) = \begin{cases}
0 & \text{if component 1 is operating and component 2} \\
  & \text{is standby at time } t, \\
1 & \text{if component 2 is operating and component 1} \\
  & \text{is standby at time } t, \\
2 & \text{if component 1 is operating and component 2} \\
  & \text{is under repair at time } t, \\
3 & \text{if component 2 is operating and component 1} \\
  & \text{is under repair or waiting for replacement at time } t, \\
4 & \text{if component 1 is under repair and component 2} \\
  & \text{is waiting for repair at time } t, \\
5 & \text{if component 2 is under repair and component 1} \\
  & \text{is waiting for repair or replacement at time } t, \\
6 & \text{if the system is under replacement at time } t.
\end{cases}
$$

Thus, the sets of working states, failure states and replacement state of process $\{I(t), t \geq 0\}$ are respectively $W = \{0, 1, 2, 3\}$, $F = \{4, 5\}$ and $R = \{6\}$. Then the state space is $S = W \cup F \cup R$. Furthermore, for $i = 1, 2$, let $I_i(t)$ be the number of failures of component i since the installation or the last replacement. It is easy to see that

$$
I_2(t) = \begin{cases}
I_1(t) & I(t) = 0, \\
I_1(t) - 1 & I(t) = 1, \\
I_1(t) & I(t) = 2, \\
I_1(t) - 1 & I(t) = 3, \\
I_1(t) & I(t) = 4, \\
I_1(t) - 1 & I(t) = 5, \\
I_1(t) & I(t) = 6.
\end{cases} \tag{7.4.1}
$$

For $i = 1, 2$, let period 1_i be the time interval between the installation or replacement of the system and the first failure of component i. In general, let period n_i be the time interval between the $(n-1)$th failure and the nth failure of component i for $n = 1, \ldots, N$. Moreover, let period A_i be the time interval between the Nth failure of component i and the completion of replacement. We say that a cycle is completed if a replacement is completed. To understand the concepts of periods and cycle better, suppose policy $N = 3$ is applied, then a possible realization of a cycle of the cold standby system may be expressed in Figure 7.4.1.

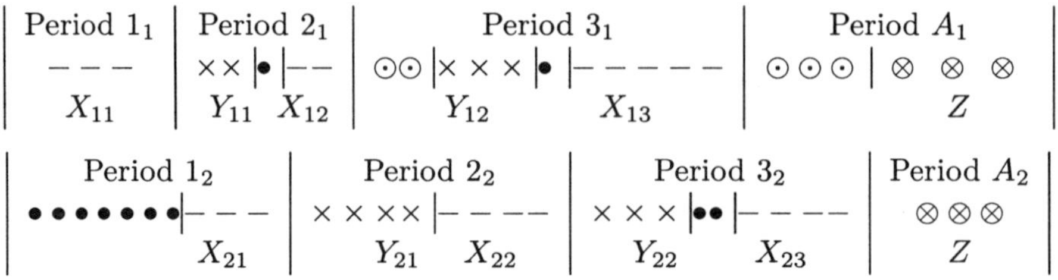

Period 1_1	Period 2_1	Period 3_1	Period A_1
$---$	$\times\times\,|\bullet\,|--$	$\odot\odot\,|\times\times\times\,|\bullet\,|------$	$\odot\,\odot\,\odot\,|\,\otimes\quad\otimes\quad\otimes$
X_{11}	$Y_{11}\quad X_{12}$	$Y_{12}\qquad\quad X_{13}$	Z

Period 1_2	Period 2_2	Period 3_2	Period A_2
$\bullet\bullet\bullet\bullet\bullet\bullet\bullet|---$	$\times\times\times\times\,|----$	$\times\times\times\,|\bullet\bullet\,|----$	$\otimes\otimes\otimes$
X_{21}	$Y_{21}\quad X_{22}$	$Y_{22}\qquad X_{23}$	Z

$--$: operating case, $\bullet\bullet$: standby case, $\times\times$: repair case,

$\odot\odot$: waiting case, $\otimes\otimes$: replacement case

Figure 7.4.1. A possible realization of a cycle

Now, we shall explain the reason why a replacement policy N is applied. In fact, under policy N, if the system is at a replacement state at time t, then $I(t) = 6$ and $I_1(t) = I_2(t) = N$. Thus both components have failed for N times. Figure 7.4.1 demonstrates that two components will operate alternatively, they play a similar role in operation. From (7.4.1), the number of failures of component 2 is at most different from that of component 1 by 1. In many practical cases, two components in a standby system are even identical. Consequently, it is reasonable to apply policy N so that we shall replace the system when both two components have failed for N times.

It follows from (7.4.1) that $I_2(t)$ is completely determined by $I_1(t)$. Therefore, we need only introduce one supplementary variable $I_1(t)$, and the system state at time t can be denoted by $(I(t), I_1(t))$. As a result, process $\{(I(t), I_1(t)), t \geq 0\}$ will be a Markov process. With the help of (7.4.1), the state

$$(I(t), I_1(t)) = (0, j), \quad j = 0, \ldots, N - 1,$$

means that at time t component 1 is operating and component 2 is standby, hence the system is up, and two components have both failed for j times. We can also explain the meaning of other state $(I(t), I_1(t))$ at time t accordingly.

Then, the probability mass function of the system state at time t is given by

$$p_{ij}(t) = P\{(I(t), I_1(t)) = (i, j) \mid (I(0), I_1(0)) = (0, 0)\},$$

$$i = 0, \quad j = 0, \ldots, N - 1;$$

$$i = 1, \quad j = 1, \ldots, N - 1;$$

$$i = 2, \quad j = 1, \ldots, N - 1;$$

$$i = 3, \quad j = 1, \ldots, N;$$
$$i = 4, \quad j = 1, \ldots, N - 1;$$
$$i = 5, \quad j = 2, \ldots, N;$$
$$i = 6, \quad j = N.$$

Consequently, by applying classical probability analysis, we can derive the following Kolmogorov forward equations.

$$(\frac{d}{dt} + \lambda_1)p_{00}(t) = \tau p_{6N}(t); \tag{7.4.2}$$

$$(\frac{d}{dt} + a_1^j \lambda_1)p_{0j}(t) = b_2^{j-1}\mu_2 p_{2j}(t), \quad j = 1, \ldots, N-1; \tag{7.4.3}$$

$$(\frac{d}{dt} + a_2^{j-1}\lambda_2)p_{1j}(t) = b_1^{j-1}\mu_1 p_{3j}(t), \quad j = 1, \ldots, N-1; \tag{7.4.4}$$

$$(\frac{d}{dt} + a_1^j \lambda_1 + b_2^{j-1}\mu_2)p_{2j}(t) = a_2^{j-1}\lambda_2 p_{1j}(t) + b_1^{j-1}\mu_1 p_{4j}(t),$$
$$j = 1, \ldots, N-1; \tag{7.4.5}$$

$$(\frac{d}{dt} + \lambda_2 + \mu_1)p_{31}(t) = \lambda_1 p_{00}(t); \tag{7.4.6}$$

$$(\frac{d}{dt} + a_2^{j-1}\lambda_2 + b_1^{j-1}\mu_1)p_{3j}(t) = a_1^{j-1}\lambda_1 p_{0j-1}(t) + b_2^{j-2}\mu_2 p_{5j}(t),$$
$$j = 2, \ldots, N-1; \tag{7.4.7}$$

$$(\frac{d}{dt} + a_2^{N-1}\lambda_2)p_{3N}(t) = a_1^{N-1}\lambda_1 p_{0N-1}(t) + b_2^{N-2}\mu_2 p_{5N}(t); \tag{7.4.8}$$

$$(\frac{d}{dt} + b_1^{j-1}\mu_1)p_{4j}(t) = a_2^{j-1}\lambda_2 p_{3j}(t), \quad j = 1, \ldots, N-1; \tag{7.4.9}$$

$$(\frac{d}{dt} + b_2^{j-2}\mu_2)p_{5j}(t) = a_1^{j-1}\lambda_1 p_{2j-1}(t), \quad j = 2, \ldots, N; \tag{7.4.10}$$

$$(\frac{d}{dt} + \tau)p_{6N}(t) = a_1^{N-1}\lambda_1 p_{3N}(t). \tag{7.4.11}$$

The initial condition is

$$p_{ij}(0) = \begin{cases} 1 & (i,j) = (0,0), \\ 0 & \text{elsewhere.} \end{cases} \tag{7.4.12}$$

Once again, the probability mass function $p_{ij}(t)$ can be determined by using the Laplace transform. To do this, let the Laplace transform of $p_{ij}(t)$ be

$$p_{ij}^*(s) = \int_0^{\infty} e^{-st}p_{ij}(t)dt.$$

Then, it follows from (7.4.2)-(7.4.12) that

$$(s + \lambda_1)p_{00}^*(s) = 1 + \tau p_{6N}^*(s); \tag{7.4.13}$$

$$(s + a_1^j\lambda_1)p_{0j}^*(s) = b_2^{j-1}\mu_2 p_{2j}^*(s), \quad j = 1,\ldots,N-1; \tag{7.4.14}$$

$$(s + a_2^{j-1}\lambda_2)p_{1j}^*(s) = b_1^{j-1}\mu_1 p_{3j}^*(s), \quad j = 1,\ldots,N-1; \tag{7.4.15}$$

$$(s + a_1^j\lambda_1 + b_2^{j-1}\mu_2)p_{2j}^*(s) = a_2^{j-1}\lambda_2 p_{1j}^*(s) + b_1^{j-1}\mu_1 p_{4j}^*(s),$$
$$j = 1,\ldots,N-1; \tag{7.4.16}$$

$$(s + \lambda_2 + \mu_1)p_{31}^*(s) = \lambda_1 p_{00}^*(s); \tag{7.4.17}$$

$$(s + a_2^{j-1}\lambda_2 + b_1^{j-1}\mu_1)p_{3j}^*(s) = a_1^{j-1}\lambda_1 p_{0j-1}^*(s) + b_2^{j-2}\mu_2 p_{5j}^*(s),$$
$$j = 2,\ldots,N-1; \tag{7.4.18}$$

$$(s + a_2^{N-1}\lambda_2)p_{3N}^*(s) = a_1^{N-1}\lambda_1 p_{0N-1}^*(s) + b_2^{N-2}\mu_2 p_{5N}^*(s); \tag{7.4.19}$$

$$(s + b_1^{j-1}\mu_1)p_{4j}^*(s) = a_2^{j-1}\lambda_2 p_{3j}^*(s), \quad j = 1,\ldots,N-1; \tag{7.4.20}$$

$$(s + b_2^{j-2}\mu_2)p_{5j}^*(s) = a_1^{j-1}\lambda_1 p_{2j-1}^*(s), \quad j = 2,\ldots,N; \tag{7.4.21}$$

$$(s + \tau)p_{6N}^*(s) = a_1^{N-1}\lambda_1 p_{3N}^*(s). \tag{7.4.22}$$

Then, we can solve (7.4.13)-(7.4.22) for $p_{ij}^*(s)$. Afterward, we could determine the probability mass function $p_{ij}(t)$ by inverting. Thus, we are also able to determine the reliability indices. The availability of the system at time t is given by

$$A(t) = P(\text{the system is up at time } t)$$
$$= \sum_{j=0}^{N-1} p_{0j}(t) + \sum_{j=1}^{N-1} p_{1j}(t) + \sum_{j=1}^{N-1} p_{2j}(t) + \sum_{j=1}^{N} p_{3j}(t). \tag{7.4.23}$$

Furthermore, let the ROCOF at time t be $m_f(t)$, then

$$m_f(t) = \sum_{j=1}^{N-1} p_{2j}(t)a_1^j\lambda_1 + \sum_{j=1}^{N} p_{3j}(t)a_2^{j-1}\lambda_2. \tag{7.4.24}$$

To determine the reliability, we may look on the failure states 4 and 5 and replacement state 6 as the absorbing states, and then denote the system state at time t by $\tilde{I}(t)$. Afterward, by introducing the same supplementary variable $I_1(t)$, a new Markov process $\{(\tilde{I}(t), I_1(t)), t \geq 0\}$ is defined. Now, let the probability mass function be

$$q_{ij}(t) = P\{(\tilde{I}(t), I_1(t)) = (i,j) \mid (\tilde{I}(0), I_1(0)) = (0,0)\},$$
$$i = 0, \quad j = 0,\ldots,N-1;$$
$$i = 1, \quad j = 1,\ldots,N-1;$$
$$i = 2, \quad j = 1,\ldots,N-1;$$
$$i = 3, \quad j = 1,\ldots,N.$$

Again, by using classical probability analysis, we have

$$(\frac{d}{dt} + \lambda_1)q_{00}(t) = 0; \tag{7.4.25}$$

$$(\frac{d}{dt} + a_1^j\lambda_1)q_{0j}(t) = b_2^{j-1}\mu_2 q_{2j}(t), \quad j = 1,\ldots,N-1; \tag{7.4.26}$$

$$(\frac{d}{dt} + a_2^{j-1}\lambda_2)q_{1j}(t) = b_1^{j-1}\mu_1 q_{3j}(t), \quad j = 1,\ldots,N-1; \tag{7.4.27}$$

$$(\frac{d}{dt} + a_1^j\lambda_1 + b_2^{j-1}\mu_2)q_{2j}(t) = a_2^{j-1}\lambda_2 q_{1j}(t), \quad j = 1,\ldots,N-1; \tag{7.4.28}$$

$$(\frac{d}{dt} + a_2^{j-1}\lambda_2 + b_1^{j-1}\mu_1)q_{3j}(t) = a_1^{j-1}\lambda_1 q_{0j-1}(t),$$
$$j = 1,\ldots,N-1; \tag{7.4.29}$$

$$(\frac{d}{dt} + a_2^{N-1}\lambda_2)q_{3N}(t) = a_1^{N-1}\lambda_1 q_{0N-1}(t). \tag{7.4.30}$$

The initial condition is

$$q_{ij}(0) = \begin{cases} 1 & (i,j) = (0,0), \\ 0 & \text{elsewhere.} \end{cases} \tag{7.4.31}$$

Taking the Laplace transform with the help of (7.4.31), it follows from (7.4.25)-(7.4.30) that

$$(s + \lambda_1)q_{00}^*(s) = 1; \tag{7.4.32}$$
$$(s + a_1^j\lambda_1)q_{0j}^*(s) = b_2^{j-1}\mu_2 q_{2j}^*(s), \quad j = 1,\ldots,N-1; \tag{7.4.33}$$
$$(s + a_2^{j-1}\lambda_2)q_{1j}^*(s) = b_1^{j-1}\mu_1 q_{3j}^*(s), \quad j = 1,\ldots,N-1; \tag{7.4.34}$$
$$(s + a_1^j\lambda_1 + b_2^{j-1}\mu_2)q_{2j}^*(s) = a_2^{j-1}\lambda_2 q_{1j}^*(s), \quad j = 1,\ldots,N-1; \tag{7.4.35}$$
$$(s + a_2^{j-1}\lambda_2 + b_1^{j-1}\mu_1)q_{3j}^*(s) = a_1^{j-1}\lambda_1 q_{0j-1}^*(s),$$
$$j = 1,\ldots,N-1; \tag{7.4.36}$$
$$(s + a_2^{N-1}\lambda_2)q_{3N}^*(s) = a_1^{N-1}\lambda_1 q_{0N-1}^*(s). \tag{7.4.37}$$

After determining $q_{ij}(t)$, the reliability of the system at time t is given by

$$R(t) = P(\text{the system operating time} > t)$$
$$= \sum_{j=0}^{N-1} q_{0j}(t) + \sum_{j=1}^{N-1} q_{1j}(t) + \sum_{j=1}^{N-1} q_{2j}(t) + \sum_{j=1}^{N} q_{3j}(t). \tag{7.4.38}$$

Then the mean time to the first failure (MTTFF) is determined by

$$MTTFF = \int_0^\infty R(t)dt = \lim_{s \to 0} R^*(s). \tag{7.4.39}$$

As an example, consider a special case, $N = 2$. Then it follows from (7.4.13)-(7.4.22), we have

$$p_{01}^*(s) = \frac{\lambda_1\lambda_2\mu_1\mu_2(2s + \lambda_2 + \mu_1)p_{00}^*(s)}{(s + a_1\lambda_1 + \mu_2)(s + \lambda_2 + \mu_1)(s + a_1\lambda_1)}$$
$$\times \frac{1}{(s + \lambda_2)(s + \mu_1)}, \tag{7.4.40}$$

$$p_{11}^*(s) = \frac{\lambda_1\mu_1 p_{00}^*(s)}{(s + \lambda_2 + \mu_1)(s + \lambda_2)}, \tag{7.4.41}$$

$$p_{21}^*(s) = \frac{\lambda_1\lambda_2\mu_1(2s + \lambda_2 + \mu_1)p_{00}^*(s)}{(s + a_1\lambda_1 + \mu_2)(s + \lambda_2 + \mu_1)(s + \lambda_2)(s + \mu_1)}, \tag{7.4.42}$$

$$p_{31}^*(s) = \frac{\lambda_1 p_{00}^*(s)}{s + \lambda_2 + \mu_1}, \tag{7.4.43}$$

$$p_{32}^*(s) = \frac{a_1\lambda_1^2\lambda_2\mu_1\mu_2(2s + a_1\lambda_1 + \mu_2)(2s + \lambda_2 + \mu_1)p_{00}^*(s)}{(s + a_1\lambda_1)(s + a_1\lambda_1 + \mu_2)(s + a_2\lambda_2)(s + \lambda_2 + \mu_1)}$$
$$\times \frac{1}{(s + \lambda_2)(s + \mu_1)(s + \mu_2)}, \tag{7.4.44}$$

$$p_{41}^*(s) = \frac{\lambda_1\lambda_2 p_{00}^*(s)}{(s + \lambda_2 + \mu_1)(s + \mu_1)}, \tag{7.4.45}$$

$$p_{52}^*(s) = \frac{a_1\lambda_1^2\lambda_2\mu_1(2s + \lambda_2 + \mu_1)p_{00}^*(s)}{(s + a_1\lambda_1 + \mu_2)(s + \lambda_2 + \mu_1)}$$
$$\times \frac{1}{(s + \lambda_2)(s + \mu_1)(s + \mu_2)}, \tag{7.4.46}$$

$$p_{62}^*(s) = -\frac{1}{\tau} + \frac{(s + \lambda_1)p_{00}^*(s)}{\tau}. \tag{7.4.47}$$

However, it follows from (7.4.22) that

$$p_{62}^*(s) = \frac{a_1^2\lambda_1^3\lambda_2\mu_1\mu_2(2s + a_1\lambda_1 + \mu_2)(2s + \lambda_2 + \mu_1)p_{00}^*(s)}{(s + a_1\lambda_1 + \mu_2)(s + a_2\lambda_2 + b_1\mu_1)(s + \lambda_2 + \mu_1)(s + a_1\lambda_1)}$$
$$\times \frac{1}{(s + \lambda_2)(s + \mu_1)(s + \mu_2)(s + \tau)}. \tag{7.4.48}$$

Thus, the combination of (7.4.47) and (7.4.48) will give an expression of $p_{00}^*(s)$. Afterward, by using partial fraction technique and inverting, we can obtain the expression of $p_{00}(t)$.

To determine the reliability, from (7.4.32) and (7.4.36), we have the following results.

$$q_{00}^*(s) = \frac{1}{s + \lambda_1}; \tag{7.4.49}$$

$$q_{31}^*(s) = \frac{A_{31}}{s + \lambda_2 + \mu_1} + \frac{B_{31}}{s + \lambda_1}. \tag{7.4.50}$$

where

$$A_{31} = \frac{\lambda_1}{\lambda_1 - \lambda_2 - \mu_1},$$

$$B_{31} = -\frac{\lambda_1}{\lambda_1 - \lambda_2 - \mu_1}.$$

Then it follows from (7.4.34) that

$$q_{11}^*(s) = \frac{A_{11}}{s + \lambda_2 + \mu_1} + \frac{B_{11}}{s + \lambda_1} + \frac{C_{11}}{s + \lambda_2}, \tag{7.4.51}$$

where

$$A_{11} = -\frac{\lambda_1}{\lambda_1 - \lambda_2 - \mu_1},$$

$$B_{11} = \frac{\lambda_1 \mu_1}{(\lambda_1 - \lambda_2 - \mu_1)(\lambda_1 - \lambda_2)},$$

$$C_{11} = \frac{\lambda_1}{\lambda_1 - \lambda_2}.$$

Thus from (7.4.35), we have

$$q_{21}^*(s) = \frac{A_{21}}{s + a_1\lambda_1 + \mu_2} + \frac{B_{21}}{s + \lambda_2 + \mu_1} + \frac{C_{21}}{s + \lambda_1} + \frac{D_{21}}{s + \lambda_2}, \tag{7.4.52}$$

where

$$A_{21} = -\frac{\lambda_1 \lambda_2 \mu_1}{(a_1\lambda_1 - \lambda_2 - \mu_1 + \mu_2)(a_1\lambda_1 - \lambda_1 + \mu_2)(a_1\lambda_1 - \lambda_2 + \mu_2)},$$

$$B_{21} = -\frac{\lambda_1 \lambda_2}{(a_1\lambda_1 - \lambda_2 - \mu_1 + \mu_2)(\lambda_1 - \lambda_2 - \mu_1)},$$

$$C_{21} = \frac{\lambda_1 \lambda_2 \mu_1}{(a_1\lambda_1 - \lambda_1 + \mu_2)(\lambda_1 - \lambda_2 - \mu_1)(\lambda_1 - \lambda_2)},$$

$$D_{21} = \frac{\lambda_1 \lambda_2}{(a_1\lambda_1 - \lambda_2 + \mu_2)(\lambda_1 - \lambda_2)}.$$

Afterward, (7.4.33) yields

$$q_{01}^*(s) = \frac{A_{01}}{s + a_1\lambda_1 + \mu_2} + \frac{B_{01}}{s + \lambda_2 + \mu_1} + \frac{C_{01}}{s + a_1\lambda_1} + \frac{D_{01}}{s + \lambda_1} + \frac{E_{01}}{s + \lambda_2},$$

$$\tag{7.4.53}$$

where

$$A_{01} = \frac{\lambda_1\lambda_2\mu_1}{(a_1\lambda_1 - \lambda_2 - \mu_1 + \mu_2)(a_1\lambda_1 - \lambda_1 + \mu_2)(a_1\lambda_1 - \lambda_2 + \mu_2)},$$

$$B_{01} = -\frac{\lambda_1\lambda_2\mu_2}{(a_1\lambda_1 - \lambda_2 - \mu_1 + \mu_2)(a_1\lambda_1 - \lambda_2 - \mu_1)(\lambda_1 - \lambda_2 - \mu_1)},$$

$$C_{01} = -\frac{\lambda_2\mu_1}{(a_1 - 1)(a_1\lambda_1 - \lambda_2 - \mu_1)(a_1\lambda_1 - \lambda_2)},$$

$$D_{01} = \frac{\lambda_2\mu_1\mu_2}{(a_1 - 1)(a_1\lambda_1 - \lambda_1 + \mu_2)(\lambda_1 - \lambda_2 - \mu_1)(\lambda_1 - \lambda_2)},$$

$$E_{01} = \frac{\lambda_1\lambda_2\mu_2}{(a_1\lambda_1 - \lambda_2 + \mu_2)(a_1\lambda_1 - \lambda_2)(\lambda_1 - \lambda_2)}.$$

Finally, (7.4.37) gives

$$q_{32}^*(s) = \frac{A_{32}}{s + a_1\lambda_1 + \mu_2} + \frac{B_{32}}{s + \lambda_2 + \mu_1} + \frac{C_{32}}{s + a_1\lambda_1} + \frac{D_{32}}{s + a_2\lambda_2}$$
$$+ \frac{E_{32}}{s + \lambda_1} + \frac{F_{32}}{s + \lambda_2}, \tag{7.4.54}$$

where

$$A_{32}$$
$$= -\frac{a_1\lambda_1^2\lambda_2\mu_1}{(a_1\lambda_1 - \lambda_2 - \mu_1 + \mu_2)(a_1\lambda_1 - a_2\lambda_2 + \mu_2)(a_1\lambda_1 - \lambda_1 + \mu_2)(a_1\lambda_1 - \lambda_2 + \mu_2)},$$

$$B_{32} = -\frac{a_1\lambda_1^2\lambda_2\mu_2}{(a_1\lambda_1 - \lambda_2 - \mu_1 + \mu_2)(a_1\lambda_1 - \lambda_2 - \mu_1)(a_2\lambda_2 - \lambda_2 - \mu_1)(\lambda_1 - \lambda_2 - \mu_1)},$$

$$C_{32} = \frac{a_1\lambda_1\lambda_2\mu_1}{(a_1 - 1)(a_1\lambda_1 - \lambda_2 - \mu_1)(a_1\lambda_1 - a_2\lambda_2)(a_1\lambda_1 - \lambda_2)},$$

$$D_{32} = -\frac{a_1\lambda_1^2\mu_1\mu_2}{(a_2 - 1)(a_1\lambda_1 - a_2\lambda_2 + \mu_2)(a_2\lambda_2 - \lambda_2 - \mu_1)(a_1\lambda_1 - a_2\lambda_2)(a_2\lambda_2 - \lambda_1)},$$

$$E_{32} = \frac{a_1\lambda_1\lambda_2\mu_1\mu_2}{(a_1 - 1)(a_1\lambda_1 - \lambda_1 + \mu_2)(\lambda_1 - \lambda_2 - \mu_1)(a_2\lambda_2 - \lambda_1)(\lambda_1 - \lambda_2)},$$

$$F_{32} = \frac{a_1\lambda_1^2\mu_2}{(a_2 - 1)(a_1\lambda_1 - \lambda_2 + \mu_2)(a_1\lambda_1 - \lambda_2)(\lambda_1 - \lambda_2)}.$$

Consequently, we can determine $q_{ij}(t)$ by inverting. Then the reliability at time t is given by

$$R(t) = \sum_{(i,j)\in W} q_{ij}(t). \tag{7.4.55}$$

It follows from (7.4.49)-(7.4.54) that the reliability at time t is given by

$$
\begin{aligned}
R(t) = {} & (A_{01} + A_{21} + A_{32})e^{-(a_1\lambda_1+\mu_2)t} \\
& + (B_{01} + A_{11} + B_{21} + A_{31} + B_{32})e^{-(\lambda_2+\mu_1)t} \\
& + (C_{01} + C_{32})e^{-a_1\lambda_1 t} + D_{32}e^{-a_2\lambda_2 t} \\
& + (1 + D_{01} + B_{11} + C_{21} + B_{31} + E_{32})e^{-\lambda_1 t} \\
& + (E_{01} + C_{11} + D_{21} + F_{32})e^{-\lambda_2 t}.
\end{aligned}
\tag{7.4.56}
$$

Then the MTTFF is given by

$$
\begin{aligned}
MTTFF = {} & \int_0^\infty R(t)dt \\
= {} & \frac{A_{01} + A_{21} + A_{32}}{a_1\lambda_1 + \mu_2} + \frac{B_{01} + A_{11} + B_{21} + A_{31} + B_{32}}{\lambda_2 + \mu_1} \\
& + \frac{C_{01} + C_{32}}{a_1\lambda_1} + \frac{D_{32}}{a_2\lambda_2} \\
& + \frac{1 + D_{01} + B_{11} + C_{21} + B_{31} + E_{32}}{\lambda_1} \\
& + \frac{E_{01} + C_{11} + D_{21} + F_{32}}{\lambda_2}.
\end{aligned}
\tag{7.4.57}
$$

7.5 A Geometric Process Maintenance Model for a Cold Standby System

A GP maintenance model for a cold standby system is introduced here by making the following assumptions.

Assumption 1. At the beginning, a new two-component system is installed in which component 1 is operating and component 2 is in cold standby. Whenever an operating component fails, it will be repaired by the repairman, and the standby component is operating. A failed component after repair will be operating if the other one fails, and standby otherwise. If one component fails when the other one is under repair it will wait for repair, and the system is down. Assume that the shift of switch is reliable and the change of system state is instantaneous. A replacement policy N is applied by which the system will be replaced by a new and identical one following the Nth failure of component 2.

Assumption 2. For $i = 1, 2$, let X_{i1} be the operating time of component i after the installation or a replacement. In general, for $n > 1$, let

X_{in} be the operating time of component i after its $(n-1)$th repair, then $\{X_{in}, n = 1, 2, \ldots\}$ form a GP with ratio $a_i \geq 1$. Assume that the distribution function of X_{i1} is F_i with mean $\lambda_i > 0$. Let Y_{in} be the repair time of component i after its nth failure. Then $\{Y_{in}, n = 1, 2, \ldots\}$ constitute a GP with ratio $0 < b_i \leq 1$. Assume that the distribution function of Y_{i1} is G_i with mean $\mu_i \geq 0$.

Assumption 3. Let Z be the replacement time of the system. Assume that the distribution of Z is H with mean τ.

Assumption 4. Assume further $\{X_{in}, n = 1, 2, \ldots\}$, $\{Y_{in}, n = 1, 2, \ldots\}$ and Z are all independent.

Assumption 5. The operating reward rate is r, the repair cost rate is c. The replacement cost comprises two parts, one part is the basic replacement cost R, and the other part is proportional to the replacement time Z at rate c_p.

As in Chapter 6, we say a cycle is completed if a replacement is completed. Then a cycle is in fact a time interval between the installation of a system and the first replacement or the time interval between two consecutive replacements.

To apply Theorem 1.3.15 for evaluation of the average cost, we should calculate the expected cost incurred in a cycle and the expected length of a cycle. With the help of Figure 7.4.1, the length of a cycle is given by

$$L = X_{11} + \sum_{j=2}^{N}(X_{1j} \vee Y_{2j-1})$$

$$+ \sum_{j=1}^{N-1}(X_{2j} \vee Y_{1j}) + X_{2N} + Z, \tag{7.5.1}$$

where $X \vee Y = \max\{X, Y\}$. To evaluate $E[L]$, we need the following lemma. The proof is straightforward.

Lemma 7.5.1. Assume that X and Y are two independent random variables with distributions F and G respectively. Then the distribution H of $Z = X \vee Y$ is given by

$$H(z) = F(z)G(z). \tag{7.5.2}$$

Now, if in addition, X and Y are nonnegative, then the expectation

$Z = X \vee Y$ is given by

$$E[Z] = E[X \vee Y] = \int_0^\infty [1 - H(z)]dz = \int_0^\infty [1 - F(z)]dz + \int_0^\infty [1 - G(z)]F(z)dz$$

$$= E[X] + \int_0^\infty [1 - G(z)]F(z)dz \tag{7.5.3}$$

$$= E[Y] + \int_0^\infty [1 - F(z)]G(z)dz. \tag{7.5.4}$$

Recall that for $i = 1, 2$, the distribution of X_{ij} is $F_i(a_i^{j-1}x)$, and the distribution of Y_{ij} is $G_i(b_i^{j-1}y)$. Then, let

$$\alpha_j = \int_0^\infty [1 - G_2(b_2^{j-2}z)]F_1(a_1^{j-1}z)dz,$$

and

$$\beta_j = \int_0^\infty [1 - G_1(b_1^{j-1}z)]F_2(a_2^{j-1}z)dz.$$

Then (7.5.3) yields that

$$E[(X_{1j} \vee Y_{2j-1})] = E[X_{1j}] + \int_0^\infty [1 - G_2(b_2^{j-2}z)]F_1(a_1^{j-1}z)dz$$

$$= \frac{\lambda_1}{a_1^{j-1}} + \alpha_j. \tag{7.5.5}$$

Similarly, we have

$$E[X_{2j} \vee Y_{1j}] = E[X_{2j}] + \int_0^\infty [1 - G_1(b_1^{j-1}z)]F_2(a_2^{j-1}z)dz$$

$$= \frac{\lambda_2}{a_2^{j-1}} + \beta_j. \tag{7.5.6}$$

Consequently, by using Theorem 1.3.15, the average cost $C(N)$ is given by

$$C(N) = \frac{E\{c \sum_1^{N-1} [Y_{1j} + Y_{2j}] - r \sum_1^N [X_{1j} + X_{2j}] + R + c_p Z\}}{E[L]}$$

$$= \frac{c \sum_{j=1}^{N-1} (\frac{\mu_1}{b_1^{j-1}} + \frac{\mu_2}{b_2^{j-1}}) - r \sum_{j=1}^N (\frac{\lambda_1}{a_1^{j-1}} + \frac{\lambda_2}{a_2^{j-1}}) + R + c_p \tau}{\sum_{j=1}^N (\frac{\lambda_1}{a_1^{j-1}} + \frac{\lambda_2}{a_2^{j-1}}) + \sum_{j=2}^N \alpha_j + \sum_{j=1}^{N-1} \beta_j + \tau}. \tag{7.5.7}$$

Thus an optimal policy can be determined from (7.5.7) numerically or analytically.

A special case of our model that two components in a cold standby system are identical and the replacement time is negligible was considered by Zhang et al. (2006), in which a numerical example with exponential distribution is provided, see there for reference.

7.6   Notes and References

It is well known that the maintenance problem and reliability analysis of a system are two important topics in reliability. The applications of GP to these two topics have been developed quickly. A number of literatures were appeared in many international journals. In Chapter 6, the application of GP to the reliability problem of a one-component system is investigated, it concentrates on the study of maintenance problem for the one-component system. In Chapter 7, the application of GP to the reliability problem of a two-component system is considered, the main concern is the reliability analysis of a two-component system, especially the determination of the reliability indices of the two-component system.

In Section 7.2, we analyze a two-component series system, this is based on the work of Lam and Zhang (1996a). On the other hand, Lam (1995) and Lam and Zhang (1996b) studied a GP model for a two-component parallel system, both papers assume that one component after repair is 'as good as new', but the other one after repair is not 'as good as new'. However, in Section 7.3, a new GP model is introduced by assuming that both components after repair are not 'as good as new'. Therefore, it is a more general model than that of Lam (1995) and Lam and Zhang (1996b).

About the same time, Zhang (1995) studied a GP model for a two-component system with a cold standby component but without considering the replacement of the system. The model we considered in Section 7.4 is also new, since it takes into account the replacement of the system. By introducing some supplementary variables, we can obtain a Markov process and then derive the Kolmogorov forward equations. Sometimes, the equations can be solved for an analytic solution by using the Laplace transform. In general, the equations could be solved by using a numerical method such as the Runge-Kutta method. Note that the idea and method developed here could be applied to more general system such as a three-component system.

Recently, Zhang et al. (2006) studied a GP maintenance model for a cold standby system by assuming that two components are identical and the replacement time is negligible. In Section 7.5, we consider a GP maintenance model for a cold standby system by assuming that two components are different and the replacement time is not negligible but a random variable. Therefore, the model studied here is a generalization of the work of Zhang et al. (2006). Besides, Zhang and Wang (2006) had also considered an optimal bivariate policy for a cold standby repairable system. Moreover, a GP model for a series system was studied by Zhang and Wang (2007). Also see Wu et al. (1994) for reference.

Chapter 8

Applications of Geometric Process to Operational Research

8.1 Introduction

In this chapter, we shall study the applications of GP to some topics of operational research.

In most of queueing systems, the service station may experience breakdown. Therefore, it is realistic to consider a queueing system with a repairable service station. As most systems are deteriorating, so are the most service stations. Thus, it is reasonable to study a GP model for a queueing system with a repairable service station. Then, we shall study a GP $M/M/1$ queueing model in Section 8.2.

A warranty can be viewed as a contractual obligation incurred by a manufacturer in connection with the sale of a product. In case the product fails to perform its intended function under normal use, it will be repaired or replaced either free or at a reduced rate. It is also an advertising tool for the manufacturer to promote the product and compete with other manufacturers. Better warranty terms convey the message that the risk of the product is lower and hence is more attractive. Therefore, a product warranty has often cost implications to both manufacturer and consumer. Actually, a warranty problem is a game with two players, the manufacturer and consumer. Then, in Section 6.3, we shall study a GP warranty model by game theory approach.

8.2 A Geometric Process M/M/1 Queueing Model

A GP model for $M/M/1$ queueing system with a repairable service station (or simply the GP $M/M/1$ model) is introduced by making the following

255

assumptions.

Assumption 1. At the beginning, a queueing system with a new service station and one repairman is installed, assume that there is m (≥ 0) customers in the system.

Assumption 2. Suppose that the number of arrivals forms a Poisson process with rate λ. Then the successive interarrival times $\{\nu_n, n = 1, 2, \ldots\}$ are i.i.d. random variables each having an exponential distribution $Exp(\lambda)$ with distribution

$$F(x) = P(\nu_n \leq x) = 1 - e^{-\lambda x} \quad x \geq 0,$$

and 0 otherwise. The customers will be served according to 'first in first out' service discipline. The consecutive service times $\{\chi_n, n = 1, 2, \ldots\}$ are also i.i.d. random variables each having an exponential distribution $Exp(\mu)$ with distribution

$$G(x) = P(\chi_n \leq x) = 1 - e^{-\mu x} \quad x \geq 0,$$

and 0 otherwise. Assume that $\mu > \lambda$.

Assumption 3. Whenever the service station fails, it is repaired immediately by the repairman. During the repair time, the system closes so that a new arrival can not join the system. In other words, no more customer will arrive, but the customers waiting in the system will remain in the system, while the service to the customer being served will be stopped. After completion the repair, the service station restarts its service to the customer whose service was stopped due to failure of the service station with the same exponentially distributed service time, and the system reopens so that a new arrival can join the system. If there is no customer in the system, the service station will remain in an operating state.

Assumption 4. Let $X_n, n = 1, 2, \ldots$, be the operating time of the service station after the $(n-1)$th repair, and let $Y_n, n = 1, 2, \ldots$, be the repair time of the service station after the nth failure. Then $\{X_n, n = 1, 2, \ldots\}$ form a GP with ratio $a \geq 1$. Assume that X_1 has an exponential distribution $Exp(\alpha)$ with $\alpha > 0$ and distribution

$$X_1(x) = P(X_1 \leq x) = 1 - e^{-\alpha x} \quad x \geq 0,$$

and 0 otherwise. On the other hand, $\{Y_n, n = 1, 2, \ldots\}$ follow a GP with ratio $0 < b \leq 1$, and Y_1 has an exponential distribution $Exp(\beta)$ with $\beta > 0$ and distribution

$$Y_1(x) = P(Y_1 \leq x) = 1 - e^{-\beta x} \quad x \geq 0,$$

and 0 otherwise.

Assumption 5. The sequences $\{\nu_n, n = 1, 2, \ldots\}, \{\chi_n, n = 1, 2, \ldots\}, \{X_n, n = 1, 2, \ldots\}$, and $\{Y_n, n = 1, 2, \ldots\}$ are independent sequences of independent random variables.

Remarks

(1) Assumption $\mu > \lambda$ makes a $M/M/1$ queueing system a real queue, otherwise the length of the queueing system might tend to infinity.

(2) Assumption 3 is reasonable. For example, consider a computer network system in which several workstations are connected together to form a local area network as a 'system', and a printer is the 'service station', a computer officer is the 'repairman'. Whenever a workstation needs to submit a printing job that will queue up as a 'customer' in the system. If the printer breaks down due to cut supply of power or short of ink or paper, it will be repaired by the computer officer, while the job being printed for an earlier part of printing time will be stopped, and the printer will close such that any new printing job will be rejected. In other words, no more 'customer' will arrive. However, the printing jobs already submitted will remain in the 'system'. After completion of the repair, the printer will resume the job, and the 'system' will reopen. Because the exponential distribution is memoryless, the later part of the printing time (the residual service time) for the job stopped when the printer breaks down will have the same distribution $Exp(\mu)$.

(3) Assumption 4 just means that the service station is deteriorating. This is a GP model considered in Section 2 of Chapter 6.

For the GP $M/M/1$ model, the system state $(I(t), J(t))$ at time t is defined in the following way: $I(t) = i$, if at time t, the number of customers in the system is $i, i = 0, 1, \ldots$; $J(t) = 0$, if at time t the service station is in operating state or in an up state, and $J(t) = 1$, if at time t the service station breaks down or in a down state. Therefore, the state space is $\Omega = \{(i, j), i = 0, 1, \ldots; j = 0, 1\}$, the set of up states is $U = \{(i, 0), i = 0, 1, \ldots\}$, and the set of down states is $D = \{(i, 1), i = 1, 2, \cdots\}$. However, stochastic process $\{(I(t), J(t)), t \geq 0\}$ is not a Markov process. Then, we shall introduce a supplementary variable $K(t)$ and define

$$K(t) = k, \quad k = 0, 1, \ldots$$

if the number of failures of the service station by time t is k. Then stochastic process $\{(I(t), J(t), K(t)), t \geq 0\}$ will be a three-dimensional continuous-time Markov process.

Since the system state at time 0 is $(I(0), J(0), K(0)) = (m, 0, 0)$, the probability mass function of the Markov process at time t is the transition probability from $(I(0), J(0), K(0)) = (m, 0, 0)$ to $(I(t), J(t), K(t)) = (i, j, k)$, it is given by

$$p_{ijk}(t, m) = P\{(I(t), J(t), K(t)) = (i, j, k) \mid (I(0), J(0), K(0)) = (m, 0, 0)\},$$

with

$$i = 0, 1, \cdots; \quad j = 0; \quad k = 0, 1, \cdots$$

or

$$i = 1, 2, \cdots; \quad j = 1; \quad k = 1, 2, \cdots$$

By applying a classical probability analysis, it is straightforward to derive the following Kolmogorov forward equations:

$$(\frac{d}{dt} + \lambda) \ p_{00k}(t, m) = \mu p_{10k}(t, m) \quad k = 0, 1, \ldots \tag{8.2.1}$$

$$(\frac{d}{dt} + \lambda + \mu + \alpha) \ p_{i00}(t, m) = \lambda p_{i-100}(t, m) + \mu p_{i+1\,00}(t, m),$$
$$i = 1, 2, \ldots; \tag{8.2.2}$$

$$(\frac{d}{dt} + \lambda + \mu + a^k \alpha) \ p_{i0k}(t, m) = \lambda p_{i-10k}(t, m) + \mu p_{i+10k}(t, m)$$
$$+ b^{k-1} \beta p_{i1k}(t, m), \quad i = 1, 2, \ldots; k = 1, 2, \ldots; \tag{8.2.3}$$

$$(\frac{d}{dt} + b^{k-1} \beta) \ p_{i1k}(t, m) = a^{k-1} \alpha p_{i0k-1}(t, m),$$
$$i = 1, 2, \ldots; k = 1, 2, \ldots \ . \tag{8.2.4}$$

The initial condition is

$$p_{ijk}(0, m) = \begin{cases} 1 & (i, j, k) = (m, 0, 0), \\ 0 & \text{elsewhere.} \end{cases} \tag{8.2.5}$$

Let δ_{ij} be the Kronecker δ defined by

$$\delta_{ij} = \begin{cases} 1 & j = i, \\ 0 & j \neq i. \end{cases}$$

Then define the Laplace transform of $p_{ijk}(t, m)$ by

$$p^*_{ijk}(s, m) = \int_0^\infty e^{-st} p_{ijk}(t, m) dt. \tag{8.2.6}$$

Now, taking the Laplace transform on the both sides of (8.2.1)-(8.2.4) and using initial condition (8.2.5) give

$$(s+\lambda)\ p^*_{00k}(s,m) = \mu p^*_{10k}(s,m) + \delta_{0k}\delta_{m0},$$
$$k = 0,1,\ldots; \tag{8.2.7}$$

$$(s+\lambda+\mu+\alpha)\ p^*_{i00}(s,m) = \lambda p^*_{i-100}(s,m) + \mu p^*_{i+100}(s,m) + \delta_{mi},$$
$$i = 1,2,\ldots; \tag{8.2.8}$$

$$(s+\lambda+\mu+a^k\alpha)\ p^*_{i0k}(s,m) = \lambda p^*_{i-10k}(s,m) + \mu p^*_{i+10k}(s,m),$$
$$+b^{k-1}\beta p^*_{i1k}(s,m),\quad i = 1,2,\ldots; k = 1,2,\ldots \tag{8.2.9}$$

$$(s+b^{k-1}\beta)\ p^*_{i1k}(s,m) = a^{k-1}\alpha p^*_{i0k-1}(s,m),$$
$$i = 1,2,\ldots; k = 1,2,\ldots \tag{8.2.10}$$

At first, we shall determine the distribution of busy period. A busy period is a time interval that starts when the number of customers in the system increases from 0 to greater than 0, and ends at the time whenever the number of customers in the system reduces to 0. In classical model for $M/M/1$ queueing system (or simply the classical $M/M/1$ model) in which the service station is not subject to failure, assume that at the beginning there is no customer or just 1 customer in the system, let $\tilde{B}_i$ be the ith busy period. Then it is well known that $\{\tilde{B}_1, \tilde{B}_2, \cdots\}$ are i.i.d. random variables each having a common distribution $\tilde{B}(x) = P(\tilde{B}_2 \leq x)$ with the Laplace-Stieltjes transform given by

$$\tilde{B}^*(s) = \int_0^\infty e^{-st}d\tilde{B}(t) = \frac{s+\lambda+\mu - \sqrt{(s+\lambda+\mu)^2 - 4\lambda\mu}}{2\lambda}, \tag{8.2.11}$$

and

$$E(\tilde{B}) = -\frac{d\tilde{B}^*(s)}{ds}\Big|_{s=0} = \frac{1}{\mu-\lambda} > 0 \tag{8.2.12}$$

(see e. g. Kleinrock (1975) for reference). Thereafter, we shall use $\tilde{B}$ to denote a busy period in classical M/M/1 model with distribution $\tilde{B}(x)$, given that at the beginning the number of customer in the system is 0 or 1.

In general, if at the beginning there are m customers in the system, then $\{\tilde{B}_1, \tilde{B}_2, \cdots\}$ are still independent but the distribution of $\tilde{B}_1$ is given by

$$\tilde{B}^{(m)}(x) = \tilde{B} * \tilde{B} * \cdots * \tilde{B}(x), \tag{8.2.13}$$

the m-fold convolution of $\tilde{B}(x)$ with itself, while the distribution of $\tilde{B}_i, i = 2,3,\ldots$ is still given by $\tilde{B}(x)$.

In the GP $M/M/1$ model, a busy period will include the total service time plus the total repair time of the service station. Let B_i be the ith busy period in the GP $M/M/1$ model. Then, each B_i will consist of two parts, the first part is the total service time of the service station corresponding to busy period $\tilde{B}_i$ in the classical $M/M/1$ model, the second part is the total repair time of the service station. Consequently, by summing up the total repair time of the service station, the sum of the first $n+1$ busy periods given that the total number of repairs is k is given by

$$\sum_{i=1}^{n+1} B_i = \sum_{i=1}^{n+1} \tilde{B}_i + \sum_{i=1}^{k} Y_i. \tag{8.2.14}$$

From (8.2.13), the distribution of $\sum_{i=1}^{n+1} \tilde{B}_i$ is given by $\tilde{B}^{(m+n)}(x)$, the $(m+n)$-fold convolution of $\tilde{B}(x)$ with itself.

Now, let the convolution of distributions $X_{n+1}(x), \ldots, X_{n+k}(x)$ be

$$X_{n+1}^{(k)}(x) = X_{n+1} * X_{n+2} * \cdots * X_{n+k}(x),$$

and the convolution of distributions $Y_{n+1}(x), \ldots, Y_{n+k}(x)$ be

$$Y_{n+1}^{(k)}(x) = Y_{n+1} * Y_{n+2} * \cdots * Y_{n+k}(x).$$

Moreover, define

$$X_{n+1}^{(0)}(x) \equiv 1, \quad Y_{n+1}^{(0)}(x) \equiv 1 \quad \text{for} \quad x \geq 0. \tag{8.2.15}$$

Afterward, we have the following theorem.

Theorem 8.2.1. The distribution of the first busy period B_1 is given by

$$B_1(x) = \sum_{k=0}^{\infty} \int_0^x Y_1^{(k)}(x-u)[X_1^{(k)}(u) - X_1^{(k+1)}(u)]d\tilde{B}^{(m\vee 1)}(u), \tag{8.2.16}$$

where $m \vee 1 = \max\{m, 1\}$.

Proof.

First assume that $m > 0$. It follows from (8.2.14) that

$$B_1(x) = P\{B_1 \leq x\}$$

$$= \sum_{k=0}^{\infty} P\left\{\tilde{B}_1 + \sum_{j=1}^{k} Y_j \leq x, \sum_{j=1}^{k} X_j < \tilde{B}_1 \leq \sum_{j=1}^{k+1} X_j\right\} \tag{8.2.17}$$

$$= \sum_{k=0}^{\infty} \int_0^x P\left\{\sum_{j=1}^{k} Y_j \leq x-u, \sum_{j=1}^{k} X_j < u \leq \sum_{j=1}^{k+1} X_j\right\} d\tilde{B}^{(m)}(u) \tag{8.2.18}$$

$$= \sum_{k=0}^{\infty} \int_0^x Y_1^{(k)}(x-u)[X_1^{(k)}(u) - X_1^{(k+1)}(u)]d\tilde{B}^{(m)}(u). \tag{8.2.19}$$

Here (8.2.17) follows from (8.2.14). Because at the beginning, there are m customers in the system, then (8.2.13) yields (8.2.18). Furthermore, (8.2.19) holds since $\{X_n, n = 1, 2, \ldots\}$ and $\{Y_n, n = 1, 2, \ldots\}$ are independent.

Second, if $m = 0$, the first busy period B_1 will start when the number of customer in the system increases to 1. Then the distribution of B_1 is the same as the case $m = 1$, and (8.2.16) follows. This completes the proof of Theorem 8.2.1.

The distribution of other busy period depends on the system state at the beginning of the busy period. Therefore, in general it is different from the distribution of B_1. Assume that a busy period B_t starts at time t with state $(I(t), J(t), K(t)) = (1, 0, k)$. Then the conditional distribution $B_t(x)$ will be given by the following result.

Theorem 8.2.2. Given that a busy period B_t starts with $(I(t), J(t), K(t)) = (1, 0, k)$, the conditional distribution of B_t is given by

$$B_t(x) = \sum_{n=k}^{\infty} \int_0^x Y_{k+1}^{(n-k)}(x - u)[X_{k+1}^{(n-k)}(u) - X_{k+1}^{(n-k+1)}(u)]d\tilde{B}(u).$$

$$(8.2.20)$$

Proof.

It follows from (8.2.14) that

$$B_t(x) = P\{B_t \le x | (I(t), J(t), K(t)) = (1, 0, k)\}$$

$$= \sum_{n=k}^{\infty} P\left\{\tilde{B} + \sum_{j=k+1}^{n} Y_j \le x, X_{k+1}^L + \sum_{j=k+2}^{n} X_j < \tilde{B} \le X_{k+1}^L + \sum_{j=k+2}^{n+1} X_j\right\}$$

$$= \sum_{n=k}^{\infty} \int_0^x P\left\{\sum_{j=k+1}^{n} Y_j \le x - u, \sum_{j=k+1}^{n} X_j < u \le \sum_{j=k+1}^{n+1} X_j\right\} d\tilde{B}(u) \quad (8.2.21)$$

$$= \sum_{n=k}^{\infty} \int_0^x Y_{k+1}^{(n-k)}(x - u)[X_{k+1}^{(n-k)}(u) - X_{k+1}^{(n-k+1)}(u)]d\tilde{B}(u). \quad (8.2.22)$$

Because the service station after the kth repair, an earlier part of operating time X_{k+1} was spent in the last busy period, but the latter part X_{k+1}^L of operating time X_{k+1} is used for the service in the present busy period. Then due to the memoryless property of exponential distribution, X_{k+1}^L will have the same distribution as X_{k+1} has. Thus (8.2.21) follows. Once again, (8.2.22) is true since $\{X_n, n = 1, 2, \ldots\}$ and $\{Y_n, n = 1, 2, \ldots\}$ are independent. This completes the proof of Theorem 8.2.2.

The following theorem will give the probability that the service station is idle.

Theorem 8.2.3. The probability that the service station is idle at time t is given by

$$P(\text{the service station is idle at time } t) = \sum_{k=0}^{\infty} p_{00k}(t, m), \quad (8.2.23)$$

where

$$p_{000}(t, m) = \sum_{n=0}^{\infty} \int_0^t [F^{(n)}(t - u) - F^{(n+1)}(t - u)]e^{-\alpha u} d\tilde{B}^{(m+n)}(u),$$

$$(8.2.24)$$

and

$$p_{00k}(t, m) = \sum_{n=0}^{\infty} \int_0^t [F^{(n)} * Y_1^{(k)}(t - u) - F^{(n+1)} * Y_1^{(k)}(t - u)]$$

$$\times [X_1^{(k)}(u) - X_1^{(k+1)}(u)]d\tilde{B}^{(m+n)}(u), \quad k = 1, 2, \ldots \quad (8.2.25)$$

with

$$F^{(n)}(x) = F * F * \cdots * F(x),$$
$$\tilde{B}^{(n)}(x) = \tilde{B} * \tilde{B} * \cdots * \tilde{B}(x),$$

respectively being the n-fold convolutions of $F(x)$ and $\tilde{B}(x)$ with themselves, and $F^{(n)} * Y_1^{(k)}(x)$ is the convolution of $F^{(n)}(x)$ and $Y_1^{(k)}(x)$.

Proof.

Result (8.2.23) is trivial. To show (8.2.24) and (8.2.25), assume first that $m > 0$ and $k > 0$. Since the idle period V_i and busy period B_i in the queueing system will occur alternatively, we have

$$p_{00k}(t, m) = P\{(I(t), J(t), K(t)) = (0, 0, k)|(I(0), J(0), K(0)) = (m, 0, 0)\}$$

$$= \sum_{n=0}^{\infty} P\left\{ B_1 + \sum_{i=1}^{n}(V_i + B_{i+1}) < t \leq B_1 + \sum_{i=1}^{n}(V_i + B_{i+1}) + V_{n+1}, \right.$$

total number of repairs on the service station by time t is $k\}$

$$= \sum_{n=0}^{\infty} P\left\{ \sum_{i=1}^{n}V_i + \sum_{i=1}^{n+1}\tilde{B}_i + \sum_{i=1}^{k}Y_i < t \leq \sum_{i=1}^{n+1}V_i + \sum_{i=1}^{n+1}\tilde{B}_i + \sum_{i=1}^{k}Y_i, \right.$$

$$\left. \sum_{i=1}^{k}X_i < \sum_{i=1}^{n+1}\tilde{B}_i \leq \sum_{i=1}^{k+1}X_i \right\}. \quad (8.2.26)$$

Therefore,

$$p_{00k}(t,m) = \sum_{n=0}^{\infty} \int_0^t P\left\{ \sum_{i=1}^{n} V_i + \sum_{i=1}^{k} Y_i < t - u \le \sum_{i=1}^{n+1} V_i + \sum_{i=1}^{k} Y_i, \right.$$

$$\left. \sum_{i=1}^{k} X_i < u \le \sum_{i=1}^{k+1} X_i \right\} d\tilde{B}^{(m+n)}(u) \tag{8.2.27}$$

$$= \sum_{n=0}^{\infty} \int_0^t [F^{(n)} * Y_1^{(k)}(t-u) - F^{(n+1)} * Y_1^{(k)}(t-u)]$$

$$\times [X_1^{(k)}(u) - X_1^{(k+1)}(u)] d\tilde{B}^{(m+n)}(u). \tag{8.2.28}$$

Here (8.2.26) follows from (8.2.14), while (8.2.27) is due to the fact that the distribution of $\sum_{i=1}^{n+1} \tilde{B}_i$ is given by $\tilde{B}^{(m+n)}(x)$. On the other hand, as $\{V_i, i = 1, 2, \ldots\}$ depends on $\{\nu_n, n = 1, 2, \cdots\}$ only, then Assumption 5 implies that $\{V_i, i = 1, 2, \ldots\}$ and $\{Y_i, i = 1, 2, \ldots\}$ are independent, hence (8.2.28) follows.

Now, assume that $m > 0$ but $k = 0$. A similar argument shows that

$$p_{000}(t,m) = P\{(I(t), J(t), K(t)) = (0,0,0) | (I(0), J(0), K(0)) = (m,0,0)\}$$

$$= \sum_{n=0}^{\infty} P\left\{ \sum_{i=1}^{n} V_i + \sum_{i=1}^{n+1} \tilde{B}_i < t \le \sum_{i=1}^{n+1} V_i + \sum_{i=1}^{n+1} \tilde{B}_i, \sum_{i=1}^{n+1} \tilde{B}_i \le X_1 \right\}$$

$$= \sum_{n=0}^{\infty} \int_0^t P\left\{ \sum_{i=1}^{n} V_i < t - u \le \sum_{i=1}^{n+1} V_i, u \le X_1 \right\} d\tilde{B}^{(m+n)}(u)$$

$$= \sum_{n=0}^{\infty} \int_0^t [F^{(n)}(t-u) - F^{(n+1)}(t-u)](1 - X_1(u)) d\tilde{B}^{(m+n)}(u)$$

$$= \sum_{n=0}^{\infty} \int_0^t [F^{(n)}(t-u) - F^{(n+1)}(t-u)] e^{-\alpha u} d\tilde{B}^{(m+n)}(u).$$

If $m = 0$, then we can think $\tilde{B}_1 = 0$, the proof is similar. This completes the proof of Theorem 8.2.3.

Especially, from (8.2.24) we have

$$p_{000}(t,0) = e^{-\lambda t} + \sum_{n=1}^{\infty} \int_0^t [F^{(n)}(t-u) - F^{(n+1)}(t-u)] e^{-\alpha u} d\tilde{B}^{(n)}(u). \tag{8.2.29}$$

Thereafter, we shall determine the Laplace transform of $p_{ijk}(t,m)$ recursively. As a result, the probability mass function $p_{ijk}(t,m)$ may be obtained by inversion. For this purpose, some well known results are reviewed here. Assume that $X_1, \ldots, X_n$ are independent and X_i has an exponential distribution $Exp(\lambda_i)$. Assume that $\lambda_i, i = 1, \ldots, n$ are different. Then the density function of $\sum_{i=1}^{n} X_i$ is given by

$$x_1^{(n)}(x) = \begin{cases} (-1)^{n-1}\lambda_1\lambda_2\cdots\lambda_n \sum_{i=1}^{n} \dfrac{e^{-\lambda_i x}}{\prod\limits_{\substack{j=1 \\ j\neq i}}^{n}(\lambda_i-\lambda_j)} & x > 0, \\ 0 & \text{elsewhere,} \end{cases} \tag{8.2.30}$$

(see Chiang (1980) for reference). In particular, if $\lambda_i = a^{i-1}\alpha$ with $a > 1$, then (8.2.30) gives

$$x_1^{(n)}(x) = \begin{cases} (-1)^{n-1}a^{\frac{n(n-1)}{2}}\alpha \sum_{i=1}^{n} \dfrac{e^{-a^{i-1}\alpha x}}{\prod\limits_{\substack{j=1 \\ j\neq i}}^{n}(a^{i-1}-a^{j-1})} & x > 0, \\ 0 & \text{elsewhere.} \end{cases} \tag{8.2.31}$$

Consequently, the distribution function of $\sum_{i=1}^{n} X_i$ is given by

$$X_1^{(n)}(x) = \begin{cases} 1 - \sum\limits_{i=1}^{n}\left(\prod\limits_{\substack{j=1 \\ j\neq i}}^{n} \dfrac{a^{j-1}}{a^{j-1}-a^{i-1}}\right)e^{-a^{i-1}\alpha x} & x > 0, \\ 0 & \text{elsewhere.} \end{cases} \tag{8.2.32}$$

However, if $a = 1$, $\sum_{i=1}^{n} X_i$ will have a gamma distribution $\Gamma(n, \alpha)$ with, respectively, the following density and distribution functions.

$$x_1^{(n)}(x) = \begin{cases} \frac{\alpha^n}{(n-1)!}x^{n-1}e^{-\alpha x} & x > 0, \\ 0 & \text{elsewhere.} \end{cases} \tag{8.2.33}$$

$$X_1^{(n)}(x) = \begin{cases} \sum\limits_{i=n}^{\infty} \frac{(\alpha x)^i}{i!}e^{-\alpha x} & x > 0, \\ 0 & \text{elsewhere.} \end{cases} \tag{8.2.34}$$

By using Theorem 8.2.3, an explicit expression for $p_{00k}^{*}(s, m)$ can be obtained from the following theorem.

Theorem 8.2.4.

$$p_{00k}^{*}(s, m) =$$

$$\begin{cases} \prod\limits_{j=1}^{k} \dfrac{b^{j-1}\beta}{s+b^{j-1}\beta} \sum\limits_{i=1}^{k+1} a^{i-k-1}\left(\prod\limits_{\substack{r=1 \\ r\neq i}}^{k+1} \dfrac{a^{r-1}}{a^{r-1}-a^{i-1}}\right)\dfrac{[\tilde{B}^{*}(s+a^{i-1}\alpha)]^m}{s+\lambda-\lambda\tilde{B}^{*}(s+a^{i-1}\alpha)} & \text{for } a > 1, \\[4ex] (-1)^k \dfrac{\alpha^k}{k!}\left\{\prod\limits_{j=1}^{k} \dfrac{b^{j-1}\beta}{s+b^{j-1}\beta}\right\}\dfrac{d^k}{d\alpha^k}\left\{\dfrac{[\tilde{B}^{*}(s+\alpha)]^m}{s+\lambda-\lambda\tilde{B}^{*}(s+\alpha)}\right\} & \text{for } a = 1. \end{cases} \tag{8.2.35}$$

Proof.

Assume first $a > 1$, then from (8.2.25) and (8.2.32), we have

$$
\begin{aligned}
p_{00k}^*(s, m) &= \int_0^\infty e^{-st} p_{00k}(t, m) dt \\
&= \int_0^\infty e^{-st} \{ \sum_{n=0}^\infty \int_0^t [F^{(n)} * Y_1^{(k)}(t-u) - F^{(n+1)} * Y_1^{(k)}(t-u)] \\
&\quad \times [X_1^{(k)}(u) - X_1^{(k+1)}(u)] d\tilde{B}^{(m+n)}(u) \} dt \\
&= \sum_{n=0}^\infty \int_0^\infty \{ \int_0^\infty e^{-sv} [F^{(n)} * Y_1^{(k)}(v) - F^{(n+1)} * Y_1^{(k)}(v)] dv \} \\
&\quad \times e^{-su} [X_1^{(k)}(u) - X_1^{(k+1)}(u)] d\tilde{B}^{(m+n)}(u).
\end{aligned}
$$

Then

$$
\begin{aligned}
&p_{00k}^*(s, m) \\
&= \sum_{n=0}^\infty \frac{1}{s} [(\frac{\lambda}{s+\lambda})^n - (\frac{\lambda}{s+\lambda})^{n+1}] (\prod_{j=1}^k \frac{b^{j-1}\beta}{s+b^{j-1}\beta}) \int_0^\infty e^{-su} \\
&\quad \times \{ \sum_{i=1}^{k+1} \prod_{\substack{r=1 \\ r \neq i}}^{k+1} \frac{a^{r-1} e^{-a^{i-1}\alpha u}}{a^{r-1} - a^{i-1}} - \sum_{i=1}^k \prod_{\substack{r=1 \\ r \neq i}}^k \frac{a^{r-1} e^{-a^{i-1}\alpha u}}{a^{r-1} - a^{i-1}} \} d\tilde{B}^{(m+n)}(u) \\
&= \sum_{n=0}^\infty \frac{\lambda^n}{(s+\lambda)^{n+1}} (\prod_{j=1}^k \frac{b^{j-1}\beta}{s+b^{j-1}\beta}) \sum_{i=1}^{k+1} a^{i-k-1} (\prod_{\substack{r=1 \\ r \neq i}}^{k+1} \frac{a^{r-1}}{a^{r-1} - a^{i-1}}) \\
&\quad \times \int_0^\infty e^{-(s+a^{i-1}\alpha)u} d\tilde{B}^{(m+n)}(u) \\
&= \prod_{j=1}^k \frac{b^{j-1}\beta}{s+b^{j-1}\beta} \sum_{i=1}^{k+1} a^{i-k-1} (\prod_{\substack{r=1 \\ r \neq i}}^{k+1} \frac{a^{r-1}}{a^{r-1} - a^{i-1}}) \\
&\quad \times \sum_{n=0}^\infty \frac{\lambda^n}{(s+\lambda)^{n+1}} [\tilde{B}^*(s + a^{i-1}\alpha)]^{m+n} \\
&= \prod_{j=1}^k \frac{b^{j-1}\beta}{s+b^{j-1}\beta} \sum_{i=1}^{k+1} a^{i-k-1} (\prod_{\substack{r=1 \\ r \neq i}}^{k+1} \frac{a^{r-1}}{a^{r-1} - a^{i-1}}) \frac{[\tilde{B}^*(s + a^{i-1}\alpha)]^m}{s + \lambda - \lambda \tilde{B}^*(s + a^{i-1}\alpha)}.
\end{aligned}
$$

For $a = 1$, by a similar approach, it follows from (8.2.25) and (8.2.34) that

$$p_{00k}^*(s, m) = \int_0^\infty e^{-st} p_{00k}(t, m) dt$$

$$= \sum_{n=0}^\infty \int_0^\infty \{ \int_0^\infty e^{-sv} [F^{(n)} * Y_1^{(k)}(v) - F^{(n+1)} * Y_1^{(k)}(v)] dv \}$$

$$\times e^{-su} [X_1^{(k)}(u) - X_1^{(k+1)}(u)] d\tilde{B}^{(m+n)}(u)$$

Consequently, we have

$$p_{00k}^*(s, m) = \sum_{n=0}^\infty \frac{1}{s} [(\frac{\lambda}{s+\lambda})^n - (\frac{\lambda}{s+\lambda})^{n+1}] (\prod_{j=1}^k \frac{b^{j-1}\beta}{s + b^{j-1}\beta})$$

$$\int_0^\infty e^{-su} \frac{(\alpha u)^k}{k!} e^{-\alpha u} d\tilde{B}^{(m+n)}(u)$$

$$= \prod_{j=1}^k \frac{b^{j-1}\beta}{s + b^{j-1}\beta} \sum_{n=0}^\infty \frac{\lambda^n}{(s+\lambda)^{n+1}} (-1)^k \frac{\alpha^k}{k!} \frac{d^k [\tilde{B}^*(s+\alpha)]^{m+n}}{d\alpha^k}$$

$$= (-1)^k \frac{\alpha^k}{k!} \{ \prod_{j=1}^k \frac{b^{j-1}\beta}{s + b^{j-1}\beta} \} \frac{d^k}{d\alpha^k} \{ \frac{[\tilde{B}^*(s+\alpha)]^m}{s + \lambda - \lambda\tilde{B}^*(s+\alpha)} \}.$$

This completes the proof of Theorem 8.2.4.

In particular, for $k = 0$ and $a \geq 1$, from (8.2.35) we obtain

$$p_{000}^*(s, m) = \frac{[\tilde{B}^*(s+\alpha)]^m}{s + \lambda - \lambda\tilde{B}^*(s+\alpha)}. \tag{8.2.36}$$

By using Theorem 8.2.4, it is straightforward to derive the Laplace transform $p_{100}^*(s, m)$ of $p_{100}(t, m)$. In fact, it follows from (8.2.7) that

$$p_{100}^*(s, m) = \frac{s+\lambda}{\mu} p_{000}^*(s, m) - \frac{\delta_{m0}}{\mu}$$

$$= \frac{(s+\lambda)[\tilde{B}^*(s+\alpha)]^m}{\mu[s + \lambda - \lambda\tilde{B}^*(s+\alpha)]} - \frac{\delta_{m0}}{\mu}. \tag{8.2.37}$$

Furthermore, we have the following theorem.

Theorem 8.2.5.

$$p_{i00}^*(s, 0) = \frac{\lambda(p^i - q^i)\tilde{B}^*(s+\alpha) + \mu(pq^i - p^i q)}{\mu(p-q)[s + \lambda - \lambda\tilde{B}^*(s+\alpha)]}, \quad i = 0, 1, \ldots$$

$$\tag{8.2.38}$$

$$p_{i00}^*(s,m) = \begin{cases} \dfrac{[(p^i-q^i)(s+\lambda)+\mu(pq^i-p^iq)][\tilde{B}^*(s+\alpha)]^m}{\mu(p-q)[s+\lambda-\lambda\tilde{B}^*(s+\alpha)]} - \dfrac{p^{i-m}-q^{i-m}}{\mu(p-q)}, \\ \quad i=2,3,\ldots;\ m=1,2,\ldots,i-1, \\ \dfrac{[(p^i-q^i)(s+\lambda)+\mu(pq^i-p^iq)][\tilde{B}^*(s+\alpha)]^m}{\mu(p-q)[s+\lambda-\lambda\tilde{B}^*(s+\alpha)]}, \\ \quad i=0,1,\ldots;\ m=i,i+1,\ldots. \end{cases}$$

(8.2.39)

where p and q are two roots of quadratic equation

$$\mu t^2 - (s+\lambda+\mu+\alpha)t + \lambda = 0. \tag{8.2.40}$$

Proof.

Assume that $m=0$, then from (8.2.8) we have

$$p_{i+100}^*(s,0) = \frac{s+\lambda+\mu+\alpha}{\mu}p_{i00}^*(s,0) - \frac{\lambda}{\mu}p_{i-100}^*(s,0),$$

$$i=1,2,\ldots \tag{8.2.41}$$

Because the discriminant of equation (8.2.40) is positive, two roots p and q are distinct and real. Then (8.2.41) becomes

$$p_{i+100}^*(s,0) - pp_{i00}^*(s,0) = q[p_{i00}^*(s,0) - pp_{i-100}^*(s,0)], \tag{8.2.42}$$

or

$$p_{i+100}^*(s,0) - qp_{i00}^*(s,0) = p[p_{i00}^*(s,0) - qp_{i-100}^*(s,0)]. \tag{8.2.43}$$

By iteration, it is straightforward that

$$p_{i00}^*(s,0) - pp_{i-100}^*(s,0) = q^{i-1}[p_{100}^*(s,0) - pp_{000}^*(s,0)],$$

and

$$p_{i00}^*(s,0) - qp_{i-100}^*(s,0) = p^{i-1}[p_{100}^*(s,0) - qp_{000}^*(s,0)].$$

Consequently,

$$\begin{aligned} p_{i00}^*(s,0) &= \frac{p^i-q^i}{p-q}p_{100}^*(s,0) + \frac{pq^i-p^iq}{p-q}p_{000}^*(s,0) \\ &= \frac{\lambda(p^i-q^i)\tilde{B}^*(s+\alpha)+\mu(pq^i-p^iq)}{\mu(p-q)(s+\lambda-\lambda\tilde{B}^*(s+\alpha))}, \quad i=0,1,\ldots \end{aligned}$$

Thus (8.2.38) follows. To prove (8.2.39), consider the case $m<i$ first. By a similar argument to (8.2.42), from (8.2.8) we have

$$\begin{aligned} & p_{i+100}^*(s,m) - pp_{i00}^*(s,m) \\ &= q^{i-m}(p_{m+100}^*(s,m) - pp_{m00}^*(s,m)) \\ &= q^{i-m}[q(p_{m00}^*(s,m) - pp_{m-100}^*(s,m)) - \frac{1}{\mu}] \\ &= q^{i-m+1}(p_{m00}^*(s,m) - pp_{m-100}^*(s,m)) - \frac{1}{\mu}q^{i-m}. \end{aligned}$$

Therefore,

$$p^*_{i00}(s,m) - pp^*_{i-100}(s,m) = q^{i-1}(p^*_{100}(s,m) - pp^*_{000}(s,m)) - \frac{1}{\mu}q^{i-m-1}.$$

$$(8.2.44)$$

Similarly

$$p^*_{i00}(s,m) - qp^*_{i-100}(s,m) = p^{i-1}(p^*_{100}(s,m) - qp^*_{000}(s,m)) - \frac{1}{\mu}p^{i-m-1}.$$

$$(8.2.45)$$

Then for the case $m < i$, (8.2.39) follows from (8.2.44) and (8.2.45) directly. On the other hand, for the case $m \geq i$, (8.2.39) is trivial. This completes the proof of Theorem 8.2.5.

As a result, on the basis of Theorems 8.2.4 and 8.2.5, the Laplace transforms $p^*_{ijk}(s,m)$ can be evaluated from equations (8.2.7)-(8.2.10) recursively. Then, we can determine the probability mass functions $p_{ijk}(t,m)$ by inversion.

As a simple application, we can determine the Laplace transform of the distribution of $I(t)$, i.e, the Laplace transform of the distribution of the number of customers in the system at time t. In fact

$$P(I(t) = i \mid (I(0), J(0), K(0)) = (m,0,0))$$

$$= \sum_{j=0}^{1}\sum_{k=0}^{\infty} P\{(I(t), J(t), K(t)) = (i,j,k) \mid (I(0), J(0), K(0)) = (m,0,0)\}$$

$$= \sum_{j=0}^{1}\sum_{k=0}^{\infty} p_{ijk}(t,m).$$

Thus, the Laplace transform of $P(I(t) = i \mid (I(0), J(0), K(0)) = (m,0,0))$ can be determined accordingly.

Now, we shall study the distribution of waiting time in the queue and in the system for a new arrival respectively.

Let Q_t be the waiting time in the queue for a new arrival at time t. Suppose that the system state at time t is $(I(t), J(t), K(t)) = (i,0,k)$. Then when the new arrival arrives, there are i customers in the queue and the first one is being served. The new arrival must wait for the time until completion of the services to these i customers. As the earlier part of the service time to the first customer was conducted before the new arrival arrives, then the total service time to these i customers since time t is given by

$$G_i = \chi_1^L + \sum_{j=2}^{i} \chi_j.$$

Therefore, G_i is the sum of the latter part χ_1^L of the service time χ_1 for the first customer plus the service time for the other $i - 1$ customers. Because of the memoryless property of exponential distribution, χ_1^L and χ_1 will have the same exponential distribution $G(x)$. Consequently, the distribution of G_i will be given by

$$G^{(i)}(x) = G * G * \cdots * G(x),$$

the i-fold convolution of G with itself. On the other hand, since $J(t) = 0$ and $K(t) = k$, the service station is in an up state and has been repaired for k times. To complete the services to i customers in the queue, the services may be completed by time X_{k+1}^L, the latter part of operating time X_{k+1} of the service station after the kth repair, since the earlier part of operating time X_{k+1} has been used for the service before the new arrival arrives. As a result, the event $Q_t \leq x$ is equivalent to the event

$$G_i \leq x, \quad G_i \leq X_{k+1}^L.$$

In general, the services may be completed after n-time more repairs on the service station, $n = 1, 2, \dots$. Thus, the real time for completion of the services to these i customers is $G_i + \sum_{j=k+1}^{k+n} Y_j$. Consequently, the event $Q_t \leq x$ now is equivalent to

$$G_i + \sum_{j=k+1}^{k+n} Y_j \leq x, \quad X_{k+1}^L + \sum_{j=k+2}^{k+n} X_j < G_i \leq X_{k+1}^L + \sum_{j=k+2}^{k+n+1} X_j.$$

Note that when the new arrival arrives, the service station has served for the earlier part of operating time X_{k+1}. The latter part X_{k+1}^L of X_{k+1} is the residual operating time of the service station. Again, due to the memoryless property of exponential distribution, X_{k+1}^L and X_{k+1} will have the same exponential distribution $X_{k+1}(x)$. According to the above explanation, we can prove the following theorem.

Theorem 8.2.6. Let the distribution of Q_t be $Q_t(x)$, then

$$Q_t(x) = Q_t(0) + \sum_{i=1}^{\infty} \sum_{k=0}^{\infty} p_{i0k}(t, m) \left\{ \int_0^x e^{-a^k \alpha u} dG^{(i)}(u) \right.$$

$$\left. + \sum_{n=1}^{\infty} \int_0^x Y_{k+1}^{(n)}(x - u)[X_{k+1}^{(n)}(u) - X_{k+1}^{(n+1)}(u)] dG^{(i)}(u) \right\} \quad \text{for} \quad x > 0,$$

$$(8.2.46)$$

with $Q_t(0) = \sum\limits_{k=0}^{\infty} p_{00k}(t,m)$.

Proof.

For $x = 0$, the result is trivial. Now consider the case $x > 0$, we have

$$Q_t(x) = P(Q_t \le x) = P(Q_t = 0) + P(0 < Q_t \le x)$$

$$= Q_t(0) + \sum_{i=1}^{\infty}\sum_{k=0}^{\infty} p_{i0k}(t,m)P\{Q_t \le x \mid (I(t), J(t), K(t)) = (i,0,k)\}$$

$$= Q_t(0) + \sum_{i=1}^{\infty}\sum_{k=0}^{\infty} p_{i0k}(t,m)\{P(G_i \le x, G_i \le X_{k+1}^L)$$

$$+ \sum_{n=1}^{\infty} P(G_i + \sum_{j=k+1}^{k+n} Y_j \le x, X_{k+1}^L + \sum_{j=k+2}^{k+n} X_j < G_i \le X_{k+1}^L + \sum_{j=k+2}^{k+n+1} X_j)\}$$

$$= Q_t(0) + \sum_{i=1}^{\infty}\sum_{k=0}^{\infty} p_{i0k}(t,m) \left\{ \int_0^x (1 - X_{k+1}(u))dG^{(i)}(u) \right.$$

$$+ \sum_{n=1}^{\infty} \int_0^x P(\sum_{j=k+1}^{k+n} Y_j \le x - u,$$

$$\left. X_{k+1}^L + \sum_{j=k+2}^{k+n} X_j < u \le X_{k+1}^L + \sum_{j=k+2}^{k+n+1} X_j)dG^{(i)}(u) \right\}$$

$$= Q_t(0) + \sum_{i=1}^{\infty}\sum_{k=0}^{\infty} p_{i0k}(t,m) \left\{ \int_0^x e^{-a^k \alpha u}dG^{(i)}(u) \right.$$

$$\left. + \sum_{n=1}^{\infty} \int_0^x Y_{k+1}^{(n)}(x-u)[X_{k+1}^{(n)}(u) - X_{k+1}^{(n+1)}(u)]dG^{(i)}(u) \right\}.$$

This completes the proof of Theorem 8.2.6.

Now, let S_t be the waiting time in the system for a new arrival at time t. It is equal to the waiting time in the queue plus the service time to the new arrival. Thus, by a similar argument, we can easily obtain the following formula.

Theorem 8.2.7. Let the distribution of S_t be $S_t(x)$, then

$$S_t(x) = \sum_{i=0}^{\infty}\sum_{k=0}^{\infty} p_{i0k}(t,m) \left\{ \int_0^x e^{-a^k \alpha u}dG^{(i+1)}(u) \right.$$

$$\left. + \sum_{n=1}^{\infty} \int_0^x Y_{k+1}^{(n)}(x-u)[X_{k+1}^{(n)}(u) - X_{k+1}^{(n+1)}(u)]dG^{(i+1)}(u) \right\}, \quad \text{for } x > 0,$$

and $S_t(x) = 0$ otherwise.

Finally, we shall discuss some reliability indices of the service station.

1. Mean time to the first failure (MTTFF)

Given that there are m customers in the system at the beginning, let T_m be the time to the first failure of the service station, and let the distribution of T_m be

$$T_m(x) = P\{T_m \le x \mid (I(0), J(0), K(0)) = (m, 0, 0)\}. \qquad (8.2.47)$$

Then we have the following theorem.

Theorem 8.2.8. The Laplace-Stieltjes transform of $T_m(x)$ is given by

$$T_m^*(s) = \frac{\alpha}{s + \alpha} - \frac{\alpha s [\tilde{B}^*(s + \alpha)]^m}{(s + \alpha)[s + \lambda - \lambda \tilde{B}^*(s + \alpha)]}. \qquad (8.2.48)$$

Proof.

Once again, let V_i be the ith idle period. Note that all busy periods prior to the first failure are the same as that in the classical $M/M/1$ model. Then it follows from (8.2.46) that

$$T_m(x) = P\{X_1 \le x, X_1 \le \tilde{B}_1 \mid (I(0), J(0), K(0)) = (m, 0, 0)\}$$

$$+ \sum_{n=1}^{\infty} P\left\{\sum_{i=1}^{n} V_i + X_1 \le x, \sum_{i=1}^{n} \tilde{B}_i < X_1 \le \sum_{i=1}^{n+1} \tilde{B}_i\right.$$

$$\left. \mid (I(0), J(0), K(0)) = (m, 0, 0)\right\} \qquad (8.2.49)$$

$$= \int_0^x P(\tilde{B}_1 \ge u) dX_1(u)$$

$$+ \sum_{n=1}^{\infty} \int_0^x P\left\{\sum_{i=1}^{n} V_i \le x - u, \sum_{i=1}^{n} \tilde{B}_i < u \le \sum_{i=1}^{n+1} \tilde{B}_i\right\} dX_1(u)$$

$$= \int_0^x (1 - \tilde{B}^{(m)}(u))\alpha e^{-\alpha u} du$$

$$+ \sum_{n=1}^{\infty} \int_0^x F^{(n)}(x - u)[\tilde{B}^{(m+n-1)}(u) - \tilde{B}^{(m+n)}(u)]\alpha e^{-\alpha u} du,$$

$$(8.2.50)$$

where (8.2.49) is obtained by conditional on the number of idle periods prior to the first failure of the service station, while (8.2.50) is due to (8.2.13).

Now, let the common density of ν_n be $f(x)$, $n = 1, 2, \ldots$, and let

$$f^{(n)}(x) = \frac{dF^{(n)}(x)}{dx} = f * f \cdots * f(x)$$

be the n-fold convolution of f with itself. Then by taking the Laplace-Stieltjes transform of $T_m(x)$, it follows from (8.2.50) that

$$T_m^*(s) = \int_0^\infty e^{-sx} dT_m(x)$$

$$= \int_0^\infty e^{-sx}(1 - \tilde{B}^{(m)}(x))\alpha e^{-\alpha x} dx + \sum_{n=1}^\infty \int_0^\infty e^{-sx}$$

$$\times \left\{ \int_0^x f^{(n)}(x - u)[\tilde{B}^{(m+n-1)}(u) - \tilde{B}^{(m+n)}(u)]\alpha e^{-\alpha u} du \right\} dx$$

$$= \frac{\alpha}{s+\alpha} - \int_0^\infty \alpha \tilde{B}^{(m)}(x)e^{-(s+\alpha)x} dx + \sum_{n=1}^\infty \int_0^\infty e^{-sv} f^{(n)}(v) dv$$

$$\times \left\{ \int_0^\infty \alpha e^{-(s+\alpha)u}[\tilde{B}^{(m+n-1)}(u) - \tilde{B}^{(m+n)}(u)] du \right\}$$

$$= \frac{\alpha}{s+\alpha} - \frac{\alpha[\tilde{B}^*(s+\alpha)]^m}{s+\alpha}$$

$$+ \sum_{n=1}^\infty (\frac{\lambda}{s+\lambda})^n \frac{\alpha}{s+\alpha} \left\{ [\tilde{B}^*(s+\alpha)]^{m+n-1} - [\tilde{B}^*(s+\alpha)]^{m+n} \right\}$$

$$= \frac{\alpha}{s+\alpha} - \frac{\alpha[\tilde{B}^*(s+\alpha)]^m}{s+\alpha} + \frac{\alpha\lambda[\tilde{B}^*(s+\alpha)]^m(1 - \tilde{B}^*(s+\alpha))}{(s+\alpha)[s+\lambda - \lambda\tilde{B}^*(s+\alpha)]}$$

$$= \frac{\alpha}{s+\alpha} - \frac{\alpha s[\tilde{B}^*(s+\alpha)]^m}{(s+\alpha)[s+\lambda - \lambda\tilde{B}^*(s+\alpha)]}.$$

Consequently, with the help of (8.2.11), the expectation of T_m, i.e. the mean time to the first failure (MTTFF) of the service station, will be given by

$$E(T_m) = \int_0^\infty t\, dT_m(t) = -\frac{dT_m^*(s)}{ds}\Big|_{s=0}$$

$$= \frac{1}{\alpha} + \frac{[\tilde{B}^*(\alpha)]^m}{\lambda[1 - \tilde{B}^*(\alpha)]}$$

$$= \frac{1}{\alpha} + \frac{\left\{\alpha + \lambda + \mu - \sqrt{(\alpha + \lambda + \mu)^2 - 4\lambda\mu}\right\}^m}{2^{m-1}\lambda^m\left\{-\alpha + \lambda - \mu + \sqrt{(\alpha + \lambda + \mu)^2 - 4\lambda\mu}\right\}}. \tag{8.2.51}$$

2. Availability of the service station

Let the availability of the service station at time t be

$$A_m(t)$$

$$= P\{\text{the service station is up at time } t \mid (I(0), J(0), K(0)) = (m, 0, 0)\}.$$

Moreover, let the probability that the service station breaks down at time t be

$$\bar{A}_m(t)$$
$$= P\{\text{the service station is down at time } t \mid (I(0), J(0), K(0)) = (m, 0, 0)\}.$$

Now, denote the Laplace transforms of $A_m(t)$ and $\bar{A}_m(t)$ by $A_m^*(s)$ and $\bar{A}_m^*(s)$ respectively. Then, the following theorem follows directly.

Theorem 8.2.9.

$$A_m^*(s) = \sum_{i=0}^{\infty} \sum_{k=0}^{\infty} p_{i0k}^*(s, m).$$

Proof.

It is clear that

$$A_m(t)$$
$$= \sum_{i=0}^{\infty} \sum_{k=0}^{\infty} P\{(I(t), J(t), K(t)) = (i, 0, k) \mid (I(0), J(0), K(0)) = (m, 0, 0)\}$$
$$= \sum_{i=0}^{\infty} \sum_{k=0}^{\infty} p_{i0k}(t, m).$$

Therefore

$$A_m^*(s) = \sum_{i=0}^{\infty} \sum_{k=0}^{\infty} p_{i0k}^*(s, m).$$

Furthermore, due to the fact $A_m(t) + \bar{A}_m(t) = 1$, we have

$$A_m^*(s) + \bar{A}_m^*(s) = \frac{1}{s}.$$

Thus

$$\bar{A}_m^*(s) = \frac{1}{s} - \sum_{i=0}^{\infty} \sum_{k=0}^{\infty} p_{i0k}^*(s, m).$$

3. The rate of occurrence of failures (ROCOF)

Let $M_f(t)$ be the expected number of failures of the service station that have occurred by time t, then its derivative $m_f(t) = M_f'(t)$ is called the rate of occurrence of failures (ROCOF). According to Theorem 1.6.5, the ROCOF can be evaluated in the following way:

$$m_f(t) = \sum_{i=1}^{\infty} \sum_{k=0}^{\infty} a^k \alpha p_{i0k}(t, m).$$

Therefore, the Laplace transform of $m_f(t)$ is given by

$$m_f^*(s) = \sum_{i=1}^{\infty} \sum_{k=0}^{\infty} a^k \alpha p_{i0k}^*(s, m).$$

As $p_{i0k}^*(s, m)$ has been determined, we can evaluate $m_f^*(s)$ and hence $m_f(t)$ accordingly.

8.3 A Geometric Process Warranty Model

A warranty can be viewed as a contractual obligation incurred by a manufacturer in connection with the sale of a product. In practice, a warranty strategy provided by a manufacturer may depend on the product nature and its usage. Two broad classes of policies have been studied so far. A free repair (replacement) warranty (FRW) provides product repair (replacement) by the manufacturer, at no cost to a consumer for a limited period of time from the time of initial purchase. A pro-rata warranty (PRW) will repair (replace) a failed product at a cost proportional to the working age of the product to a consumer. Whenever a replacement is conducted, a failed product is replaced by a new and identical one.

In marketing research, the sales volume of a product is one of main concerns of a manufacturer. In general, the sales volume depends on national economic performance, it also depends on the quality and price of product as well as the warranty offered. However, the national economic performance is an external factor, the manufacturer has no control over this factor. The manufacturer could try to improve the quality of the product and reduce the price. Nevertheless, for a specific product the quality is specified by the national standard bureau and the price is more or less the same in market because of the free competition in market. As a result, to promote the sales volume of a product, a manufacturer should offer more attractive warranty condition to consumer and apply his optimal strategy for maximizing his revenues or minimizing his cost. This is not an easy problem, as a warranty that is favourable to consumer will be unfavourable to manufacturer, and vice versa. In this section, a GP warranty model is considered by making the following assumptions.

Assumption 1. At the beginning, a manufacturer sells a product to a consumer with FRW on time period $[0, W)$. Let X_1 be the operating time of the product after sale or replacement and $X_i, i = 2, 3, \ldots$ be the operating time of the product following the $(i-1)$th repair. Assume that the successive operating times of the product form a GP with ratio $a \geq 1$, and X_1 has an exponential distribution $Exp(1/\lambda)$ with density function

$$f(x) = \begin{cases} \frac{1}{\lambda} \exp(-x/\lambda) & x > 0, \\ 0 & x \leq 0, \end{cases}$$

and $\lambda > 0$. The repair time and replacement time are negligible.

Assumption 2. According to FRW, the manufacturer agrees to repair the product free of charge when it fails in $[0, W)$. By applying an extended warranty policy, at time W, the manufacturer will offer an option to the consumer either to renew or not to renew the warranty. A consumer can renew the warranty at time W by paying a renewal cost c_N to the manufacturer, in return the manufacturer will provide free repair service to the consumer for an extended warranty period $[W, W + L)$. The consumer can further renew the warranty at time $W + L$ again for a period $[W + L, W + 2L)$ by paying the same renewal cost c_N to the manufacturer, and so on.

Assumption 3. A consumer can adopt the following k-renewal strategy: for $i = 0, \ldots, k - 1$, renew the warranty contract at times $W + iL$ for an extended period $[W + iL, W + (i + 1)L)$. After renewing the warranty for k times, the consumer replaces the product by a new and identical one following the first failure after time $W + kL$, where k satisfies

$$W + kL \leq \frac{a\lambda}{a - 1}, \tag{8.3.1}$$

or $k = 0, 1, \ldots, K$ with

$$K = [\frac{a\lambda - (a - 1)W}{(a - 1)L}], \tag{8.3.2}$$

and $[x]$ is the integer part of x. Assume that once the consumer chooses a k-renewal strategy, he will choose the same strategy forever.

Assumption 4. Let the initial purchase price or the replacement cost to a consumer be c_R and the production cost per product to the manufacturer is c_r. The renewal cost of the warranty contract incurred by a consumer is c_N. The repair cost incurred by the manufacturer is c_e. The manufacturer will adopt a c_N strategy by choosing a renewal cost c_N satisfying the following constraint:

$$c_N < c_R. \tag{8.3.3}$$

Remarks.

In Assumption 1, the successive operating times $\{X_n, n = 1, 2, \ldots\}$ form a GP with ratio $a \geq 1$. This means that the product is a deteriorating product.

By Theorem 2.2.3, inequality (8.3.1) is reasonable. To show this, assume that $a > 1$. As the operating time of a product after the $(i-1)$th repair is X_i, the expected total operating time of the product will be

$$\sum_{i=1}^{\infty} E(X_i) = \frac{a\lambda}{a-1}. \tag{8.3.4}$$

Because the end time of the last warranty period $W + kL$ should be no more than the expected total operating time $\frac{a\lambda}{a-1}$, then (8.3.1) is a natural assumption.

In practice, the market values of c_R, c_r and c_e are more or less the same in the commercial market due to a similar real expenses and free competition. Moreover, W and L are also more or less the same in the market because of general custom or tacit understanding among manufacturers. Therefore, the manufacturer can only change these values a little by his discretion. Thus the manufacturer can apply a c_N strategy by choosing a cost c_N flexibly.

Constraint (8.3.3) is also reasonable since the renewal cost c_N to a consumer should be cheaper than the initial purchase cost or replacement cost c_R. Otherwise, the consumer will simply buy a new and identical one rather than renew for an extended warranty period.

As the manufacturer can choose cost c_N and a consumer can choose a k-renewal strategy, then the manufacturer and consumer form two players of a game. This game may be called as a warranty game.

Now, we study the extensive form representation of the warranty game. First of all, we say a cycle is completed if a replacement is completed. Thus, a cycle is actually the time interval between the sale of a product and the first replacement or a time interval between two successive replacements. Clearly, the successive cycles and the costs incurred in each cycle form a renewal reward process. Then, let

$$M(t) = E[N(t)],$$

be the expected number of failures that have occurred by time t. In fact, $M(t)$ is the geometric function $M(t, a)$ defined in Chapter 3.

Assume that the manufacturer has adopted a c_N strategy by choosing value c_N, and a consumer applies a k-renewal strategy. The cost incurred in a cycle to the consumer is clearly given by $c_R + kc_N$. As the number of failures of the product by time $W + kL$ is $N(W + kL)$, the product will be replaced by a new and identical one at time $S_{N(W+kL)+1}$. Consequently, the expected length of a cycle is $E(S_{N(W+kL)+1})$. Then from Corollary 2.5.2, the average cost to a consumer is given by

$$
\begin{aligned}
A_c(c_N, k) &= \frac{c_R + kc_N}{E(S_{N(W+kL)+1})} \\
&= \frac{a(c_R + kc_N)}{a\lambda + W + kL} \quad \text{for} \quad k = 0, 1, \ldots, K.
\end{aligned}
\tag{8.3.5}
$$

To determine the average cost to the manufacturer, as a consumer adopts a k-renewal strategy and the manufacturer is responsible for repairs on the product up to time $W + kL$, the expected cost incurred in a cycle is

$$
c_e E(N(W + kL)) + c_r - c_R - kc_N.
$$

The expected length of a cycle is also $E(S_{N(W+kL)+1})$. Therefore, the average cost to the manufacturer is given by

$$
\begin{aligned}
A_m(c_N, k) &= \frac{c_e E(N(W + kL)) + c_r - c_R - kc_N}{E(S_{N(W+kL)+1})} \\
&= \frac{a\{c_e M(W + kL) + c_r - c_R - kc_N\}}{a\lambda + W + kL}.
\end{aligned}
\tag{8.3.6}
$$

Thus the warranty game is a two-stage dynamic game of complete and perfect information. The extensive form representation of the warranty game (see e.g. Gibbons (1992)) can be described in the following way.

1. The manufacturer and a consumer are two players in the game.

2. At the beginning, the manufacturer chooses a strategy or action, a cost c_N, from the feasible set $A_1 = [0, c_R)$. Afterward, the consumer chooses a strategy or action, a k-renewal strategy, from the feasible set $A_2 = \{0, 1, \ldots, K\}$. They know each other the full history of the play of the game. Moreover, both players are rational players, so that they will always choose their own optimal strategies or actions.

3. Given a combination of strategies or actions (c_N, k), i. e. the manufacturer chooses a c_N strategy and the consumer chooses a k-renewal strategy, the average costs incurred by the two players are given by $A_m(c_N, k)$ and $A_c(c_N, k)$ respectively.

In the extensive form representation of the warranty game, rather than the payoff received by a player, we study the average cost incurred by a player. Therefore, instead of maximizing the payoff, we should minimize the average cost in the warranty game.

Now, we study the optimal strategies for manufacturer and consumer. To do this, we shall apply the backward induction (see e.g. Gibbons (1992)) to determine the optimal strategies for the two players respectively. For this purpose, assume that the manufacturer has chosen cost c_N, we shall derive the following lemma.

Lemma 8.3.1. For $k \geq 0$, $A_c(c_N, k)$ is nondecreasing (nonincreasing) in k, if and only if

$$c_N \geq (\leq) \, \alpha c_R,$$

where $\alpha = L/(a\lambda + W)$.

Proof.

Consider the first difference of $A_c(c_N, k)$.

$$
\begin{aligned}
A_c(c_N, k+1) - A_c(c_N, k) &= \frac{a[c_R + (k+1)c_N]}{a\lambda + W + (k+1)L} - \frac{a(c_R + kc_N)}{a\lambda + W + kL} \\
&= \frac{a[(a\lambda + W)c_N - Lc_R]}{(a\lambda + W + kL)[a\lambda + W + (k+1)L]}.
\end{aligned}
$$

Then Lemma 8.3.1 is trivial.

To determine the optimal strategies for the manufacturer and consumer respectively, we consider two cases.

Case 1. $\alpha < 1$.

Then by Lemma 8.3.1 and (8.3.5), for given c_N, the minimum of $A_c(c_N, k)$ is given by

$$
\min_{k \in A_2} A_c(c_N, k)
= \begin{cases} A_c(c_N, 0) = ac_R/(a\lambda + W) & \alpha c_R \leq c_N < c_R, \\ A_c(c_N, K) & c_N \leq \alpha c_R. \end{cases} \tag{8.3.7}
$$

where K is given by (8.3.2) and is independent of c_N. Accordingly, if the manufacturer chooses a strategy c_N, the optimal k^*-renewal strategy for a consumer is either 0-renewal strategy or K-renewal strategy, namely

$$
k^* = \begin{cases} 0 & \alpha c_R \leq c_N < c_R, \\ K & c_N \leq \alpha c_R. \end{cases} \tag{8.3.8}
$$

To determine an optimal strategy for the manufacturer, as a consumer is a rational player and the game is of complete and perfect information, the manufacturer could assume that the consumer always plays his optimal

k^*-renewal strategy. Then from (8.3.6), the minimum average cost to the manufacturer is given by

$$A_m(c_N^*, k^*) = \min_{c_N \in A_1} A_m(c_N, k^*)$$

$$= \min\{\min_{\alpha c_R \leq c_N < c_R} A_m(c_N, 0), \min_{c_N \leq \alpha c_R} A_m(c_N, K)\}$$

$$= \min\{\frac{a[c_e M(W) + c_r - c_R]}{a\lambda + W}, \frac{a[c_e M(W + KL) + c_r - c_R - K\alpha c_R]}{a\lambda + W + KL}\}$$

$$= \min\{c_1, c_2\}, \tag{8.3.9}$$

where

$$c_1 = \frac{a\{c_e M(W) + c_r - c_R\}}{a\lambda + W} \tag{8.3.10}$$

and

$$c_2 = \frac{a\{c_e M(W + KL) + c_r - c_R - K\alpha c_R\}}{a\lambda + W + KL}. \tag{8.3.11}$$

Consequently, we can determine an optimal strategy for the manufacturer by simply making a comparison between c_1 and c_2. Nevertheless, in general, an optimal strategy for the manufacturer may not exist. For example, if $c_1 > c_2$, it seems that an optimal strategy for the manufacturer is $c_N^* = \alpha c_R$. Unfortunately, this is not true. In fact in this case, (8.3.7) shows that the optimal strategy for the consumer is either 0-renewal strategy or K-renewal strategy. If the consumer applies K-renewal strategy, the average cost to the manufacturer is $A_m(c_N^*, K) = c_2$, this is the minimum. Nevertheless, if the consumer applies 0-renewal strategy, the average cost to the manufacturer now will be $A_m(c_N^*, 0) = c_1$ that is not the minimum. Consequently, the strategy $c_N^* = \alpha c_R$ is not an optimal strategy for the manufacturer. In other words, if $c_1 > c_2$, the optimal strategy for the manufacturer does not exist! To overcome this difficulty, we shall introduce the concept of ϵ-optimal strategy here.

Definition 8.3.2. A strategy c_N^ϵ is an ϵ-optimal strategy for the manufacturer, if

$$A_m(c_N^\epsilon, k^*) \leq \min_{c_N \in A_1} A_m(c_N, k^*) + \epsilon. \tag{8.3.12}$$

If $\min_{c_N \in A_1} A_m(c_N, k^*)$ does not exist, then (8.3.12) is replaced by

$$A_m(c_N^\epsilon, k^*) \leq \inf_{c_N \in A_1} A_m(c_N, k^*) + \epsilon. \tag{8.3.13}$$

The ϵ-optimal strategy had been studied by Thomas (1979) and Lam (1987) in inventory control and queueing control problems respectively. Afterward, it was also applied to a warranty model by Lam and Lam (2001).

Theorem 8.3.3. For $\alpha < 1$, an optimal strategy c_N^* or ϵ-optimal strategy c_N^ϵ for the manufacturer is determined in the following way.

(1) If $c_1 < c_2$, then any strategy c_N^* satisfying

$$\alpha c_R < c_N^* < c_R$$

is an optimal strategy for the manufacturer.

(2) If $c_1 = c_2$, then any strategy c_N^* satisfying

$$\alpha c_R \leq c_N^* < c_R$$

is an optimal strategy for the manufacturer.

(3) If $c_1 > c_2$, then the strategy c_N^ϵ with

$$c_N^\epsilon = \alpha c_R - \frac{\epsilon(a\lambda + W + KL)}{aK}$$

is an ϵ-optimal strategy for the manufacturer.

Proof.

1. Assume that $c_1 < c_2$. Now for any strategy c_N^* satisfying $\alpha c_R < c_N^* < c_R$, according to (8.3.8), the optimal strategy for a consumer is 0-renewal strategy. As a result, from (8.3.6) the average cost to the manufacturer is given by

$$A_m(c_N^*, k^*) = A_m(c_N^*, 0) = c_1 = \min\{c_1, c_2\}. \tag{8.3.14}$$

Therefore, strategy c_N^* is an optimal strategy for the manufacturer.

2. Assume that $c_1 = c_2$. Then for strategy $c_N^* = \alpha c_R$, it follows from (8.3.8) that an optimal strategy for a consumer is either 0-renewal or K-renewal strategy. If the consumer adopts 0-renewal strategy, then by (8.3.10), we have

$$A_m(c_N^*, k^*) = A_m(c_N^*, 0) = c_1 = \min\{c_1, c_2\};$$

if the consumer applies K-renewal strategy, then from (8.3.11),

$$A_m(c_N^*, k^*) = A_m(c_N^*, K) = c_2 = \min\{c_1, c_2\}.$$

In both cases, $A_m(c_N^*, k^*)$ takes its minimum. Thus, strategy $c_N^* = \alpha c_R$ is an optimal strategy for the manufacturer.

Moreover, for any strategy $\alpha c_R < c_N^* < c_R$, the optimal strategy for a consumer is 0-renewal strategy. Then

$$A_m(c_N^*, k^*) = A_m(c_N^*, 0) = c_1 = \min\{c_1, c_2\}.$$

Thus, any strategy c_N^* satisfying

$$\alpha c_R \leq c_N^* < c_R$$

is an optimal strategy for the manufacturer.

3. Now assume that $c_1 > c_2$. For strategy c_N^ϵ

$$c_N^\epsilon = \alpha c_R - \frac{\epsilon(a\lambda + W + KL)}{aK},$$

because $c_N^\epsilon < \alpha c_R$, K-renewal strategy is the optimal strategy for a consumer. As a result, from (8.3.6) we have

$$
\begin{aligned}
A_m(c_N^\epsilon, k^*) &= A_m(c_N^\epsilon, K) \\
&= \frac{a\{c_e M(W + KL) + c_r - c_R - K\alpha c_R\}}{a\lambda + W + KL} + \epsilon \\
&= c_2 + \epsilon = \min\{c_1, c_2\} + \epsilon \\
&= \min_{c_N \in A_1} A_m(c_N, k^*) + \epsilon.
\end{aligned}
$$

Therefore, strategy c_N^ϵ is an ϵ-optimal strategy for the manufacturer.

This completes the proof of Theorem 8.3.3.

Case 2. $\alpha \geq 1$.

In this case, it follows from (8.3.3) that $c_N < c_R \leq \alpha c_R$. Then (8.3.7) yields that

$$\min_{k \in A_2} A_c(c_N, k) = A_c(c_N, K). \tag{8.3.15}$$

Hence, K-renewal strategy is the optimal strategy for a consumer, i. e. $k^* = K$.

On the other hand, (8.3.6) shows that $\min_{c_N \in A_1} A_m(c_N, k^*)$ does not exist, then we define

$$
\begin{aligned}
A_m(c_N^*, k^*) &= \inf_{c_N < c_R} A_m(c_N, K) \\
&= \inf_{c_N < c_R} \frac{a\{c_e M(W + KL) + c_r - c_R - K c_N\}}{a\lambda + W + KL} = c_3, \tag{8.3.16}
\end{aligned}
$$

where

$$c_3 = \frac{a\{c_e M(W + KL) + c_r - c_R - K c_R\}}{a\lambda + W + KL}. \tag{8.3.17}$$

Thus, we have the following result.

Theorem 8.3.4. For $\alpha \geq 1$, an ϵ-optimal strategy c_N^ϵ for the manufacturer is given by

$$c_N^\epsilon = c_R - \frac{\epsilon(a\lambda + W + KL)}{aK}.$$

Proof.

It is easy to check that

$$
\begin{aligned}
A_m(c_N^\epsilon, k^*) &= A_m(c_N^\epsilon, K) \\
&= \frac{a\{c_e M(W + KL) + c_r - c_R - Kc_R\}}{a\lambda + W + KL} + \epsilon \\
&= c_3 + \epsilon = \inf_{c_N < c_R} A_m(c_N, k^*) + \epsilon.
\end{aligned}
$$

Therefore, c_N^ϵ is an ϵ-optimal strategy for the manufacturer.

In application of Theorems 8.3.3 and 8.3.4, we should evaluate values c_1, c_2 or c_3 first. To do this, several methods for calculating the values $M(W)$ and $M(W + KL)$ have been developed in Chapter 3. In many practical situations, $L \leq W$, then $\alpha < 1$ and Theorem 8.3.3 is applicable.

Remarks.

Because both players are rational players and the warranty game has complete and perfect information, the manufacturer and a consumer will always apply their optimal strategies respectively. By using backward induction, we can determine the optimal strategies for a consumer and the manufacturer respectively.

Note that the optimal strategy for the manufacturer is a conservative strategy. In fact, if a consumer does not use his optimal strategy, the consumer will incur a higher average cost, while the manufacturer will incur even lower average cost and gain a higher profits.

If $\alpha < 1$ and $c_1 \leq c_2$, Theorem 8.3.3 concludes that the optimal strategies c_N^* and k^* for the manufacturer and consumer respectively will exist, then any combination of strategies (c_N^*, k^*) is a backward induction outcome of the warranty game. Now, the backward induction outcome is not unique. However, if $\alpha < 1$ and $c_1 > c_2$, as there exists no optimal strategy for the manufacturer, the backward induction outcome does not exist!

If $\alpha \geq 1$, we can see from Theorem 8.3.4 that K-renewal strategy will be the unique optimal strategy for a consumer. Because $\min_{c_N < c_R} A_m(c_N, K)$ does not exist, so does not an optimal strategy for the manufacturer. Thus the backward induction outcome also does not exist. This is the motivation to introduce an ϵ-optimal strategy c_N^ϵ. Accordingly, we may call (c_N^ϵ, k^*)

as the ϵ backward induction outcome of the warranty game.

In the GP warranty model, if $\alpha < 1$, there exists an indifferent point αc_R. Whenever $c_N = \alpha c_R$, a consumer can apply either 0-renewal or K-renewal strategy. Both strategies are optimal for a consumer. Although a consumer will incur the same average cost, the manufacturer will incur different average costs. Then if $c_1 > c_2$, the manufacturer will have no optimal strategy, i. e. an optimal strategy for the manufacturer does not exist! In other words, there is no backward induction outcome in the warranty game. Therefore, an indifferent point may cause the nonexistence of an optimal strategy for the manufacturer, and hence the nonexistence of an backward induction outcome in the warranty game.

In the GP warranty model, we can see that the manufacturer can always apply either an optimal strategy c_N^* or an ϵ-optimal strategy c_N^ϵ, so that a consumer will fall into a 'trap'. Consequently, the warranty game will be beneficial to the manufacturer but not to a consumer. We are not surprising with this result, because the warranty contract is proposed by the manufacturer, nobody wishes to be caught in his own 'trap'.

Of course, for promotion of marketing sales, the manufacturer should apply an optimal strategy and make the strategy as attractive as possible to a consumer. To do this, the manufacturer should choose the value c_N^* as small as possible, while the cost $A_m(c_N^*, k^*)$ will keep unchanged. This will be possible if the optimal strategy for the manufacturer is not unique. For example, if $\alpha < 1$ and $c_1 = c_2$, then by Theorem 8.3.3, any strategy c_N^* satisfying $\alpha c_R \leq c_N^* < c_R$ is an optimal strategy for the manufacturer. Clearly, the optimal strategy $c_N^* = \alpha c_R$ is the most attractive one to consumers, it should be adopted by the manufacturer. Nevertheless, if $\alpha < 1$ and $c_1 < c_2$, then any strategy c_N^* satisfying $\alpha c_R < c_N^* < c_R$ is optimal for the manufacturer. To choose a more attractive policy to a consumer, the manufacturer may adopt an optimal policy

$$c_N^* = \alpha c_R + \delta,$$

where δ is a small positive number. If $\alpha < 1$ and $c_1 > c_2$, by Theorem 8.3.3, the manufacturer can choose an appropriate ϵ-optimal strategy c_N^ϵ to compromise the profits and the attraction to consumers.

The GP warranty model should have much potential application to promotion of marketing sales. According to Theorems 8.3.3 and 8.3.4, a manufacturer can put forward a warranty term, and make more profits while the warranty term is still attractive. On the other hand, a consumer will also benefit from Theorems 8.3.3 and 8.3.4 by using the optimal strategy to reduce his cost and avoid more expenses.

8.4   Notes and References

In Chapter 5, we consider the application of GP to data analysis. Through the analysis of many real data sets, it has been shown that a GP model is a good and simple model for analysis of data with a single trend or multiple trends. In Chapters 6 and 7, we study the application of GP to the maintenance problem and reliability analysis of a system. It is the application of GP to reliability and electronic engineering. In fact, the applications of GP to data analysis and maintenance problem both are the initial motivations of introducing GP. In Chapter 8, we consider more applications of GP to operational research and management science. Here, we have just focused on queueing theory and warranty problem.

Section 8.2 concerns an application of GP to a $M/M/1$ queueing system, it is based on Lam et al. (2006). As $M/M/1$ queue system is a simple queueing system, more research work is expected in the future.

Warranty problem is an interesting topic in operational research and management science and has important application in marketing research. For reference, see Blischke and Scheuer (1975, 1981), Mamer (1982), Nguyen and Murthy (1984), Balcer and Sahin (1986), Blischke and Murthy (1994), among others. Section 8.3 is due to Lam's work (2007c), it studies a GP warranty model. We deal with the GP warranty model using game theory approach. A warranty problem is treated as a game between two players, a manufacturer and a consumer. It is interesting not only in warranty problem, but also in game theory.

As GP is a simple monotone process, we can expect it will have more and more applications in different topics of statistics, operational research, management science, engineering and other subjects.

Appendix A

SARS Data Sets

A.1   Hong Kong SARS Daily Infected Case Data

N	Date	Daily Cases	N	Date	Daily Cases	N	Date	Daily Cases	N	Date	Daily Cases
1	12/3	10	24	4/4	27	47	27/4	16	70	20/5	4
2	13/3	14	25	5/4	39	48	28/4	14	71	21/5	1
3	14/3	5	26	6/4	42	49	29/4	15	72	22/5	3
4	15/3	8	27	7/4	41	50	30/4	17	73	23/5	2
5	16/3	5	28	8/4	45	51	1/5	11	74	24/5	0
6	17/3	53	29	9/4	42	52	2/5	11	75	25/5	1
7	18/3	28	30	10/4	28	53	3/5	10	76	26/5	1
8	19/3	27	31	11/4	61	54	4/5	8	77	27/5	2
9	20/3	23	32	12/4	49	55	5/5	8	78	28/5	2
10	21/3	30	33	13/4	42	56	6/5	9	79	29/5	2
11	22/3	19	34	14/4	40	57	7/5	8	80	30/5	4
12	23/3	20	35	15/4	42	58	8/5	7	81	31/5	3
13	24/3	18	36	16/4	36	59	9/5	6	82	1/6	3
14	25/3	26	37	17/4	29	60	10/5	7	83	2/6	4
15	26/3	30	38	18/4	30	61	11/5	4	84	3/6	1
16	27/3	51	39	19/4	31	62	12/5	5	85	4/6	1
17	28/3	58	40	20/4	22	63	13/5	6	86	5/6	0
18	29/3	45	41	21/4	22	64	14/5	9	87	6/6	2
19	30/3	60	42	22/4	32	65	15/5	5	88	7/6	2
20	31/3	80	43	23/4	24	66	16/5	3	89	8/6	0
21	1/4	75	44	24/4	30	67	17/5	4	90	9/6	1
22	2/4	23	45	25/4	22	68	18/5	3	91	10/6	1
23	3/4	26	46	26/4	17	69	19/5	1	92	11/6	1

http://www.info.gov.hk/info/sars/eindex.htm

A.2 Singapore SARS Daily Infected Case Data

N	Date	Daily Cases	N	Date	Daily Cases	N	Date	Daily Cases	N	Date	Daily Cases
1	13/3	3	18	30/3	2	35	16/4	5	52	3/5	0
2	14/3	6	19	31/3	1	36	17/4	4	53	4/5	0
3	15/3	7	20	1/4	3	37	18/4	1	54	5/5	1
4	16/3	4	21	2/4	3	38	19/4	5	55	6/5	0
5	17/3	1	22	3/4	2	39	20/4	1	56	7/5	0
6	18/3	2	23	4/4	1	40	21/4	6	57	8/5	0
7	19/3	8	24	5/4	2	41	22/4	2	58	9/5	1
8	20/3	3	25	6/4	3	42	23/4	3	59	10/5	0
9	21/3	5	26	7/4	7	43	24/4	3	60	11/5	0
10	22/3	5	27	8/4	5	44	25/4	3	61	12/5	0
11	23/3	7	28	9/4	8	45	26/4	3	62	13/5	0
12	24/3	14	29	10/4	7	46	27/4	1	63	14/5	0
13	25/3	4	30	11/4	7	47	28/4	0	64	15/5	0
14	26/3	5	31	12/4	7	48	29/4	2	65	16/5	0
15	27/3	4	32	13/4	4	49	30/4	0	66	17/5	0
16	28/3	8	33	14/4	7	50	1/5	0	67	18/5	0
17	29/3	3	34	15/4	4	51	2/5	2	68	19/5	1

http://www.moh.gov.sg/sars/

A.3   Ontario SARS Daily Infected Case Data

N	Date	Daily Cases	N	Date	Daily Cases	N	Date	Daily Cases	N	Date	Daily Cases
1	18/3	7	32	18/4	0	63	19/5	0	94	19/6	0
2	19/3	1	33	19/4	2	64	20/5	0	95	20/6	1
3	20/3	0	34	20/4	2	65	21/5	0	96	21/6	0
4	21/3	0	35	21/4	3	66	22/5	0	97	22/6	3
5	22/3	2	36	22/4	1	67	23/5	0	98	23/6	0
6	23/3	0	37	23/4	0	68	24/5	0	99	24/6	0
7	24/3	0	38	24/4	0	69	25/5	0	100	25/6	0
8	25/3	8	39	25/4	2	70	26/5	0	101	26/6	0
9	26/3	9	40	26/4	0	71	27/5	0	102	27/6	0
10	27/3	1	41	27/4	1	72	28/5	2	103	28/6	0
11	28/3	7	42	28/4	3	73	29/5	22	104	29/6	0
12	29/3	0	43	29/4	1	74	30/5	15	105	30/6	0
13	30/3	7	44	30/4	0	75	31/5	2	106	1/7	0
14	31/3	9	45	1/5	2	76	1/6	8	107	2/7	0
15	1/4	5	46	2/5	0	77	2/6	15	108	3/7	0
16	2/4	4	47	3/5	0	78	3/6	3	109	4/7	0
17	3/4	7	48	4/5	0	79	4/6	2	110	5/7	0
18	4/4	4	49	5/5	0	80	5/6	0	111	6/7	0
19	5/4	8	50	6/5	0	81	6/6	4	112	7/7	0
20	6/4	8	51	7/5	0	82	7/6	7	113	8/7	0
21	7/4	1	52	8/5	0	83	8/6	0	114	9/7	1
22	8/4	3	53	9/5	0	84	9/6	0	115	10/7	0
23	9/4	3	54	10/5	0	85	10/6	1	116	11/7	0
24	10/4	1	55	11/5	0	86	11/6	8	117	12/7	0
25	11/4	2	56	12/5	0	87	12/6	4	118	13/7	0
26	12/4	0	57	13/5	0	88	13/6	1	119	14/7	0
27	13/4	0	58	14/5	0	89	14/6	0	120	15/7	0
28	14/4	3	59	15/5	0	90	15/6	2	121	16/7	0
29	15/4	5	60	16/5	0	91	16/6	0	122	17/7	1
30	16/4	17	61	17/5	0	92	17/6	0			
31	17/4	6	62	18/5	0	93	18/6	0			

http://www.health.gov.on.ca/english/public/updates/archives/
hu03/husars.html

A.4 Taiwan SARS Daily Infected Case Data

N	Date	Daily Cases	N	Date	Daily Cases	N	Date	Daily Cases	N	Date	Daily Cases
1	1/3	1	28	28/3	2	55	24/4	18	82	21/5	11
2	2/3	0	29	29/3	0	56	25/4	12	83	22/5	9
3	3/3	0	30	30/3	2	57	26/4	25	84	23/5	9
4	4/3	0	31	31/3	1	58	27/4	10	85	24/5	9
5	5/3	0	32	1/4	1	59	28/4	17	86	25/5	3
6	6/3	0	33	2/4	1	60	29/4	13	87	26/5	6
7	7/3	2	34	3/4	2	61	30/4	17	88	27/5	7
8	8/3	0	35	4/4	2	62	1/5	21	89	28/5	0
9	9/3	0	36	5/4	1	63	2/5	14	90	29/5	5
10	10/3	0	37	6/4	3	64	3/5	18	91	30/5	3
11	11/3	0	38	7/4	0	65	4/5	15	92	31/5	0
12	12/3	0	39	8/4	0	66	5/5	14	93	1/6	3
13	13/3	2	40	9/4	0	67	6/5	16	94	2/6	3
14	14/3	1	41	10/4	0	68	7/5	20	95	3/6	1
15	15/3	0	42	11/4	0	69	8/5	18	96	4/6	2
16	16/3	1	43	12/4	1	70	9/5	18	97	5/6	2
17	17/3	2	44	13/4	1	71	10/5	18	98	6/6	2
18	18/3	1	45	14/4	1	72	11/5	24	99	7/6	4
19	19/3	2	46	15/4	2	73	12/5	24	100	8/6	2
20	20/3	0	47	16/4	2	74	13/5	25	101	9/6	3
21	21/3	1	48	17/4	4	75	14/5	18	102	10/6	0
22	22/3	1	49	18/4	4	76	15/5	20	103	11/6	0
23	23/3	1	50	19/4	4	77	16/5	17	104	12/6	1
24	24/3	0	51	20/4	10	78	17/5	13	105	13/6	0
25	25/3	0	52	21/4	25	79	18/5	12	106	14/6	0
26	26/3	0	53	22/4	10	80	19/5	15	107	15/6	1
27	27/3	2	54	23/4	24	81	20/5	16			

http://www.cdc.gov.tw/sarsen/

Remark: The url at the end of each table is the source of data.

Bibliography

Andrews, D. F. and Herzberg, A. M. (1985) *Data.* Springer-Verlag, New York.

Arnold, S. F. (1990) *Mathematical Statistics.* Prentice-Hall, New Jersey.

Aronszajn, N. (1950) Theory of reproducing kernels, *Transactions of American Mathematical Society*, 68, 337-404.

Ascher, H. and Feingold, H. (1969) 'Bad-as-old' analysis of system failure data, in *Annals of Assurance Sciences,* Gordon and Breach, New York.

Ascher, H. and Feingold, H. (1984) *Repairable Systems Reliability* . Marcel Dekker, New York.

Balcer, Y. and Sahin, I. (1986) Replacement costs under warranty: cost moments and time variability, *Operations Research*, 34, 554–559.

Barlow, R. E. and Proschan, F. (1965) *Mathematical Theory of Reliability.* Wiley, New York.

Barlow, R. E. and Proschan, F. (1975) *Statistical Theory, Reliability and Life Testing.* Holt, Reinehart and Winston, New York.

Blischke, W. R. and Murthy, D. N. P. (1994) *Warranty Cost Analysis.* Marcel Dekker, Inc., New York.

Blischke, W. R. and Scheuer, E. M. (1975) Calculation of the cost of warranty policies as a function of estimated life distributions, *Naval Research Logistics Quarterly*, 22, 681–696.

Blischke, W. R. and Scheuer, E. M. (1981) Applications of renewal theory in analysis of the free-replacement warranty, *Naval Research Logistics Quarterly*, 28, 193–205.

Braun, W. J., Li, W. and Zhao, Y. Q. (2005) Properties of the geometric and related processes, *Naval Research Logistics*, 52, 607-616.

Chan, J. S. K., Lam, Y. and Leung, D. Y. P. (2004) Statistical inference for geometric processes with gamma distributions. *Computational Statistics and Data Analysis*, 47, 565-581.

Chan, J. S. K., Yu, P. L. H., Lam, Y. and Ho, A. P. K. (2006) Modelling SARS data using threshold geometric process. *Statistics in Medicine*, 25, 1826-1839.

Cheng, K. (1999) *Life Distribution Classes and Mathematical Theory of Reliability*, Science Press, Beijing (in Chinese).

Chiang, C. L. (1980) *An Introduction to Stochastic Processes and Their Applications* Robert E. Krieger Publishing Co, New York.

Choi, B. C. K. and Pak, A. W. P. (2003) A simple approximate mathematical model to predict the number of severe acute respiratory syndrome cases and deaths, *Journal of Epidemiology and Community Health*, 57, 831-835.

Cox, D. R. and Lewis, P. A. (1966) *The Statistical Analysis of Series of Events*, Mathuen, London.

Dellacherie, C. and Meyer, P. A. (1982) *Probabilities and Potential B Theory of Martingales*, North Holland, Amsterdam New York.

Feller, W. (1970) *An Introduction to Probability Theory and Its Applications, Vol. II*, Wiley, New York.

Gibbons, R. (1992) *A Primer in Game Theory*, Harvester Wheatsheaf, New York.

Hand, D. J., Daly, F., Lunn, A. D., McConway, K. J. and Ostrowski, E. (edited) (1994) *A Hand Book of Small Data Sets*, Chapman and Hall, London.

He, S. W., Wang, J. G. and Yan, J. A. (1995) *Semimartingale and Stochastic Calculus*, Science Press, Beijing (in Chinese).

Hill, V. L. and Blischke, W. R. (1987) An assessment of alternative models in warranty analysis, *Journal of Information and Optimization Sciences*, 8, 33–55.

Hsieh, Y. H., Chen, C. W. S. and Hsu, S. B. (2004) SARS outbreak, Taiwan, 2003, *Emerging Infectious Diseases* 10, 2, 201-206.

Jarrett, R. G. (1979) A Note on the Intervals between coal-mining disasters, *Biometrika*, 66, 191-193.

Kang, Z. R. and Yan, Y. B. (2000) The analytic solution of the integral equation of the renewal theory, *Chinese Journal of Applied Probability and Statistics*, 16, 125-132 (in Chinese).

Kleinrock, L (1975) *Queueing System Theory*, Vol. 1. Wiley, New York.

Lam, Y. (1987) A Finite algorithm for ϵ-optimal solutions of adaptive queueing control, *Journal of Mathematical Analysis and Applications*, 125, 218-233.

Lam, Y. (1988a) A note on the optimal replacement problem. *Advances in Applied Probability*, 20, 479-782.

Lam, Y. (1988b) Geometric processes and replacement problem. *Acta Mathematicae Applicatae Sinica*, 4, 366-377.

Lam, Y. (1991a) A replacement model for repairable deteriorating system. *Asia-Pacific Journal of Operational Research* 8, 119-127.

Lam, Y. (1991b) Optimal policy for a general repair replacement: average reward case. *IMA Journal of Mathematics Applied in Business & Industry* 3, 117-129.

Lam, Y. (1991c) An optimal repairable replacement model for deteriorating systems. *Journal of Applied Probability*, 28, 843-851.

Lam, Y. (1992a) Optimal geometric process replacement model. *Acta Mathematicae Applicatae Sinica*, 8, 73-81.

Lam, Y. (1992b) Nonparametric inference for geometric processes. *Communications in Statistics-Theory and Methods*, 21, 2083-2105.

Lam, Y. (1992c) Optimal policy for a general repair replacement model: Discounted Reward Case. *Communications in Statistics-Stochastic Models*, 8,

245-267.

Lam, Y. (1995) Calculating the rate of occurrence of failure for continuous-time Markov chains with application to a two-component parallel system. *Journal of Operational Research Society*, 45, 528-536.

Lam, Y. (1997) The rate of occurrence of failures. *Journal of Applied Probability*, 34, 234-247.

Lam, Y. (2003) A geometric process maintenance model. *Southeast Asian Bulletin of Mathematics*, 27, 1-11.

Lam, Y. (2005a) A monotone Process maintenance model for a multistate system. *Journal of Applied Probability*, 42, 1-14.

Lam, Y. (2005b) The expected number of events for a geometric process. Research report, No.420, Department of Statistics and Actuarial Science, The University of Hong Kong.

Lam, Y. (2006) Geometric process. *Encyclopedia of Statistical Sciences, 2nd edition*, Balakrishnan, N., Read, C. B., Kotz, S. and Vidakovic, B., ed. John Wiley & Sons, Inc., New York, to appear.

Lam, Y. (2007a) A geometric process maintenance model with preventive repair, *European Journal of Operational Research*, 182, 806-819.

Lam, Y. (2007b) A threshold geometric process maintenance model. Research report, No.449, Department of Statistics and Actuarial Science, The University of Hong Kong.

Lam, Y. (2007c) A geometric process warranty model-game theory approach. Research report, No.450, Department of Statistics and Actuarial Science, The University of Hong Kong.

Lam, Y. and Chan, S. K. (1998) Statistical inference for geometric processes with lognormal distribution. *Computational Statistics and Data Analysis*, 27, 99-112.

Lam, Y. and Tang, Y. Y. (2007) The analytic solution of an integral equation in geometric process. Research report, No.451, Department of Statistics and Actuarial Science, The University of Hong Kong.

Lam, Y. and Tse, Y. K. (2003) Optimal maintenance model for a multistate deteriorating system-a geometric process approach. *International Journal of Systems Science*, 34, 303-308.

Lam, Y. and Zhang, Y. L. (1996a) Analysis of a two-component series system with a geometric process model. *Naval Research Logistics*, 43, 491-502.

Lam, Y. and Zhang, Y. L. (1996b) Analysis of a parallel system with two different units. *Acta Mathematicae Applicatae Sinica*, 12, 408-417.

Lam, Y. and Zhang, Y. L. (2003) A geometric process maintenance model for a deteriorating system under a random environment. *IEEE Transactions on Reliability*, 53, 83-89.

Lam, Y. and Zhang, Y. L. (2004) A shock model for the maintenance problem of a repairable system. *Computers and Operations Research*, 31, 1807-1820.

Lam, Y., Zhang, Y. L. and Liu, Q. (2006) A geometric process model for $M/M/1$ queueing system with a repairable service station. *European Journal of Operational Research*, 168, 100-121.

Lam, Y., Zhang, Y. L. and Zheng, Y. H. (2002) A geometric process equivalent model for a multistate degenerative system. *European Journal of Operational Research*, 142, 21-29.

Lam, Y., Zheng, Y. H. and Zhang, Y. L. (2003) Some limit theorems in geometric processes. *Acta Mathematicae Applicatae Sinica*, 19, 405-416.

Lam, Y., Zhu, L. X., Chan, S. K. and Liu, Q. (2004) Analysis of data from a series of events by a geometric process model. *Acta Mathematicae Applicatae Sinica*, 20, 263-282.

Lee, L. (1980a) Testing adequacy of the Weibull and loglinear rate models for a Poisson process. *Technometrics*, 22, 195-199.

Lee, L. (1980b) Comparing rates of several independent Weibull processes, *Technometrics*, 22, 427-430.

Lehmann, E. L. (1983) *Theory of Point Estimation*, Wiley, New York.

Leung, F. K. N. (2001) A note on "a bivariate optimal replacement policy for a repairable system", Research Report, Department of Management Science, City University of Hong Kong.

Leung, F. K. N. (2005) Statistically inferential analogies between arithmetic and geometric processes. *International Journal of Reliability, Quality and Safety Engineering*, 12, 323-335.

Leung, F. K. N. and Lee Y. M. (1998) Using geometric processes to study maintenance problems for engines. *Int. J. Industrial Engineering*, 5, 316-323.

Lewis, T. and the M345 Course Team (1986) M345 *Statistical Methods, Unit 2: Basic Methods: Testing and Estimation, Milton Keynes:* The Open University, 16.

Li, Z. H. (1984) Some distributions related to Poisson processes and their application in solving the problem of traffic jam, *Journal of Lanzhou University*, 20 127-136.

Li, Z. H., Chan, L. Y. and Yuan, Z. H. (1999) Failure time distribution under a δ-shock model and its application to economic design of systems. *International Journal of Reliability, Quality and Safety Engineering*, 6, 237-247.

Liang, X. L., Li, Z. H. and Lam, Y. (2006) Optimal replacement policy subject to a general δ-shock, submitted for publication.

Maguire, B. A., Pearson, E. S. and Wynn, A. H. A. (1952) The time intervals between industrial accidents, *Biometrika*, 39, 168-180.

Mamer, J. W. (1982) Cost analysis of pro rata and free-replacement warranties, *Naval Research Logistics Quarterly*, 29, 345-356.

Morimura, H. (1970) On some preventive maintenance policies for IFR, *Journal of Operational Research Society of Japan*, 12, 94-124.

Musa, B. D. (1979) *Software Reliability Data*, Data Analysis Center for Software, Rome Air Development Center, Rome, NY.

Musa, J. D., Iannino, A. and Okumoto, K. (1987) *Software Reliability: Measurement, Prediction, Application*, McGraw-Hill, New York.

Nguyen, D. G. and Murthy, D. N. P. (1984) Cost analysis of warranty policies, *Naval Research Logistics Quarterly*, 31, 525-541.

Park, K. S. (1979) Optimal number of minimal repairsbefore replacement, *IEEE Transactions on Reliability*, 28, 137-140.

Péréz-Ocon, R. and Torres-Castro, I. (2002) A reliability semi-Markov model involving geometric processes. *Applied Stochastic Models In Business And Industry*, 18, 157-170.

Rangan, A. and Esther Grace, R. (1989) Optimal replacement policies for a deteriorating system with imperfect maintenance. *Advances in Applied Probability*, 21, 949-951.

Ross, S. M. (1996) *Stochastic Processes*, second edition, Wiley, New York.

Seber, G. A. F. (1977) *Linear Regression Analysis*, Wiley, New York.

Shanthikumar, J. G. and Sumita, U. (1983) General shock models associated with correlated renewal sequences. *Journal of Applied Probability*, 20, 600-614.

Shanthikumar, J. G. and Sumita, U. (1984) Distribution properties of the system failure time in a general shock model. *Advances in Applied Probability*, 16, 363-377.

Sheu, S. H. (1999) Extended optimal replacement model for deteriorating systems. *European Journal of Operational Research*, 112, 503-516.

Shi, D. (1985) A new method for calculating the mean failure numbers of a repairable system during $(0, t]$. *Acta Mathematicae Applicatae Sinica*, 8, 101-110.

Stadje, W. and Zuckerman, D. (1990) Optimal strategies for some repair replacement models. *Advances in Applied Probability*, 22, 641-656.

Stoer, J. and Bulirsch, R. (1980) *Introduction to Numerical Analysis*, Springer-Verlag, New York.

Stout, W. F. (1974) *Almost Sure Convergence*. Academic Press, New York.

Tang, Y. Y. and Lam, Y. (2006a) A lognormal δ-shock model for a deteriorating system, *Journal of Sichuan University*, 43, 59-65.

Tang, Y. Y. and Lam, Y. (2006b) A δ-shock maintenance model for a deteriorating system. *European Journal of Operational Research*, 168, 541-556.

Tang, Y. Y. and Lam, Y. (2007) A numerical solution to the integral equation in geometric process. *Journal of Statistical computation and Simulation*, to appear.

Thomas, L. C. (1979) A finite algorithm for ϵ-optimal solutions of the infinite horizon adaptive inventory model. Notes in Decision Theory, No. 71, Department of Decision Theory, University of Manchester.

Tuen, W.N., Turinici, G. and Danchin, A. (2003) A double epidemic model for the SARS propagation. *BMC Infectious Diseases*, 3, 19.

Wang, G. J. and Zhang, Y. L. (2005) A shock model with two-type failures and optimal replacement policy. *International Journal of Systems Science*, 36, 209-214.

Wang, G. J. and Zhang, Y. L. (2006) Optimal periodic preventive repair and replacement policy assuming geometric process repair. *IEEE Transactions on Reliability*, 55,118-122.

Wu, S. M., Huang, R. and Wan, D. J. (1994) Reliability analysis of a repairable system without being repaired "as good as new". *Microelectronics and Reliability*, 34, 357-360.

Zhang, Y. L. (1991) Excess life of the geometric process and its distribution, *Journal of Southeast University*, 21, 27-34.

Zhang, Y. L. (1994) A bivariate optimal replacement policy for a repairable system. *Journal of Applied Probability*, 31, 1123-1127.

Zhang, Y. L. (1995) The reliability analysis of a two-unit cold standby repairable system and its optimal replacement policy. *Applied Mathematics - A Journal of Chinese Universities*, 10, 1-11 (in Chinese).

Zhang, Y. L. (1999) An optimal geometric process model for a cold standby repairable system. *Reliability Engineering and Systems Safety*, 63, 107-110.

Zhang, Y. L. (2002) A geometric-process repair-model with good-as-new preventive repair. *IEEE Transactions on Reliability*, 51, 223-228.

Zhang, Y. L. (2007) A discussion on "A bivariate optimal replacement policy for a repairable system". *European Journal of Operational Research*, 179, 275-276.

Zhang, Y. L. and Wang, G. J. (2006) A bivariate optimal repair-replacement model using geometric process for cold standy repairable system. *Engineering Optimization*, 38, 609-619.

Zhang, Y. L. and Wang, G. J. (2007) A geometric process repair model for a series repairable system with dissimilar components. *Applied Mathematical Modelling*, 31,1997-2007.

Zhang, Y. L., Wang, G. J. and Ji, Z. C. (2006) Replacement problems for a cold standby repairable system. *International Journal of Systems Science*, 37, 17-25.

Zhang, Y. L., Yam, R. C. M. and Zuo, M. J. (2001) Optimal replacement policy for a deteriorating production system with preventive maintenance. *International Journal of Systems Science*, 32, 1193-1198.

Zhang, Y. L., Yam, R. C. M. and Zuo, M. J. (2002) Optimal replacement policy for a multistate repairable system. *Journal of Operational Research Society*, 53, 336-341.

Zhang, Y. L., Yam, R. C. M. and Zuo, M. J. (2007) A bivariate optimal replacement policy for a multistate repairable system. *Reliability Engineering and System Safety*, 92, 535-542,

Index